PROGRESS IN
AEROSPACE SCIENCES

(Incorporating Progress in Aeronautical Sciences)

VOLUME 14

PROGRESS IN AEROSPACE SCIENCES

VOLUME 14

EDITED BY

D. KÜCHEMANN

Royal Aircraft Establishment,
Farnborough, England

AND

P. CARRIÈRE, *Paris* B. ETKIN, *Toronto* W. FISZDON, *Warsaw*
N. ROTT, *Zürich* J. SMOLDEREN, *Brussels* I. TANI, *Tokyo*
W. WUEST, *Göttingen*

PERGAMON PRESS

Oxford · New York · Toronto
Sydney · Braunschweig

Pergamon Press Ltd., Headington Hill Hall, Oxford
Pergamon Press Inc., Maxwell House, Fairview Park, Elmsford, New York 10523
Pergamon of Canada Ltd., 207 Queen's Quay West, Toronto 1
Pergamon Press (Aust.) Pty. Ltd., 19a Boundary Street, Rushcutters Bay, N.S.W. 2011, Australia
Vieweg & Sohn GmbH, Burgplatz 1, Braunschweig

First edition 1973

Library of Congress Catalog Card No. 74-618347

Printed in Great Britain by A. Wheaton & Co., Exeter

ISBN 0 08 017138 9

CONTENTS

PREFACE

In view of the vital importance that experimental facilities have in the activities in aerospace sciences, this volume contains two articles which are specifically concerned with such tools. The remaining three articles deal with aerodynamic problems of particular topical interest.

The first article is a critical review of the development of aerodynamic facilities for testing at high speeds during the past 25 years, in which the author himself has played an outstanding part. This is stocktaking of a most useful kind, which also provides a basis for looking into the possibilities of further progress to satisfy future needs.

The second article gives an exhaustive account of the technique for suspending models magnetically in wind tunnels, which has been developed in recent years and now presents a practical and powerful tool for the experimentalist. The team of the authors has been foremost in this development, and this is the first comprehensive account in which the many aspects and pieces of information are presented together.

The third article deals with the problem of generating aerodynamic lift either for sustained hypersonic flight or during re-entry in the atmosphere. Rapid advances have been made in recent years in the theoretical treatment of novel shapes of bodies in which the provision of volume and the production of lift are combined. This is a case where theoretical considerations have led to the evolution of an entirely new concept and a new type of aircraft, and an authoritative review of this work should be timely and welcome.

The fourth article gives a review of methods for predicting the base pressure in two-dimensional steady flows. This problem is of practical importance in aircraft design and is still not solved satisfactorily. This review informs the readers of the present state of knowledge.

The last article deals with the subject of aerodynamic noise, which is of great importance and interest not only to aeronautical engineers but also to the population at large. The article takes the form of four lectures which have been given at the University of Tennessee and at the Technische Hochschule, Aachen. Between them, these lectures give a concise and lucid introduction to the theory of aerodynamic sound generation and present in a unified form the various and often very complex theoretical approaches which have been put forward. This summary should be welcome not only to the specialist but also to the general reader.

D. K.

1

A CRITICAL REVIEW OF DEVELOPMENT OF EXPERIMENTAL METHODS IN HIGH-SPEED AERODYNAMICS*

J. LUKASIEWICZ†

Carleton University, Ottawa, Canada

INTRODUCTION

This paper is concerned with two general aspects of development of high-speed aerodynamic test facilities during the past 25 years. First, the efficiency with which aerodynamic test technology was developed is assessed through examination of several shortcomings which may have caused delays in the availability of urgently needed equipment and data, and excessive costs. Secondly, from the technical standpoint, an overall trend in the development of high-speed aerodynamic test facilities is noted and possibilities of future progress are discussed.

DEFICIENCIES IN THE DEVELOPMENT OF HIGH-SPEED AERODYNAMIC TESTING

A review of the development of high-speed aerodynamic test facilities reveals a number of deficiencies which can be grouped under the following headings:

(i) *Ad-hoc development*. Potentially applicable techniques have been seldom evaluated; instead, the supposedly coordinated effort proceeded in an *ad-hoc* manner, each group promoting its own device.

(ii) *Technical judgement*. Traditional outlook or lack of detailed understanding of the problem at hand often delayed introduction of successful design solutions. On the other hand, excessive performance has been often claimed for new types of facilities, particularly in the hypersonic regime.

(iii) *Information transfer lag*. There have been cases of significant achievements by one group unknown for a long time to other groups working in the same field.

(iv) *Lack of a system for support of continuing and integrated research into development of test techniques and for reaching timely decisions on construction of major test facilities.* Examination of the past experience suggests that research support was provided for specific solutions rather than toward well-defined performance objectives and that decisions have

* Based on an invited lecture delivered at the XXIInd International Astronautical Congress, Brussels, 20–25 September 1971.

† Professor of Engineering.

been often taken somewhat late relative to the development of flight hardware, as a result of outside pressures rather than recognition of technical needs.

Ad-hoc Development

Although already around 1930 (Ackeret, 1927, 1935; Anon., 1936; Busemann, 1931, 1971) the basic information (one-dimensional, steady and unsteady, compressible flow theory) was available to visualize shock tube, shock tunnel, the expansion, compression and Ludwieg tube cycles, the light gas gun and the free-piston compressor applications, the development of these techniques followed a haphazard path. Applications based on the shock tube were pursued relatively early, from the late 1940s, but have not matured until about 1960 (Hertzberg *et al.*, 1962) after the first generation of the U.S. ballistic missiles was already deployed. The expansion and compression tube cycles were not investigated until 1962 (Trimpi, 1962). It was only in 1955 that the Ludwieg tube cycle, the simplest of all, was proposed, and it was not until 1968 that it was applied to hypersonic tunnels, for which it was eminently suitable (Hottner, 1968). The application of the Ludwieg tube drive to large, high Reynolds number supersonic tunnels was first proposed in 1963 (Falk, 1963) but the development of large-scale transonic and supersonic Ludwieg tube-type facilities has been only recently initiated (AGARD, 1971; Lukasiewicz, 1971; Whitfield *et al.*, 1971). Another conceptually simple solution to the problem of high Reynolds number transonic tunnel design, which derives from the Ludwieg tube cycle and involves the use of a piston, has been proposed only in 1971 by Evans. This scheme, as well as an even simpler reservoir and tube tunnel (RTT) design proposed by the writer, are discussed later in this section.

The light gas gun was developed already in 1949, but several years passed before it was used in hypervelocity ranges. Although it achieved high performance, much effort went into the development of various "augmentation schemes", none of which were truly successful (cf. next section). The possibility of application of the multi-stage rocket propulsion to launch large models at high hypersonic speeds into an aeroballistic range was only examined in the late 1960s (Lukasiewicz, 1970). The free-piston compressor, as a basic technique applicable to a wide variety of hypersonic test facilities, was never systematically investigated and assessed. Chronologically, it was first applied to light gas guns in 1947 (Crozier and Hume, 1957, in the U.S.), and later to gun tunnels (Cox and Winter, 1957, in the U.K.), shock tubes (Stalker, 1961, in Canada; Greif and Bryson, 1965, in the U.S.), adiabatic compressor hypersonic tunnels (Leuchter, 1964, in France), longshot tunnels (Perry, 1964, in the U.S.) and shock tunnels (Stalker, 1967, in Australia). Thus the development of free-piston applications extended over a span of 20 years and was pursued in five countries.

The development of heaters for hypersonic tunnels was equally erratic and resulted eventually, over a period of some 15 years, in several practical designs, including the electric (for air and/or nitrogen, using nichrome, Kanthal, graphite or tungsten elements) and the refractory (storage) types. In general, techniques were worked on as individuals and organizations were attracted to different schemes (for example, shock tunnels at the Cornell Aeronautical Laboratory, hotshot tunnels at the AEDC, light gas guns at NACA's Ames Aeronautical Laboratory). It would appear that much time and effort could have been saved had a comprehensive morphological analysis been performed (although much research would be needed to provide a positive proof of this assertion).

A rudimentary morphology of hypersonic test facilities is shown in Table 1. Some of the fundamental gas dynamic processes are given at the top of the table, separately for the

driver and the driven (working) gases.* The list is not complete: other processes, such as diffusion and cryogenic pumping, heating by radio frequency discharge, etc., could be included.

In the first column the more important facility types are listed and grouped into wind tunnels, ranges and sleds, and counter-flow devices. Since in shock tubes the model is stationary in the laboratory frame of reference, they have been included in the wind tunnel group. In the case of two-stage guns and rockets, the propellant which acts on the model or projectile is considered as the working gas.

The processes that may be applicable to the particular facilities listed are indicated in Table 1 by dots. Of course, other combinations than those indicated could be envisaged. For example, a number of different permutations could be indicated for the shock tubes and shock tunnels, depending on whether heating of the driver gas is used, or whether MHD augmentation is applied. Also, cycles which so far have not been fully investigated, or new ones, could be shown, such as adiabatic compression and constant pressure heating cycle (which could be regarded as an arc heated gun or shock tunnel) or MHD augmentation of light gas guns.

Table 1 indicates several options for various classes of hypersonic facilities. For example, wind tunnels which rely on steady, isentropic expansion for acceleration of the working gas to the desired Mach number, may employ drives ranging from mechanical compression of the working gas to shock compression. Although Table 1 contains retrospective data, it suggests that the various gas dynamic options (including the corresponding design solutions) could have been identified and evaluated when the development of hypersonic facilities was being initiated, a course of action which was not followed. As a result of such approach, if the potential of the Ludwieg tube and of the slow-piston compressor were appreciated in the late 1940s, hypersonic tunnels could have been developed much earlier, at a small cost. Their aerodynamic performance would have compared well with that of more conventional, and more expensive designs. ONERA's slow piston compressor tunnel (Leuchter, 1964) achieved stagnation conditions of 1700°K, 3000 psi in air (or $M = 14$ at saturation) with a 120-msec useful run duration.

A similar relationship is apparent in the development of hotshot and longshot tunnels. Except for the method of drive, the longshot cycle, in which the working gas is expanded from a constant volume reservoir, is the same as the hotshot cycle and results in similar aerodynamic performance. Compared to the hotshot and its electrical supply, the longshot technique appears to offer much more economical means of achieving relatively long duration, high Reynolds number hypersonic flow. While in the 1950s and the early 1960s the main effort was directed toward the development of shock tunnels (Hertzberg *et al.*, 1962) and hotshots (Lukasiewicz *et al.*, 1961) in the hope of achieving hypervelocities, the longshot tunnel was not developed until 1968 (Richards and Enkenhus, 1969). As regards the recently proposed, multi-stage rocket launchers for hypervelocity range testing (Lukasiewicz, 1970), the provision of this technique in the early phases of the ICBM development might have reduced the need for extremely costly atmospheric flight testing and effected considerable savings.

It is pertinent to note that the unavailability of sophisticated instrumentation has been sometimes mentioned as the cause of delay in the development of hypersonic test techniques. It is probably more accurate to state that the development of facilities and instrumentation

* Except for conventional wind tunnels and single-stage guns, an auxiliary driver gas is usually employed in hypersonic facilities for acceleration of the driven (working) gas or model.

usually occurred simultaneously, since the instrumentation needs could not have been determined without a detailed knowledge of the facility characteristics, and the determination of the latter required appropriate instruments. The relative progress in the two areas was not necessarily always equal, as illustrated by the history of development of shock and hotshot tunnels. Although both techniques required about a decade of development, the progress of the shock tunnel, with its short run time (in the 1-msec range), was paced by the instrumentation, whereas development of the hotshot type, with its much longer run time (in the 100-msec range), depended on improvements in the arc chamber design aimed at reduction of flow contamination.

Development of facilities other than hypersonic could have also benefited from a morphological approach. This is apparent, for example, in the area of high Reynolds number, transonic and supersonic wind tunnel designs which are being currently investigated.

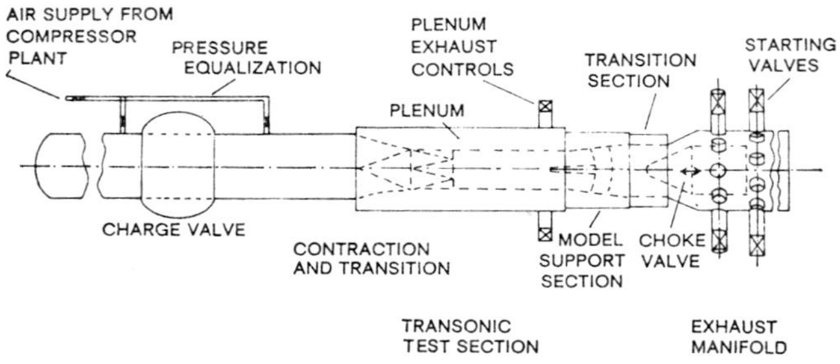

FIG. 1. High Reynolds number, transonic Ludwieg-tube tunnel (Lukasiewicz, 1971).

From a morphological viewpoint, the already mentioned Ludwieg (1955) and Evans (1971) alternatives, in view of their conceptual simplicity, are particularly interesting; they are shown schematically in Figs. 1 and 2.

In the Evans design (Fig. 2) the air is stored in a long tube in which a movable piston is located upstream of the test section. At the diffuser end a plug-valve is situated and connected by a return circuit to the tube on the upstream side of the piston. Before a test run, the valve is closed, the piston is in the extreme upstream position and the tube-test section-diffuser assembly is filled with air at the desired pressure. The return circuit contains air at a lower pressure. The flow is initiated by controlled opening of the plug valve, which causes an expansion to propagate toward the piston and the air to escape into the return circuit. The piston is started when the expansion reaches the piston face, the opening of the valve and the piston motion being matched so as to cancel the expansion wave. Following the acceleration phase, the piston, moving at a constant velocity, maintains a constant flow through the test section. The piston is driven by a number of counter-balancing pistons located in the return circuit and connected by cables and pulleys to the primary piston. The secondary pistons are loaded with the return circuit air and operate in evacuated cylinders, so that the pull exerted by them on the primary piston is proportional to the tunnel stagnation pressure. After a run, the tunnel is recharged by closing the valve at the diffuser exit and pumping the air from the return circuit into the barrel downstream of the piston. Provided the cancellation of the expansion wave is accomplished satisfactorily, the proposed drive should result in an exceptionally quiet tunnel flow, of the quality achieved in a simple Ludwieg tube. For this reason, the design has been referred to as "Evans Clean Tunnel" or ECT.

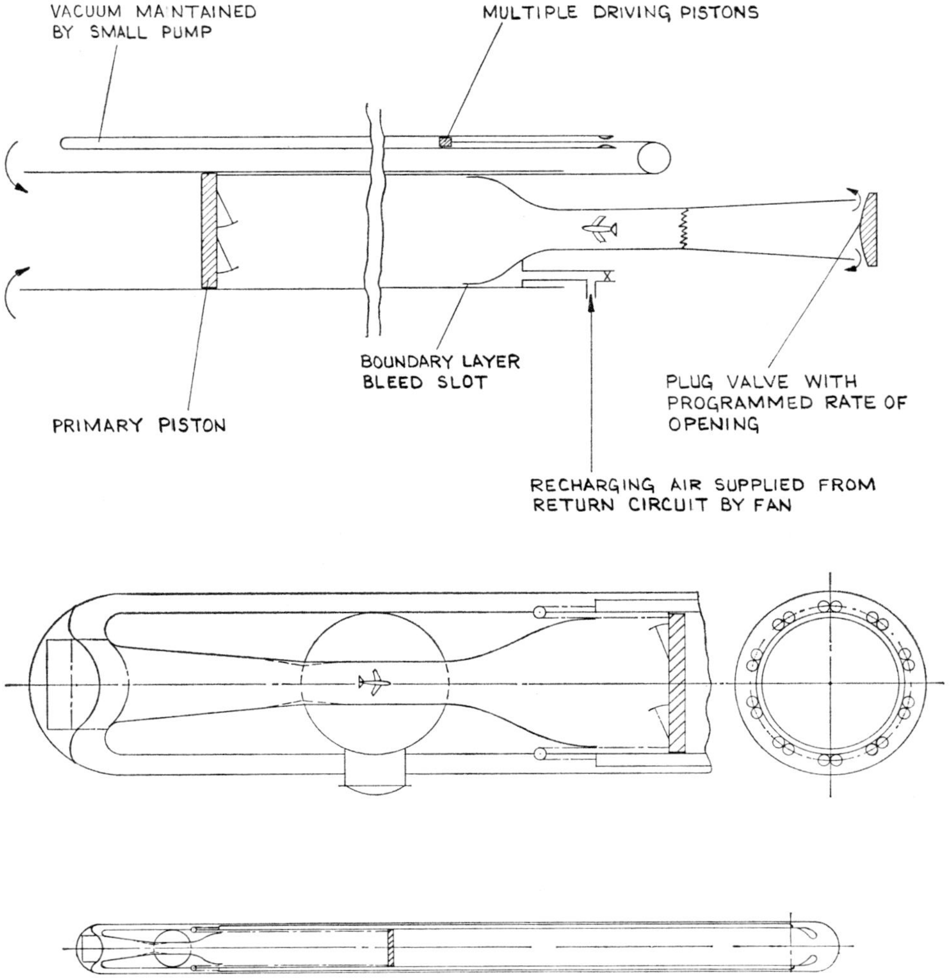

FIG. 2. The ECT design (Evans, 1971). *Top*: schematic layout of drive system. *Bottom*: general arrangement of wind tunnel.

In both the Ludwieg tube and the ECT the working fluid is accelerated by means of an unsteady expansion. In a Ludwieg cycle, the "prime" test time is terminated by arrival of the expansion wave after its reflection from the closed tube end; subsequent reflection cycles may provide test periods at lower stagnation pressures. The Ludwieg tube "prime" test time, being determined by the reflection cycle, is short (on the order of 1 sec), with only a small fraction of the working gas stored in the tube being used. The ECT cycle, in which a piston is used to cancel the expansion wave and to subsequently push through the test section all of the gas initially stored in the tube, achieves a much longer run (of the order of 10 sec). Since the ECT cycle is based on acceleration through unsteady expansion, it could be regarded as a modification of the Ludwieg cycle toward maximization of run time. Alternatively, it could be thought of as derived from a conventional, Prandtl-type, closed-circuit, continuous wind tunnel concept applied to the case of a short test time. This was the route followed by Evans (1971), who stated that his proposed "... design takes advantage of the fact that a closed circuit of a conventional transonic tunnel ... would be so large that, if the tunnel could be started and stopped quickly, the air would not complete one full circuit during 10 seconds of operation". Indeed, the Evans scheme retains some of the essential features of a closed circuit, continuous tunnel, including high driving efficiency and a return path for the working gas.

From the physical point of view, the Evans design is a derivative of the Ludwieg cycle; yet, in the absence of a systematic, morphological analysis of the available options, the ECT concept was developed as an extension of closed circuit, continuous wind tunnel technology and experience.

Taking a morphological viewpoint, one should ask what other means are available for lengthening the duration of Ludwieg tube test cycle? The use of a fluid, rather than solid, piston, offers yet another option, which will be referred to as the Reservoir and Tube Tunnel (RTT). In the RTT the piston is formed by admission of gas at the upstream end of the tube, at a velocity and pressure which match those produced by unsteady expansion. In the simplest design, the reservoir would be initially at the same pressure as the tube, and would be connected to it by means of a fixed opening (such as a perforated plate) smaller than the tube cross-sectional area.* Since at Mach numbers smaller than $4/(3 - \gamma) = 2.5$ for $\gamma = 1.4$, at a given pressure ratio the mass flux and velocity are smaller in unsteady than in steady isentropic expansion (Lukasiewicz, 1950), the pressure difference generated by the expansion wave would suffice to induce the required flow from the reservoir. Indeed, such a design would be analogous to that of the transonic test section perforated walls which, because of their mixed, open-solid geometry, are effective in attenuating reflections of shock-expansion disturbances caused by the model (see, e.g., Lukasiewicz, 1961).

With the initial reservoir pressure equal to the tube pressure, the reservoir volume required to attenuate the reflection of the expansion wave is comparable to the extra tube volume which would be necessary to achieve the same test time without a reservoir. With a more sophisticated design, employing a reservoir pressure higher than the initial tube pressure, and an active flow control valve, large test time increase is possible with relatively much smaller reservoir volume. The limit of the fluid piston technique is reached when all of the gas initially stored in the tube is used up during the test time, as in the ECT design. Thus, depending on the relative tube/test section contraction ratio, an arbitrarily long test time may be obtained. For a contraction ratio of about 1.5, considered close to the minimum for a transonic tube tunnel, the limiting test time amounts to more than double the simple Ludwieg-tube cycle test duration.

Compared to the ECT scheme, the fluid piston RTT design would appear to be mechanically simpler, but would have a lower drive efficiency and would be subject to noise caused by admission of gas from the reservoir. The significance of the latter would have to be assessed relative to the disturbances present in a transonic, mixed-boundary wall test section.

The propagation of disturbances from the fluid piston face could be impeded by inclusion of a diaphragm at the piston–working gas interface, which would slide as a light-weight, sound absorbing piston inside the tube.

It is apparent from the above review that systematic examination of the possibilities

* The reservoir-perforated plate arrangement has been used successfully with shock tunnels (Henshall *et al.*, 1962) and has resulted in a substantial increase in the available test time (from 1 msec to 4.4 msec, with a 1/3 open perforated plate, using cold hydrogen drive), at the expense of total reservoir and driver volume.

based on the Ludwieg tube cycle was not undertaken when options for high Reynolds number, transonic wind tunnel designs were being considered, and that the approaches pursued do not necessarily reflect a process of optimization.

Technical Judgement

In retrospect, it appears that traditional conservative outlook or lack of detailed understanding of the problem at hand often delayed the introduction of a successful test technique. On the other hand, excessive performance has been frequently claimed for new types of facilities.

Continuous versus intermittent, trisonic wind tunnels.

The history of development of large, transonic and supersonic wind tunnels in the West after 1945 provides an example of excessive engineering conservatism.*

For reasons of power economy and design simplicity, supersonic tunnels developed in Germany before 1945 were of the intermittent, atmosphere-to-vacuum-storage type. They provided Mach numbers up to 4.4 and run times on the order of 15 sec, in test sections up to 40×40-cm size; suitable force, pressure and heat transfer instrumentation was concurrently developed (Owen, 1945).

The basic concept of short duration wind tunnel operation, with its attendant advantages of low cost and high performance (wide Mach number range at large Reynolds numbers), was not immediately seized by the West. Instead, the tradition of continuous wind tunnel design was followed in the major national programs of development of new aerodynamic test facilities, in the U.S. (Unitary Wind Tunnel Program of 1949), U.K. (post-war facilities at RAE, Bedford) and France (ONERA facilities near Paris and Modane). The design and construction of these large (on the order of 10×10-ft test section) wind tunnels occupied many years, and was not completed until mid-1950s and early 1960s. The major design and fabrication difficulties, delays and costs stemmed from the continuous operation principle and were related to the basic incompatibility of compressor pressure ratio–volume flow characteristics versus wind tunnel requirements over a wide Mach number range (as noted by Ackeret already in 1935), the large power and the complexity of the drive (comprising motors, compressors and coolers). In fact, with the continuous tunnels the major share of the design and construction effort was devoted to the drive, rather than to the test section and instrumentation.

The development of large, intermittent tunnels, bypassed by the programs of the major nations of the West, was later undertaken by the countries and organizations with more limited resources, and resulted in the early availability (in mid and late 1950s) of high performance (large Reynolds number) facilities. A large, trisonic, intermittent wind tunnel design was probably first proposed in 1950, in Canada (4×4-ft test section, 10-sec run duration, m.a.c. Re. No. 2 to 4 million, blowdown to the atmosphere; Lukasiewicz and Pruden, 1950), and several large facilities of this type were subsequently built by the U.S. and British aircraft manufacturers; one such wind tunnel was built in Canada (5-ft NAE wind tunnel; Tupper *et al.*, 1961). The relative merits of the continuous and intermittent installations were discussed by Lukasiewicz (1955).

* This discussion does not pertain to propulsion testing, for which relatively long duration flows may be required.

The development of subsonic and transonic–supersonic (trisonic) wind tunnels, at a standstill during the past decade, is now receiving some attention in view of the widening wind tunnel–flight Reynolds number gap, for current and future aircraft and spacecraft (AGARD, 1971; Lukasiewicz, 1971). The growth of this gap over the years is illustrated in Figs. 3, 4, and 5. The increasing size of subsonic jet transports, the critical dependence of supersonic transport economic viability on aerodynamic performance, the transonic maneuverability of new fighters call for wind tunnel Reynolds numbers which exceed those now available by a factor of 5 or more; the deficiency is even bigger in the case of large launch vehicles and recoverable orbital transport vehicles (space shuttle). In these cases the power requirements alone practically preclude consideration of continuous wind tunnels, while the intermittent designs appear capable to meet the needs at a reasonable cost.

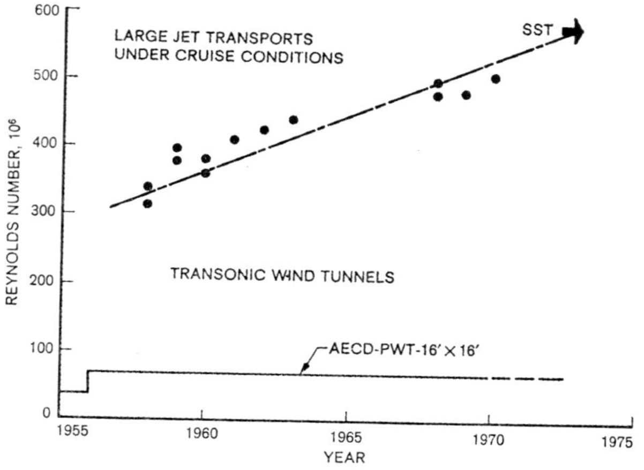

FIG. 3. Chronology of Reynolds number of large jet transports (cruise conditions) and of transonic wind tunnels. Reynolds number based on fuselage length for jet transports and on 0.7 × test section width for wind tunnels.

The tube type drives, previously discussed, with run times in the range of 1 to 10 sec, may offer the highest practical Reynolds number (up to 150 × 10^6/ft) at transonic speeds. A high-pressure, blowdown tunnel could operate at 25 × 10^6/ft Reynolds number with a run time on the order of 10 sec (AGARD, 1971; Evans, 1971; Lukasiewicz, 1971; Whitfield *et al.*, 1971).

It is interesting to note that, at present, the highest transonic unit Reynolds number (of 40 × 10^6/ft), is available in an intermittent, blowdown wind tunnel which was designed in the 1950s (Ohman and Brown, 1971). It appears that, as a result of performance requirements and the experience gained with large, trisonic intermittent tunnels, the conservative, continuous wind tunnel tradition is being discarded in favour of the more modern and superior techniques.

*Hypersonic wind tunnel nozzles: from two-dimensional to axisymmetric design.**

The development of hypersonic wind tunnels was critically dependent on the nozzle as the component responsible for acceleration of flow to the desired Mach number and the

* For an earlier discussion of this topic see Lukasiewicz (1959).

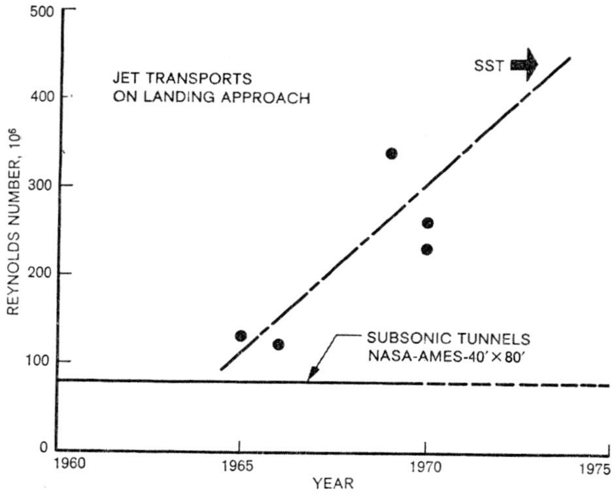

FIG. 4. Chronology of Reynolds number of jet transports (landing approach conditions) and wind tunnels. See Fig. 3 for basis of Reynolds number.

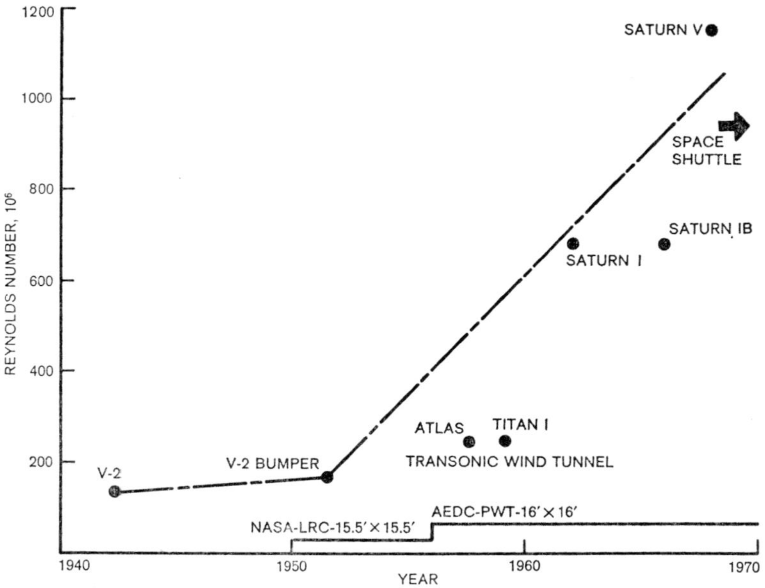

FIG. 5. Chronology of Reynolds number of large rockets and of transonic tunnels. Reynolds number based on rocket length at $1 \lesssim M \lesssim 1.2$, and on $0.7 \times$ test section width.

production of uniform conditions in the test section. In fact, the hypersonic nozzle design was one of the key factors which paced the construction of hypersonic tunnels. The examination of the technical views and arguments which governed hypersonic nozzle development is particularly interesting since many of the generally accepted premises were found to be inapplicable or mistaken.

The development of nozzles for hypersonic Mach numbers originated in Germany during World War II (Owen, 1945) and was then continued in the U.S. and elsewhere. The early, German design was the same used for the two-dimensional, supersonic nozzles. In order to minimize the boundary layer effects, the nozzles were made of the shortest possible length. In the standard design, a circular throat profile was used up to the point of inflexion at which the maximum deflection, equal to one-half of the Prandtl–Meyer angle for the nozzle exit Mach number, was attained. The boundary layer displacement thickness along the nozzle walls was calculated on the assumption of subsonic, flat plate laws, and the inviscid, characteristic profile was corrected for the boundary layer displacement thickness on both plane and curved nozzle walls. The emphasis on the minimum nozzle length was carried to the limit with the design of nozzles with sharp-edge throats. The conditions in the immediate neighbourhood of the sharp-edge correspond exactly to a Prandtl–Meyer expansion round a corner through one-half of the angle for the exit Mach number. The viscous effects in the flow around the sharp throat edge were neglected and the boundary layer correction was applied as for the circular-throat profile nozzles.

The flow uniformity which was experimentally achieved with the short, circular profile throat nozzles was rather poor, with Mach number variations from $\pm 0.5\%$ to $\pm 2.5\%$ in the supersonic range ($1.22 \leq M \leq 4.38$). The sharp throat nozzles were built for $M = 5.18$ and 8.83 for a 40×40 cm tunnel, but their tests, made with undried and unheated air, were inconclusive.

In the search for a supersonic nozzle design which would allow still smaller length and more convenient interchangeability, development of axisymmetric, multi-nozzle grids was attempted in Germany. Subsequent tests of such nozzles (Royle *et al.*, 1947) indicated large pressure losses and flow non-uniformities.

Paradoxically, the insistence on the shortest nozzle length as a means of prevention of detrimental viscous effects has served, in fact, to amplify these effects and resulted in poor flow uniformity. It was later pointed out (Lukasiewicz and Royle, 1952; Lukasiewicz, 1952) that the short, circular throat profiles, characterized by a discontinuity of curvature at the inflexion, gave rise to an abrupt change in the pressure gradient and, due to the boundary layer interaction, produced strong expansion–compression disturbances.

Although the early, German experience with the short, two-dimensional nozzles was not altogether favourable, nevertheless the development of two-dimensional nozzles for hypersonic tunnels* was continued for some ten years before the axisymmetric designs were adopted.

It was reasoned that the two-dimensional nozzles are required to provide Mach number variation, and that, with adjustable but initially highly flat flexible plates, a highly uniform flow could be obtained in the test section. Additional reasons, such as convenient installation of windows in the flat side walls and, if flexible plates were used, the ease of applying a correction for the boundary layer, were also mentioned.

On the other hand, it was feared that the focusing of disturbances in axisymmetric nozzles may cause flow non-uniformity (although there was no experimental evidence to support these fears).

It became apparent in the mid-1950s that the above arguments reflected the traditional practice of two-dimensional nozzle design rather than a sound engineering judgement. The requirement for Mach number variability in the hypersonic range appeared unwarranted, since the sensitivity of aerodynamic characteristics to Mach number decreases with Mach

* Three-dimensional, rectangular and two-dimensional nozzles were also used in the early shock tunnels.

number increasing. Moreover, a two-dimensional design in general, and the use of flexible plates in particular, posed extremely serious sealing and throat design problems; at high Reynolds numbers, the conventional throat fabrication and cooling techniques were found to be inadequate (Smelt, 1955). From the point of view of flow uniformity, the two-dimensional design suffered from (i) extreme sensitivity of the flow uniformity to the throat region configuration and (ii) a highly non-uniform distribution of the boundary layer between the flat and the curved walls.

The second factor, concerning boundary layer flow in two-dimensional nozzles, was the more fundamental one, in that it could have precluded altogether the possibility of obtaining a highly uniform flow. In two-dimensional hypersonic nozzles, due to the transverse pressure gradients, strong secondary boundary layer flows develop which tend to reduce the boundary layer thickness on the curved walls and increase it towards the centre of the flat walls, as observed experimentally already in 1950 (McLellan *et al.*, 1950). The boundary layer growth on the flat walls causes the flow to converge in planes normal to the flat walls and parallel to the centreline. Inasmuch as the contoured walls are usually over-corrected for their own boundary layer, the flow in the planes parallel to the flat walls diverges.

Since the boundary layer thickness increases with the Mach number, these effects were accentuated in the hypersonic range.

It thus became evident that an axisymmetric design was ideally suited to assure dimensional stability of the throat region and a uniform boundary layer, whereas the two-dimensional one was the least desirable.

The other supposed advantages of the two-dimensional nozzles also appeared questionable. For example, it was doubtful whether sufficient control could be exerted over the deflection of flexible plates to take full advantage of their initial flatness, and it was not certain whether such extreme precision was actually required. As regards flush mounting of the windows, this seemed unnecessary since, at high Mach numbers, the disturbances caused at the test section station would not reach the model. As regards computational procedures, these, although more complex than in the two-dimensional case, were certainly not insurmountable.

The significance of the focusing effects in internal, axisymmetric flow was more difficult to assess, particularly in view of the dearth of experimental data. It was expected that, if such effects were present, they could be reduced to acceptable limits through development of both design and fabrication techniques, just as the perfection of supersonic, two-dimensional nozzles has been attained.

The first axisymmetric nozzles for higher supersonic Mach numbers have been probably developed for the low-density wind tunnels, as part of research initiated in 1947 at the University of California. This was done mainly to reduce the boundary layer effects at the very low Reynolds numbers of tests. Axisymmetric nozzles were used with hot air in the Bofors supersonic blowdown tunnel in Sweden in 1954.

It was at the Princeton University that axisymmetric nozzles were first applied successfully to hypersonic testing at Mach numbers up to about 16 in a helium wind tunnel, operated since 1950 (Bogdonoff and Hammitt, 1954); they were adopted following unsatisfactory flow quality obtained with two-dimensional designs. At the Polytechnic Institute of Brooklyn (Ferri *et al.*, 1955) and at the Ohio State University axisymmetric nozzles were first applied to hypersonic air tunnels. In a paper published in 1955, Ferri *et al.* first presented a cogent argument for the use of hypersonic, axisymmetric nozzles in preference to the two-dimensional ones.

Following the success achieved at PIB and OSU and as a result of the growing realization of the difficulties encountered with the two-dimensional nozzles, many hypersonic tunnels have been equipped, after 1957, with axisymmetric designs. The only notable exception has been the JPL 21-in. hypersonic, continuous tunnel, which employs a flexible-plate, two dimensional nozzle and a slightly divergent (0.5 deg) plane side walls (to compensate for the non-uniformity of the boundary layer thickness) and covers $4 \leq M \leq 10$. The most ambitious two-dimensional nozzle, hypersonic tunnel project ever undertaken was that of a Mach 10, 40-in. square tunnel for the AEDC (Smelt and Sivells, 1957). Although fabricated at great expense, this tunnel was scrapped before installation and replaced by axisymmetric designs covering Mach 6 to 12 range. The two-dimensional design, which included 22-ft long, flexible, stainless steel nozzle plates (tapered in thickness from 0.3 in. near the throat to 2 in. at the test section) and tungsten throat blocks, while exceedingly complex, was limited in performance and was expected to require considerable development work.

Tube-tunnel development

The design of the first, relatively large Ludwieg tube tunnel provides another striking example of pure engineering conservatism. As already mentioned, with a Ludwieg-type drive the working fluid is accelerated from its original, quiescent state by a centered rarefaction wave—a process which must result in a highly uniform, quiet flow at the entrance to the test section. Indeed, it would be difficult to conceive of any means by which the quality (temporal and spatial uniformity) of such a flow could be improved. In particular the settling chamber (with or without flow straightening devices and turbulance screens), a standard component of conventional wind tunnels, could not enhance the flow quality and would be thus superfluous. Yet the high Reynolds number tube tunnels completed in the late 1960s at NASA's Marshall Space Flight Center (the 2.6-in. dia. pilot and the 32-in. dia. test section facility; Davis, 1968; Warmbrod, 1969; Felix, 1971) were both built with settling chambers. In the larger tunnel, provisions have been even made for the use of screens or honeycomb. Moreover, in order to achieve the contraction ratio commonly used in conventional tunnels (10:1 in the 32-in. facility), an area expansion (1:2.25) had to be provided at the tube exit, at the settling chamber entrance. It was argued that the purpose of the settling chamber is "to reduce the effects of the boundary layer, which grows with time along the supply tube, and to slow the flow before it enters the nozzle and test section" (Warmbrod, 1969). While only a moderate area *contraction*, of about 1.5, is necessary to control the unsteady boundary layer effect, the flow deceleration must be considered undesirable. Not surprisingly, a settling chamber is not a part of the more recent (Whitfield *et al.*, 1971) tube tunnel designs.

Hypersonic launchers

A fundamental improvement in the gun performance was first realized by Crozier and Hume (1957) who, in 1947, developed for the U.S. Navy a high-velocity, two-stage, light-gas gun. In the Crozier–Hume gun hydrogen,* the lightest and the most efficient propellant, was substituted for the powder and was heated by means of an adiabatic, free-piston, powder-driven compressor, used as the first stage of the gun. A maximum velocity of 12.33 kft/sec was attained in a 0.39-in. calibre gun with a relatively heavy projectile (3.9 g), and a non-

* Helium was also used.

evacuated barrel, in the free atmosphere; a velocity of 17 kft/sec could have been anticipated with an evacuated barrel.

The potential for aerodynamic testing of Crozier–Hume invention was not appreciated for several years. When in 1949 Crozier's project was transferred to another laboratory, the emphasis changed to the study of equation of state of light gases at high pressure (Rinehart, 1960) and shock-heated guns (Kurzweg, 1959; Slawsky, 1959). It was only in the mid-1950s that further work on the light-gas gun was initiated at NACA Ames Aeronautical Laboratory by Charters *et al.* (1957) and at a number of other establishments.

Although hydrogen, the highest performance propellant, was used successfully in the first light-gas gun, in subsequent development helium and other, inferior, propellant mixtures were employed for many years, before the advantages and practicality of the hydrogen were generally acknowledged in the latter part of the last decade.

The design of guns has been an old engineering art and has abounded, before and after the event of Crozier–Hume invention, in schemes which never led to a performance higher than that attainable with a conventional, two-stage, light-gas gun, developed into a reliable tool for model speeds up to 15 to 35 kft/sec. This failure has been largely due to a lack of understanding of the basic gas-dynamic processes involved in the gun cycle and of their limitations. As an illustration a few augmentation techniques will be mentioned.

(i) In the *distributed charge* designs the propellant, distributed along the barrel length, was sequentially ignited behind the projectile in the course of its travel; in the analogous *distributed energy* designs, the electrical energy was sequentially added. Probably the most ambitious in the *distributed charge* category was the German V3 weapon project (known also as *Hochdruckpumpe* or *Tausendfussler*; Irving, 1965; Kutterer and Pohl, 1960), built toward the end of World War II in Mimoyecques near Calais, France, but never completed. The V3 installation, intended for the bombardment of London, was to consist of fifty 150-mm calibre, 416-ft. long launch tubes, each capable of firing a 300-lb shell every 5 min, giving a total rate of 600 shells per hour. The powder propellant was distributed in breech chambers along the barrels at 8-ft intervals. A muzzle velocity of 5 kft/sec was to have been achieved. After the war, the U.S. General Electric Company experimented with a gun in which a light propellant (lithium hydride) was vaporized behind the projectile through sequential, electrical discharge (Yoler and Cobine, 1956; Anon., 1957; Bengson *et al.*, 1960). Later, experiments with augmentation of a light-gas gun through electrical discharge behind the projectile were made by Volpe (1960) and Howell and Orr (1967). Recently, Rodenberger (1968) proposed a uniformly distributed charge design, in which the barrel is lined with the propellant. All of these distributed charge or energy designs suffered from a common fault: they did not provide a mechanism through which the propellant could expand to the required velocity; not surprisingly, they have all failed to achieve a superior performance.

(ii) The *travelling charge* designs in which some propellant was carried in the base of the projectile, resulted in small and erratic (hence useless for artillery applications) increases in muzzle velocity (Vest, 1951; Vinti, 1952; Baer, 1960). No throat or nozzle was provided to accelerate the propellant gases and maintain combustion pressure, which was progressively decreasing due to the rarefaction accompanying the projectile travel. The *travelling charge* scheme was later extended to the consideration (Kutterer and Pohl, 1960) and use of single-stage and multi-stage, gun-fired, rocket assisted projectiles (RAP: Plattner, 1968; deGruchy, 1969), and some success (at low muzzle velocities) was achieved with single-stage RAPs. In one instance (Kutterer and Pohl, 1960) a two-stage gun design, in which a 10.6-mm calibre gun firing a 3.5-gm projectile was itself fired from a 78-mm calibre gun, was tried, and achieved an unimpressive velocity of 4 kft/sec.

(iii) In the early 1960s, a hypervelocity launcher design utilizing a *spherical implosion* to produce high pressure and temperature propellant was proposed by the University of Toronto Institute for Aerospace Studies, and has been since undergoing development (UTIAS, 1969; Glass, 1970). Theoretical estimates of velocity performance ranged into the 50 to 100 kft/sec regime, whereas actually a maximum velocity of only 11.8 kft/sec has been achieved with an intact projectile (UTIAS, 1969); 17.65 kft/sec was recorded with a distorted projectile (Glass, 1970). In evaluating the applicability of a spherical implosion to gun propulsion, apparently no account was taken of the effects of extremely high pressures on the projectile and gun barrel (resulting in projectile break-up, barrel distortion, erosion and propellant leakage), and of the very short duration of the propulsive pulse—two fundamental aspects of any practical launch cycle.

Magnetic model suspension

Ideally suited to base and wake flow studies, and drag measurements at low densities, was pioneered by ONERA in France in the mid-1950s (Tournier and Laurenceau, 1957); in spite of the then obviously advantageous application to investigations of the ballistic missiles' "signatures" during atmospheric entry and of the drag at high altitudes, this technique was not developed for several years, and has seen only limited exploitation in the 1960s. In fact, it was only in 1969 that the results of fairly extensive drag and wake measurements, made with the Princeton University magnetic suspension and balance system, started to appear in the literature (Peterson and Bogdonoff, 1969; Keel *et al.*, 1971); the first large magnetic suspension system (for a 12-in. dia. Mach 8 wind tunnel at VKF-AEDC) was completed only in 1970 (Matthews *et al.*, 1970).

Excessive performance claims

While conservatism often marked the design of hypersonic test facilities, excessive optimism was quite common as regards their performance.* For example, in 1958, on the basis of calculations, the performance of hotshot tunnels was predicted to extend from 10 kft/sec at 100 kft density altitude to 20 kft/sec at 250 kft (Lukasiewicz, 1959). In the same year, still higher performance was claimed (Stollenwerk and Perry, 1959) when it was stated that ". . . in 1957 the air supply for one of the 'Hotshot' tunnels at AEDC was heated to over 20,000°K by an electric spark, and then the air was expanded through a Laval nozzle to

* Although technology was probably more often underestimated than not, nevertheless there are other instances of overoptimistic hopes. The gunpowder and the thermonuclear fusion provide two prominent examples, both concerned with conversion of technology from military to peaceful use. In 1680 Huygens, the Dutch physicist, hoped to develop an engine driven by controlled gunpowder combustion. This proved impractical, but led instead to the development, in 1690, by Papin, Huygens' assistant, of the expanding–condensing stream cycle (Derry and Williams, 1960). As regards fusion, following the first uncontrolled release of fusion energy in 1952, there were great hopes of achieving soon a controlled nuclear fusion reaction. In 1958, at the Second United Nations Conference on the Peaceful Uses of Atomic Energy in Geneva, "essentially every proposed solution still seemed a good one which merely required adequate development at the technological end" (Furth, 1965). The result of these efforts "was one of the greatest disappointments in the history of science and engineering. The confident predictions of the scientists were proved wrong by sometimes as much as a factor of a million, and all surprises were against the inventor." According to a 1969 poll of experts, laboratory demonstration of controlled fusion is not likely to be achieved before 1980 (Gabor, 1970).

a velocity of 13 km/sec. Similar methods have also been used at the AEDC in an attempt to heat hydrogen or helium to temperatures of this magnitude or greater. A properly designed gun supplied with such a hot and light propellant gas would be expected to be capable of accelerating projectiles to speeds exceeding even the escape velocity."

Four years later, in 1962, the situation was discussed by Cox (1964) who, with the aid of a diagram (Fig. 6) commented that, in 1957, "... we still hoped that it might be possible to achieve 10,000°K in a gun tunnel although we recognized that a limitation would be imposed by heat transfer, particularly by radiation losses. In fact the maximum temperature normally obtained in gun tunnels is around 2000°K, although we are in the process of increasing this at RARDE by preheating the compressor barrel.

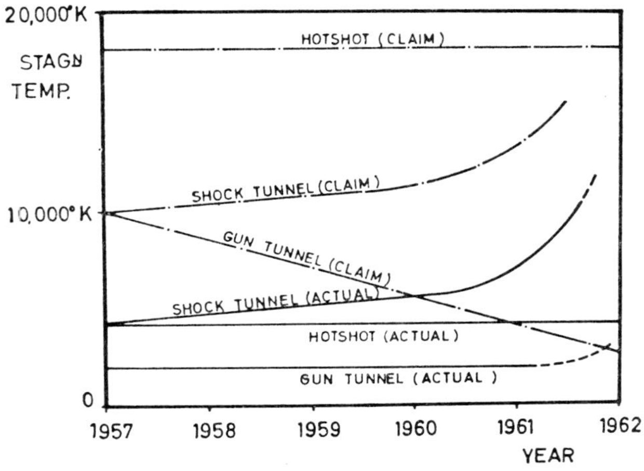

FIG. 6. Claimed and actual temperatures for various facilities during the period 1957–62.

"The various hotshot groups were still talking until recently about temperatures of 18,000°K or thereabouts. It has, however, only proved possible to use the hotshot for serious testing for temperatures up to 4000°K, and we heard yesterday from Dr. Lukasiewicz that even if a higher initial temperature is obtained it is rapidly lost by heat transfer.

"In shock tubes, which have much shorter running times, higher temperatures can be achieved. However, most tunnels operate at temperatures below 5000°K, although there have been a few shock tube tunnels—such as that of Dr. Nagamatsu—in which tests at higher temperatures have been reported. In general the higher the temperature the shorter the running time available in shock tubes."

Dr. Cox's figures of actual performance of gun and shock tunnels, and hotshot tunnels are essentially correct today, after an interval of ten years. As for the electric-arc heated, light-gas gun, this technique has not led to worthwhile performance (Lukasiewicz *et al.*, 1961) and was soon abandoned; the Earth's escape velocity was achieved by the ordinary, light-gas guns.

Even more extreme claims then quoted by Cox were made only five years ago on the basis of experimentation with an electrodeless magnetohydrodynamic accelerator. Seeman *et al.* (1967) estimated from rather meagre measurements that, with the accelerator turned on, and operating with argon at 7.5 mm Hg stagnation pressure, the Mach number at the accelerator exit was increased from 1.5 to 1.8, and a velocity of 7.5 kft/sec was attained, corresponding to a gain of 1.9 kft/sec. Extrapolating these unimpressive and uncertain data, Seeman *et al.* "concluded on the basis of experimental results . . . that future electrodeless MHD accelerator facilities . . . would operate over the altitude–velocity range of 100–350 kft and 14–40 kft/sec. . ." and stated that "In fact . . . [the] major areas of doubt and uncertainty have now been proved and resolved. No problem of a fundamental nature remains. Nor do any other basic problems remain such as might require the use of some unobtainable material, or such as might require the use of techniques that have not yet been developed or applied. The path to a successful re-entry simulator, therefore, is now well defined." These expectations were not fulfilled and the electrodeless accelerator has not seen practical wind tunnel application.

More recently, the development of free piston-driven shock tube and shock tunnel has followed the tradition of high performance expectations. Shock tunnel operation at enthalpies equivalent to 31 kft/sec (Stalker, 1967) and 40 kft/sec (Stalker and Hornung, 1970; Stalker, 1970) has been postulated although only a minimum of test section flow probing was accomplished. In 1970, it was stated that ". . . the stream kinetic energy achieved in the free-piston shock tunnel . . . exceeds that available in other shock tunnels by a factor of 3 or 4. . . . By a considerable factor, it is the fastest wind tunnel available anywhere . . ." (Stalker, 1970).

Clearly, at least in some instances enthusiasm and wishful thinking coupled with instrumentation difficulties which were not appreciated, swayed the objective judgement and led to unsubstantiated and excessive performance claims. At one time or another, probably all groups engaged in hypersonics were guilty in this respect. Perhaps a better approach, in most engineering endeavours, would be boldness and innovation in concept and design, coupled with a conservative evaluation of performance. However, to recommend such an approach may be also to underestimate the exigencies of the political environment which provides support necessary for all new work.

Diffusion of Information

Extensive research would be required to establish the importance and frequency of occurrence of cases in which progress was thwarted due to lack of awareness of developments in the field. Nevertheless, even a cursory review indicates a number of interesting cases of information transfer lag.

The early shock tube history shows that re-invention may be quite common.* The shock tube was invented in France by Vieille who published the results of his experiments in 1899 (Vieille, 1899). He did not provide a theoretical interpretation of his observations. An analysis applicable to shock tube flow was first given in 1910 by Kobes in Austria, and the shock tube equations were first formulated by Schardin in 1932 in Germany. Payman and Shepherd in England, since about 1925, were—like Vieille—concerned with mine safety and studied, in this connection, detonation and shock waves. In 1940, they submitted a paper for publication in the *Proceedings of the Royal Society*, in which they reported a comprehensive experimental investigation of shock tube flow.† Payman and Shepherd were aware of Vieille's contribution, but not of the theory developed by Kobes and Schardin. Neither was G. I. Taylor, whose shock tube theory, derived independently, was verified experimentally by Payman and Shepherd and published by them in 1946.

* For a more detailed account see Emrich (1965) and Griffith (1971).
† Because of World War II, this paper was published only in 1946.

The activity which led eventually to a spectacular development of shock tube technique for investigation of compressible flow and high temperature gas phenomena originated in Princeton in about 1942, independently of the earlier French, Austrian, German and English work. Under the leadership of Bleakney, the Princeton group, concerned with the measurement of blast pressures in air, developed a shock tube for calibration of pressure gauges. The shock tube theory was formulated in Princeton by Taub, in 1942. After these initial developments were accomplished, the English work became known at Princeton through the early reports of Payman and Shepherd. These referred to Vieille's experiments, but did not include Taylor's theory, contained in the 1946 Royal Society publication (Payman and Shepherd, 1946). Kobes' and Schardin's contributions remained unknown to Princeton workers for a number of years.

Summarizing, it is apparent that shock tube was re-invented in Princeton after it was used in France and England; the shock tube theory was independently formulated, on three occasions, in Germany, England and the U.S.A., over a span of some ten years.

In the case of the shock tube, the inefficiency of information transfer had no detrimental effects since the need for application of shock tube to compressible flow studies appeared only later. In other instances, the diffusion of information did not necessarily follow the shortest path. For example, it was a paper by Longwell *et al.* (1958) which suggested to Stalker (1961) the use of a free piston for shock tube drive, rather than the earlier work of Crozier and Hume (1957), whose light-gas gun corresponded, in the limiting case of zero-mass projectile, to the piston driven shock tube.

It should be noted that in general the diffusion of specialized information is often quite slow, as indicated by the following example. Although during the 1950s the shock tube became one of the standard tools of fluid mechanics and physics of high temperature gases, the index to the 1968 edition of *Encyclopedia Britannica* does not list shock tube. Description of the shock tube flow is to be found in an article on "Traffic Waves" (vol. 23, pp. 315–316) where the analogy of shock tube flow to traffic flow is demonstrated.

*Criteria for Decisions on Major Facility Developments**

Over the last thirty years, wind-tunnel facilities have often lagged the needs of flight hardware. During the latter stages of World War II, the Allies had inadequate high-speed, subsonic tunnels, while the Germans lacked high-Mach-number, supersonic tunnels required for support of long-range rocket developments. Transonic and large supersonic wind tunnels became available in the 1950s, after many supersonic aircraft types were already flying. Adequate hypersonic facilities were not ready during the early phases of ICBM development; atmospheric-entry heat transfer data had to be obtained from laboratory-scale shock tubes. More recently, as already mentioned, a large gap has developed between the Reynolds number available in subsonic, transonic and supersonic wind tunnels, and flight hardware (cf. Figs. 3, 4 and 5). Why have we failed to follow the "systems approach" of the Wright Brothers, who developed and instrumented a wind tunnel before undertaking the construction of the first successful aircraft?

Undoubtedly some of the factors already discussed, such as conservative or unsound technical judgement and *ad hoc* approach, contributed to the delay in the provision of aerodynamic test facilities. However, at least in some cases, the insistence of the organiza-

* The considerations presented here may also apply to decision processes in other areas of research and development.

tions responsible for the major aeronautical and aerospace projects on absolute and complete proof—in technical, military, cost-effectiveness, or economic terms—of the necessity to provide a new test capability, has been the major cause. This has been often the requirement in spite of small cost of facility relative to over-all aerospace funding, a cost often comparable to unit aircraft or spacecraft price. The advocates of the "incontrovertible proof" course have not heeded Samuel Johnson's (1709–1784) warning, who wrote that "Nothing will ever be attempted if all possible objections must be first overcome." Obviously, it is entirely preposterous to suggest that the benefits of experimental equipment, such as wind tunnels, required to uncover and investigate new phenomena, could be evaluated *precisely* a priori. Were this the case, such equipment would be unnecessary. At best, a "positive proof" can only come too late, in the shape of a serious performance deficiency, or discovery of technology lag relative to other competing groups.

The "positive proof" approach has not worked in the past and has not resulted in the development of a major, new test capability. Instead, powerful outside pressures were largely responsible for the development after 1945 of some of the large test facilities in the West. The Unitary Wind Tunnel Program of 1949 (see, e.g., Hartman, 1970, p. 150), which was to provide eventually the major United States transonic and supersonic wind tunnels, the development of large hypersonic tunnels in the United States, and the construction of large aerodynamic laboratories in the United Kingdom and France, grew in the wake of German accomplishments in World War II, as evaluated by von Kármán, Dryden, and others, and new appreciation of aircraft and missile potential. The next boost, mostly in the rocket and space tests, came when Soviet Russia provided the ballistic missile and space-flight competition.

A further evidence of the inherent limitations of projecting detailed justifications is provided by the early histories of test facilities. Consider two examples. Transonic wind tunnels, when they became available in the United States did much work on the early ballistic missile nose cones—a task that was not originally envisaged for them. A new, large air-breathing Engine Test Facility (at the Arnold Engineering Development Center, USAF) was quickly converted to high-altitude rocket testing to handle hardware and problems unanticipated at the time it was planned and built* (it reverted recently to its original name and role).

More successful—albeit inadequate—approaches have been those of construction or modification of test facilities for support of specific test projects or with the funds allocated for the development of major aerospace systems. Although these courses of action may result in a relatively fast acquisition of the required test capability and data, they clearly fall short of the optimum timing, the facilities becoming available too late to be of maximum usefulness.†

* The history of the famous system of fortifications built by the French near the German border after World War I furnishes perhaps the ultimate example of the inability of professionals to foresee the uses of technology. Although the tank was originally developed by the British and the French during World War I, its future impact was not appreciated by its inventors. Instead, the French relied on the traditional concepts of stationary warfare and constructed the Maginot (later dubbed "Imaginot") Line which played no role in World War II; eventually local farmers found these fortifications ideal for the growing of mushrooms. Recently, a cooperative Franco-German program was proposed for studies of blast wave propagation using the Maginot Line galleries (Auriol, 1971), and some of the forts have been auctioned by the French government (Kamm, 1971).

† A recent investigation of the procedures current in the United States for acquisition and modernization of major aerodynamic test facilities confirms the deficiencies noted above (Mitchell, 1971; see also Lukasiewicz, 1972, 1972a).

Another contributing factor has been the lack of well integrated, continuing programs of research directed toward attainment of higher or novel aerodynamic performance. Indeed, what support has been provided toward reaching these objectives (usually through small-scale pilot developments) was often applied to specific solutions proposed by individual groups and seldom coordinated within a long-range plan.

Possibilities of Improvements in the Future

While it would be unrealistic to believe that the deficiencies discussed above could be completely eliminated, it is nevertheless suggested that some improvements could be made in the future. The occurrence of overconservative or unsound technical judgement is probably the most difficult aspect to deal with, but its incidence could be reduced through establishment of comprehensive programs of research into development of test facilities. Such coordinated programs should be oriented toward *specific technical objectives* (e.g., attainment of high Reynolds numbers at hypersonic Mach numbers, or of hypersonic velocities) rather than toward particular solutions, and should generate freely disseminated information. This approach would lead to selection of promising concepts for pilot developments and would provide a sound basis for construction of major test facilities.

The decision-making mechanism, which governs the initiation of major programs, is perhaps the most important factor of all since it may cause long delays that cannot be recouped.* Here the recognition of the inherent limitations of our ability to project future needs is essential. Indeed, the modernization of existing facilities or provision of the new ones may be best based on the latest "state of the art" extrapolations of pilot developments rather than on the theoretically desirable, but perhaps unattainable, performance requirements. In principle, such an approach would insure the availability, at any point in time, of the most advanced test equipment that can be developed.

TECHNICAL TRENDS AND FUTURE PROSPECTS

During the past 25 years the development of high speed aerodynamic test facilities exhibited one overriding trend: the achievement of high performance in wind tunnels invariably required the test time to be reduced. Given the practical engineering constraints, the inverse relationship between aerodynamic performance and test duration was to be expected. The attainment of higher Mach and Reynolds numbers, flow velocities and enthalpies, meant increases in test flow energy flux, test section size, stagnation pressure and temperature. In all instances these could be realized through a reduction of test time. Exceedingly large test flow power levels could be achieved for very short periods of time through applications of energy storage, in thermal, chemical, kinetic or electrical form. The attainment of high temperatures was concomitant with the limitation of heat losses, and these could be minimized by operation over a small time interval. Containment of a high-temperature gas at high pressure could be achieved for time periods small enough so that the pressure vessel strength did not deteriorate through wall heating. In the limit, expendable equipment could be used, the pressure pulse being controlled by the inertia of the confining mass.

The trend toward short test times has been apparent with both the conventional and impulse type wind tunnels.

Climbing up the performance ladder, the test times of conventional hypersonic tunnels

---

* It is pertinent to note that the time advantage is a realistic measure of technical superiority.

have been shrinking from intervals measured in hours down to fractions of a minute, as with blow-down, storage heater type installations. The newest, large hypersonic tunnels will use pulse-heated nitrogen, for test runs on the order of 1-sec duration (Zakkay, 1968; Glowacki et al., 1971). The highest Mach and Reynolds numbers to date have been achieved with impulse-type facilities, such as hotshot, longshot and shock tunnels operating in the 1 to 100-msec range.

As regards wind tunnels aimed at attainment of high flow velocities or enthalpies, these have been also running for progressively shorter times. The latest high enthalpy arc-jets rely on sink cooling and are limited to durations of ~ 1 sec (Shepard, 1971), while the high performance shock tubes operate in the microsecond range. Should application of unsteady compression and expansion cycles ever meet with practical success, they will offer exceedingly short test times. The same is true of the techniques involving detonation of explosives and magnetohydrodynamic acceleration.

It is probably correct to state that in general the instrumentation techniques have kept pace with the ever-diminishing run times, and that we should look elsewhere for the lower practical limit to test time duration.* Since one is usually interested in the investigation of flight phenomena under steady, constant velocity and constant ambient conditions, the minimum absolute value of test time must correspond to the time required for the establishment of an essentially steady flow† around the model. Because the model is already immersed in the flow during the starting period or before steady conditions prevail in the test section, the establishment of steady flow around the model reflects the starting history, and the time in which it will achieve the final, steady state is difficult to estimate. Compared to the ideal situation in which the flow is started impulsively from rest, a shorter time for the establishment of the model flow field could be envisaged. In any given case, the minimum required time would also depend on the model size and shape. It is a matter of speculation how closely the minimum has already been approached with some techniques, e.g., in shock tunnels and in the high performance shock tubes. One recent set of shock tunnel results (Holden, 1971) indicates that about 30 and 60 body lengths of flow are required for the pressure and the heat transfer rate, respectively, to stabilize in a separated base flow region. Considering that a 1-ft long model requires 20 μsec to travel its own length when moving at 50 kft/sec, shock tube run durations of 10^{-5} to 10^{-6} sec could already be in the minimum practical range for flight speeds of interest. Also, under these conditions, due to a high rate of heat loss, the shock tube starting time may become the limiting factor.

In cases in which a short test time was not the consequence of augmented aerodynamic performance, test time reduction was nevertheless advantageous from the point of view of operational convenience and test productivity. This has been clearly demonstrated by the history of development of conventional supersonic and hypersonic wind tunnels. Because of drive power considerations the first large supersonic tunnels, developed in Germany in the 1930s and 1940s, were of the intermittent, vacuum-driven type. This technique was, for a time, abandoned in the West, where—continuing the low-speed tunnel tradition—large, continuous transonic, supersonic, and hypersonic facilities were built after 1945. The intermittent wind tunnel concept was revived only in the mid-1950s when it became apparent that much higher Reynolds numbers over a wide Mach number range and at a much lower

---

* These considerations apply to aerodynamic testing under cold-wall conditions ($T_w \ll T_0$); they are not pertinent to testing of materials when subjected to aerodynamic heating, such tests requiring relatively long run times.

† Including both inviscid and viscous, or boundary layer, components.

cost could be realized with large blowdown tunnels. At that time the technique of driving models through a complete angle-of-attack range during a ~15-sec tunnel run was introduced, and was later also used with continuously running, supersonic and hypersonic tunnels. The final step in increasing their productivity came with the development of model injection into the test section. Moreover, this method eliminated the occurrence of model starting loads and associated difficulties, facilitated heat transfer testing, model changes, and operation of the tunnel drive.

Thus, starting with an intermittent supersonic wind tunnel, through continuous and blowdown tunnels equipped with models driven over the angle-of-attack range during one run, and ending with an injected, "intermittent" model, the development of conventional tunnels came full circle.

As has been already noted, the performance and economic gains associated with short running times are being now recognized also at lower speeds. In fact, as mentioned above, the gap existing at transonic speeds between the wind tunnel and flight Reynolds numbers may be bridged by high performance, tube-type wind tunnels, with run times in the 1 to 10-sec range.

With the mounting demand for high performance, test convenience, and productivity, the leisurely pace of wind tunnel experimentation vanished as modern high-speed, supersonic and hypersonic tunnels came into use. In effect, the development of aerodynamic testing, particularly in the hypersonic range, has contradicted the Parkinson's law (Parkinson, 1957). Instead of "work expanding to fill the time available for its completion", in this case work was being accomplished in continually decreasing time intervals.

This trend also proved significant in economic terms. With test times of 100 msec or less, inexpensive hypersonic facilities, within the reach of university laboratories, could be built. This has been particularly true of free piston-driven (including gun tunnels, slow-compression, free piston tunnels, and longshot tunnels)* and Ludwieg-tube-type tunnels. Their cost was a fraction of the cost of conventional and hotshot tunnels of similar aerodynamic performance.

After the initial phase of space flight came to a close with the deployment of highly perfected ballistic missiles and the first landing of man on the Moon, the emphasis has been shifting in the West toward more mundane applications of technology, including environmental problems and ground transportation. As regards hypersonic speeds, the major current effort in the United States concerns the development of recoverable, Earth-to-orbit transports, or the so-called space shuttle (Newbauer, 1971). Other active projects are concerned with the exploration of the solar system with unmanned, instrumented probes (Newbauer, 1970). The interest in hypersonic travel between points on Earth has been marginal. The economic and environmental viability of even a supersonic transport has yet to be demonstrated, and the substitution of superior electronic communications for human mobility may be desirable on our crowded planet.

In any event, the maximum speeds applicable to terrestrial travel by man are likely to be limited by the acceleration level consistent with passenger's comfort. For example, if an acceleration and deceleration of three-tenths of the gravitational acceleration† at sea level were not to be exceeded, a velocity of no more than 14 kft/sec (4.3 km/sec; $M \approx 15$) could

* It may also apply to free piston-driven, high-performance shock tubes, and shock tunnels, undergoing development since the early 1960s.

† Corresponding to acceleration experienced in an automobile which attains a speed of 60 mph (100 km/hr) in about 10 sec from rest. A maximum acceleration of 0.5 g was suggested by Miller (1971) for passenger hypersonic transport.

be attained over a distance of 4000 miles traversed in about 50 min. For the maximum global range (12,500 miles) the corresponding values are 25 kft/sec ($M \approx 25$) and 3 hr.

Hypersonic test facilities and free-flight techniques developed over the past 25 years have been adequate in general for investigation of flight aerodynamics at speeds up to the orbital velocity (26 kft/sec), and, to a lesser extent, the Earth's escape velocity (37 kft/sec). It is thus apparent that the performance of space shuttle and hypersonic transport falls within the range of existing aerodynamic test equipment.*

Higher speeds are of interest in relation to interplanetary, meteroid, and deep-space flights. In considering the future gains in the performance of hypersonic test facilities required to support these developments, we should note that the accuracy of such forecasts has been unimpressive in the past and may not be better today. As discussed above, we may already be reaching the limit of the test time–performance trade-off with some techniques, and further improvements in performance may have to come from more basic advances in the techniques and equipment used. For example, applications of magnetohydrodynamic acceleration and explosive drives may lead to performance gains in terms of *both* velocity and ambient density. However, it is probable that these could be only realized through a significant research and development effort and large-scale experimentation. This latter requirement relates to the predominance of detrimental viscous effects in small-scale apparatus, and the equivalence of run time to physical length in, for example, shock-tube-type devices.

Gains could be also envisaged with the application of rocket propulsion to free-flight ranges; again, large physical scale is a necessity, in this case as a condition for attainment of large velocity.

Considerable resources would have to be allocated to carry out such large-scale developments. In the present ordering of national priorities, at least in the Western countries, this is not to be expected in the near future.

The prospects appear much brighter as regards development of major new subsonic, transonic and supersonic wind tunnels. Intensive design and experimental studies, currently underway in Europe and in the United States, should result in closing of the wind tunnel–flight Reynolds number gap in these speed regimes before the end of this decade.

In Europe, these objectives are pursued by the Large Wind Tunnels (LAWs) Group under the auspices of AGARD's Fluid Dynamics Panel. In the United States, the Aeronautics and Astronautics Coordinating Board recently recommended (Lukasiewicz, 1972) construction of three large national facilities: the NASA Full Scale Subsonic Wind Tunnel, the AEDC High Reynolds Number Tunnel (Ludwieg-tube type) and the AEDC Aero Propulsion Systems Test Facility.

* This is not necessarily true with regard to the propulsion test facilities.

REFERENCES

Note: The following abbreviations are used in the listing of references and in the text:

AEDC	Arnold Engineering Development Center (US Air Force), Tullahoma, Tennessee
AGARD	Advisory Group for Aerospace Research and Development, a division of the North Atlantic Treaty Organization, Paris, France
AIAA	American Institute of Aeronautics and Astronautics, New York
ARC	Aeronautical Research Council, London, England
ARL	Aerospace Research Laboratory, Wright Patterson Air Force Base, USAF, Ohio
AVA	Aerodynamische Versuchsanstalt, Göttingen, Germany
BRL	Ballistics Research Laboratory, U.S. Army, Aberdeen Proving Ground, Maryland
CARDE	Canadian Armament Research and Development Establishment, Valcartier, Que.
HMSO	Her Majesty's Stationery Office, London, England
HVIS	Hypervelocity Impact Symposium

HVTS	Hypervelocity Techniques Symposium
ISTS	International Shock Tube Symposium
NACA	National Advisory Committee for Aeronautics, Washington, D.C.
NASA	National Aeronautics and Space Administration, Washington, D.C.
NOL	U.S. Naval Ordnance Laboratory, White Oaks, Maryland
ONERA	Office National d'Etudes et de Recherches Aeronautiques, Paris, France
PIBAL	Polytechnic Institute of Brooklyn Aerodynamics Laboratory, Brooklyn, N.Y.
RAE	Royal Aircraft Establishment, Farnborough, Hants, England
RARDE	Royal Armament Research and Development Establishment, Fort Halstead, Kent, England
UTIAS	University of Toronto Institute for Aerospace Studies
WADC	Wright Air Development Center, WPAFB
WPAFB	Wright–Patterson Air Force Base, USAF, Ohio

ACKERET, J. (1927) Gasdynamik, *Handbuch der Physik*, H. Geiger and K. Scheel, eds., vol. 7, ch. 5, pp. 289–342, Springer, Berlin.

ACKERET, J. (1935) High speed wind tunnels, *Proceedings of the Fifth Volta Congress*, Italy; NACA-TM-808, 1936.

AGARD (1971) *Report of the High Reynolds Number Wind Tunnel Study Group of the Fluid Dynamics Panel*, AGARD Advisory Report No. 35, pp. 17; also: AGARD study of high Reynolds number wind tunnel requirements for the North Atlantic Treaty Organization nations, *Facilities and Techniques for Aerodynamic Testing at Transonic Speeds and High Reynolds Number*, AGARD Conference Pre-print No. 83, pp. 32–1 to 32–9, 1971.

ANON. (1936) *Le alte velocità in aviatione*, Convegno 30 Settembre-6 Ottobre 1935-XIII, Reale Accademia d'Italia, Fondatione Alessandro Volta, Rome, 694 pp.

ANON. (1957) Hypersonic rubs rough edges off space flight, *Jet Propulsion*, vol. 27, no. 5, pp. 548–550.

AURIOL, A. (1971) *The Experience of International Cooperation at the ISL in the Field of Specific Aerodynamic Facilities*, Institut Franco-Allemand de Recherches de Saint-Louis, Rapport 4/71.

BAER, P. G. (1960) The travelling charge gun as a hypervelocity launching device, 4 *HVIS*, Eglin A.F. Base, Florida.

BENGSON, M. H., SLAWECKI, T. K. and WILLIG, F. J. (1960) A multistage hypervelocity projector, 4 *HVIS*, Eglin A.F. Base, Florida.

BOGDONOFF, S. M. and HAMMITT, A. G. (1954) *The Princeton Helium Hypersonic Tunnel and Preliminary Results above* $M = 11$, Rept. No. 260, The James Forrestal Research Center, Princeton University.

BUSEMANN, A. (1931) Gasdynamik, *Handbuch der Experimentalphysik*, W. Wien and F. Harms, eds., vol. 4, pp. 343–453, Akademische Verlagsgesellschaft, Leipzig.

BUSEMANN, A. (1971) Compressible flow in the thirties, *Annual Review of Fluid Mechanics*, vol. 3, pp. 1–12, Annual Reviews, Inc., Palo Alto, California.

CHARTERS, A. C., DENARDO, B. P. and ROSSOW, V. J. (1957) *Development of a Piston–Compressor Type Light-gas Gun for the Launching of Free-flight Models at High Velocity*, NACA-TN-4143.

COX, R. N. (1964) Introduction, An assessment of our present status and future requirements for high temperature hypersonic facilities, *The High Temperature Aspects of Hypersonic Flow*, ed. W. C. Nelson, Pergamon Press, p. 745.

COX, R. N. and WINTER, D. F. T. (1957) *The Light Gas Hypersonic Gun Tunnel at ARDE, Fort Halstead, Kent*, AGARD Report No. 139, July 1957.

CROZIER, W. D. and HUME, W. (1957) High-velocity, light-gas gun, *Journal of Applied Physics*, vol. 28, no. 8, pp. 892–894.

DAVIS, J. W. (1968) A high Reynolds number wind tunnel and its operating concept, *Journal of Spacecraft and Rockets*, vol. 5, no. 10, pp. 1225–1227.

DEGRUCHY, D. C. (1969) Rocket-assisted projectiles promise increased artillery range, payload, *Space/Aeronautics*, vol. 51, May 1969, pp. 90–92.

DERRY, T. K. and WILLIAMS, T. I. (1960) *A Short History of Technology*, Oxford, Clarendon Press, pp. 314–315.

EMRICH, R. J. (1965) Early development of the shock tube and its role in current research, 5 *ISTS*, U.S.N.O.L., White Oak, Maryland.

EVANS, J. Y. G. (1971) *A Scheme for a Quiet Transonic Flow suitable for Model Testing at High Reynolds Number*, RAE Tech. Rept. 71, 112. Abstract given in *Facilities and Techniques for Aerodynamics Testing at Transonic Speeds and High Reynolds Number*, AGARD-CP-83-71, pp. 35–1 to 35–5.

FALK, T. J. (1963) *An SST Aerodynamic Test Facility using Short-duration Flow Techniques*, Cornell Aeronautical Laboratory, Inc., SST Memo No. 133, 19 November 1963; also FALK, T. J., *A Tube Wind Tunnel for High Reynolds Number Supersonic Testing*, ARL-68-0031, February 1968.

FELIX, A. R. (1971) MSFC high Reynolds number tube tunnel, in *Facilities and Techniques for Aerodynamic Testing at Transonic Speeds and High Reynolds Number*, AGARD-CP-83-71, pp. 30–1 to 30–10.

FERRI, A., LIBBY, P., BLOOM, M. and ZAKKAY, V. (1955) *Development of the Polytechnic Institute of Brooklyn Hypersonic Facility*, WADC TN-55-695.

FURTH, H. P. (1965) The status of world fusion, *Nucleonics*, vol. 23, no. 12, pp. 64–69.

GABOR, D. (1970) *Innovations: Scientific, Technological and Social*, Oxford University Press, 113 pp.

GLASS, I. I. (1970) *Appraisal of UTIAS Implosion-driven Hypervelocity Launchers and Shock Tubes*, UTIAS Review No. 31.

GLOWACKI, W. J., HARRIS, E. L., LOBB, R. K. and SCHLESINGER, M. I. (1971) *The NOL Hypervelocity Wind Tunnel*, AIAA Paper No. 71-253.

GREIF, R. and BRYSON, A. E. (1965) Measurements in a free piston shock tube, *AIAA J.*, vol. 3, no. 1, p. 183.

GRIFFITH, W. C. (1971) Research with the shock tube: a case history of science relevant to the needs of society, 8 *ISTS*, Imperial College of Science and Technology, London.

HARTMAN, E. P. (1970) *Adventures in Research: a History of Ames Research Center* 1940–1965, NASA-SP-4302, 555 pp.

HENSHALL, B. D., TENG, R. N. and WOOD, A. D. (1962) A driver-sphere technique for increasing the steady state test time of hypersonic shock tunnels, *Advances in Hypervelocity Techniques*, 2 HVTS, University of Denver, Colorado, Plenum Press, pp. 453–481.

HERTZBERG, A., WITTLIFF, C. D. and HALL, J. G. (1962) Development of the shock tunnel and its application to hypersonic flight, *Hypersonic Flow Research*, F. R. Riddell, ed., Academic Press, pp. 701–758.

HOLDEN, M. W. (1971) Establishment time for laminar-separated flows, *AIAA J.*, vol. 9, no. 11, pp. 2296–2298.

HOTTNER, T. (1968) *Der Rohrwindkanal der Aerodynamischen Versuchsanstalt Göttingen*, Report 68A77, AVA, Göttingen.

HOWELL, W. G. and ORR, W. R. (1967) Results of developmental research on an augmentation technique for a light gas gun, 5 *HVTS*, vol. 2, pp. 103–157.

IRVING, D. (1965) *The Mare's Nest*, Little, Brown & Co., Boston and Toronto.

KAMM, H. (1971) The fortresses of the Maginot Line fall to the highest bidders, *The New York Times*, 19 November 1971.

KEEL, A. G., KRAIGE, L. G., PASSMORE, R. D. and ZAPATA, R. N. (1971) *Hypersonic Low Density Cone Drag*, AIAA Paper No. 71-133.

KOBES, K. (1910) Die durchschlagsgeschwindigkeit bei den Luftsauge und Drucklift-bremsen, *Zeitschrift des Osterreichischen Ingenieur und Architekten-Vereines*, vol. 62, p. 558, Vienna.

KURZWEG, H. H. (1959) Special ballistic ranges and gas guns, *Selected Topics in Ballistics*, W. C. Nelson, ed., Pergamon Press, pp. 183–199.

KUTTERER, R. E. and POHL, J. (1960) Contribution pour atteindre des vitesses absolues élevées avec des maquettes, *Proceedings of the Tenth AGARD General Assembly*, Istanbul, Turkey, pp. 61–72.

LEUCHTER, O. (1964) *Slow Compression Heater for Intermittent Hypersonic Wind Tunnels reaching Mach* 15, ONERA T.P. 171.

LONGWELL, P. A., REAMER, H. H., WILBURN, N. P. and SAGE, B. H. (1958) Ballistic piston for investigating gas phase reactions, *Industrial and Engineering Chemistry*, vol. 50, p. 603.

LUDWIEG, H. (1955) Der Rohrwindkanal, *Z. für Flugwissenschaften*, vol. 3, no. 7, pp. 206–216.

LUKASIEWICZ, J. (1950) *Shock Tube Theory and Applications*, National Research Council Rept. MT-10, Ottawa, Canada. Also Rept. 15, National Aeronautical Establishment, Ottawa, Canada, 1952.

LUKASIEWICZ, J. (1952) *Design and Calibration Tests of a 5.5-inch square Supersonic Wind Tunnel*, R&M No. 2745 (13,425), ARC Tech. Rept. HMSO, London.

LUKASIEWICZ, J. (1955) Development of large, intermittent wind tunnels, *J. Roy. Aero. Soc.*, pp. 259–278, April 1955.

LUKASIEWICZ, J. (1959) Experimental investigation of hypervelocity flight, *Advances in Aeronautical Sciences*, von Kármán, ed., Pergamon Press, vol. 1, pp. 127–186.

LUKASIEWICZ, J. (1961) Effects of boundary layer and geometry on characteristics of perforated walls for transonic wind tunnels, *Aerospace Engineering*, vol. 20, no. 4, pp. 22–23, 62–68.

LUKASIEWICZ, J. (1970) Atmospheric entry test facilities: basic limitations and proposal for a new technique, *J. Spacecraft & Rockets*, vol. 7, no. 6, pp. 741–747.

LUKASIEWICZ, J. (1971) The need for developing a high Reynolds Number transonic wind tunnel in the U.S., *Astronautics & Aeronautics*, vol. 9, no. 4, pp. 64–70.

LUKASIEWICZ, J., ed. (1972) *Aerodynamic Test Simulation: Lessons from the Past and Future Prospects*, AGARD Rept. (in press).

LUKASIEWICZ, J. (1972a) Toward a viable system for superior test capability, *Astronautics & Aeronautics*, vol. 10, no. 8, pp. 18–20.

LUKASIEWICZ, J. and PRUDEN, F. W. (1950) *An Economical High-speed Wind Tunnel of High Performance*, National Research Council (internal lab. memo), Ottawa, 22 September 1950.

LUKASIEWICZ, J. and ROYLE, J. K. (1952) *Boundary Layer and Wake Investigation in Supersonic Flow*, R&M No. 2613 (12,130), ARC Tech. Rept. HMSO, London.

LUKASIEWICZ, J., STEPHENSON, W. B., CLEMENS, P. L. and ANDERSON, D. E. (1961) *Development of Hypervelocity Range Techniques at Arnold Engineering Development Center*, AEDC-TR-61-9; also CARDE Tech. Memo Q-646/61.

MATTHEWS, R. K., BROWN, M. D. and LANGFORD, J. M. (1970) *Description and Initial Operation of the AEDC Magnetic Model Suspension Facility—Hypersonic Wind Tunnel E*, AEDC-TR-70-80.

MCLELLAN, C. H., WILLIAMS, T. W. and BECKWITH, I. E. (1950) *Investigation of the Flow through a Single-stage, Two-dimensional Nozzle in the Langley* 11-*inch Hypersonic Tunnel*, NACA-TN-2223.

MILLER, R. H. (1971) Thinking "hypersonic", *Astronautics & Aeronautics*, vol. 9, no. 8, pp. 40–44.

MITCHELL, J. G. (1971) *The Test Facility's Role in the Effective Development of Aerospace Systems*, U.S. Air Force Systems Command Rept. AFSC-TR-71-01.

NEWBAUER, J., ed. (1970) Planetary navigation—the new challenges, Special Section, *Astronautics & Aeronautics*, vol. 8, no. 5, pp. 26–70.

NEWBAUER, J., ed. (1971) Space shuttle, Special A&A Report, *Astronautics & Aeronautics*, vol. 9, no. 2, pp. 22–67.

OHMAN, L. H. and BROWN, D. (1971) *The NAE High Reynolds Number* 15″ × 60″ *Two-dimensional Test Facility; Description, Operating Experiences and Some Representative Results*, AIAA Paper No. 71-293.

OWEN, P. R. (1945) *Note on the Apparatus and Work of the WVA Supersonic Institute at Kochel, S. Germany*, RAE Tech. Notes Aero 1711, 1712.

PARKINSON, C. N. (1957) *Parkinson's Law*, Houghton-Mifflin Co., Boston, 113 pp.

PAYMAN, W. and SHEPHERD, F. (1946) Explosion waves and shock waves. VI. The disturbance produced by bursting diaphragms with compressed air. *Proc. Royal Society of London*, Series A, vol. 186, pp. 243–321.

PERRY, R. W. (1964) The "longshot" type of high Reynolds number hypersonic tunnel, 3 *HVTS*, pp. 395–422, University of Denver, Colorado.

PETERSON, C. W. and BOGDONOFF, S. M. (1969) *An Experimental Study of Laminar Hypersonic Blunt Cone Wakes*, AIAA Paper 69-714.

PLATTNER, C. M. (1968) Liquid-filled cases stretch rocket range, *Aviation Week and Space Technology*, March 25, 1968, pp. 57–61.

RICHARDS, D. E. and ENKENHUS, K. R. (1969) *Hypersonic Testing in the VKI Longshot Free-piston Tunnel*, AIAA Paper No. 69–333.

RINEHART, J. S. (1960) Some historical highlights of hypervelocity research, 1 *HVTS*, pp. 4–11, University of Denver, Colorado.

RODENBERGER, C. A. (1968) *The Feasibility of Obtaining Hypervelocity Acceleration using Propellant Lined Launch Tubes*, Texas A&M University, College Station, Texas.

ROYLE, J. K., BOWLING, A. G. and LUKASIEWICZ, J. (1947) *Calibration of Two-dimensional and Conical Supersonic Multi-nozzles*, RAE Rept. No. AERO 2221, SD23.

SCHARDIN, H. (1932) Bemerkungen zum Druckausgleichsvorgang in einer Rohrleitung, *Phys. Zeits*., vol. 33, p. 60.

SEEMAN, G. R., THORNTON, J. A. and PENFOLD, A. S. (1967) Development of electrodeless MHD accelerator technology, 5 *HVTS*, University of Denver, Colorado, pp. 374–410.

SEIFF, A. (1955) *A Free-flight Wind Tunnel for Aerodynamic Testing at Hypersonic Speeds*, NACA-TR-1222.

SHEPARD, C. E. (1971) *Advanced High-power Arc Heaters for Simulating Entries into the Atmospheres of the Outer Planets*, AIAA Paper No. 71–263.

SLAWSKY, Z. I. (1959) Survey of NOL hyperballistic research, *NOL Aeroballistic Research Facilities Dedication and Decennial*, NOLR 1238, pp. 106–123.

SMELT, R. (1955) Test facilities for ultra-high-speed aerodynamics, *Proceedings of the Conference on High-Speed Aeronautics*, PIB.

SMELT, R. and SIVELLS, J. C. (1957) *Design and Operation of Hypersonic Wind Tunnels*, AGARD Rept. 135.

STALKER, R. J. (1961) *An Investigation of Free-piston Compression of Shock Tube Driver Gas*, National Research Council, Ottawa, Canada, Rept. MT-44.

STALKER, R. J. (1967) A study of the free-piston shock tunnel, *AIAA J*., vol. 5, no. 12, pp. 2160–2165.

STALKER, R. J. (1970) Shock tube developments at the Australian National University, *The Australian Physicist*, vol. 7, no. 7, pp. 99–102.

STALKER, R. J. and HORNUNG, H. G. (1970) Two developments with free-piston drivers, 7 *ISTS*, pp. 242–258, University of Toronto.

STOLLENWERK, E. J. and PERRY, R. W. (1959) Preliminary planning for a hypervelocity aeroballistic range at AEDC, *Selected Topics in Ballistics*, W. C. Nelson, ed., Pergamon Press, pp. 200–212.

TOURNIER, M. and LAURENCEAU, P. (1957) Suspension magnetique d'une maquette en soufflerie, *La Recherche Aeronautique*, no. 59, pp. 21–26.

TRIMPI, R. L. (1962) *A Preliminary Theoretical Study of the Expansion Tube, a New Device for Producing High-enthalpy Short-duration Hypersonic Gas Flow*, NASA-TR-R-133.

TUPPER, K. F., DILWORTH, P. B. and JENKINS, L. A. (1961) *The N.A.E. Five-foot Supersonic Wind Tunnel*, Engineering Institute of Canada, 1961 Annual General Meeting, Paper No. 40.

UTIAS (1969) *Annual Progress Report* 1969, Institute for Aerospace Studies, University of Toronto.

VEST, D. C. (1951) *An Experimental Travelling Charge Gun*, BRL Rept. 773.

VIEILLE, P. (1899) Sur les discontinuités produites par la détente brusque de gas comprimés, *Comptes Rendus*, vol. 129, pp. 1228–1229, Paris; reproduced in 7 *ISTS*, pp. 9–10, with English translation, pp. 6–8, University of Toronto.

VINTI, J. P. (1952) *Theory of the Rapid Burning of Propellants*, BRL Rept. 841.

VOLPE, V. F. (1960) An experimental evaluation of a sequential electrical discharge light gas gun, 1 *HVTS*, pp. 12–15, University of Denver, Colorado.

WARMBROD, J. D. (1969) *A Theoretical and Experimental Study of Unsteady Flow Processes in a Ludwieg-tube Tunnel*, NASA-TN-D-5469.

WHITFIELD, J. D., SCHUELER, C. J. and STARR, R. F. (1971) High Reynolds number transonic wind tunnels—blowdown or Ludwieg tube? *Facilities and Techniques for Aerodynamic Testing at Transonic Speeds and High Reynolds Number*, AGARD-CP-83-71, pp. 29–1 to 29–17.

YOLER, Y. A. and COBINE, J. D. (1956) U.S. Patent No. 2790354.

ZAKKAY, V. (1968) *The Design and Operation of a High Mach Number–Reynolds Number Facility*, ARL-68-0213.

2

MAGNETIC BALANCE AND SUSPENSION SYSTEMS FOR USE WITH WIND TUNNELS

EUGENE E. COVERT and MORTON FINSTON

Massachusetts Institute of Technology

and

MILAN VLAJINAC* and TIMOTHY STEPHENS*

M.I.T. Aerophysics Laboratory

I. INTRODUCTION

The idea of a magnetic suspension system has a certain appeal in its own right. In seeming to overcome the so-called law of gravity, such a suspension is a source of amusement.† However, magnetic suspensions are practical for use as bearings of gyroscopes, as plasma containers in proposed nuclear fusion electric power generators, as supports for high-speed ground transportation systems, and as interference-free model support systems for use with wind tunnels. In the latter application merely suspending an object magnetically is inadequate. This application requires that information be generated and made available in useful form at the same time. Hence the idea of suspension is generalized to include this additional aspect where necessary.

Only the application of magnetic suspension systems as a wind tunnel instrument will be considered in this review. Certain elements of magnetic suspension systems may have additional uses in wind tunnel testing. These other uses will be discussed in Section V.

Even in this specialized case the suspension may or may not be combined with direct measurement. When the suspension has the latter function the term magnetic balance will be used. Here the word balance is taken in the sense of a chemical analytical balance. An unknown but measurable quantity is balanced against something known. In the case of a wind tunnel instrument, unknown aerodynamic forces and torques applied to a model by the relative velocity of the wind in the wind tunnel are balanced against gravitational and inertial forces and torques which can be known, and by magnetic forces and torques which are known in terms of electric currents. The currents are measurable, so the aerodynamic forces and torques may be determined within the limits of experimental accuracy. Whether the system under discussion is either a suspension or a balance system the problem of determining the magnitude and direction of the unknown forces and torques must be solved either implicitly or explicitly.

* Now at Lincoln Laboratory, M.I.T.

† To the ancients suspension by means of invisible magnetic field lines was thought to imply omnipotence.[1]

Assessment of advantages and disadvantages is difficult, and is subjective at best. However, it is virtually impossible to make steady-state interference-free measurements of wake flow field without a magnetic system. The higher accuracy of this type of balance is compatible with the more recent requirements in applied aerodynamics. Further, magnetic suspension is attractive from the standpoint of investigating more subtle aerodynamic details and as a means of improving techniques for studying vehicle stability. These reasons seem sufficient to justify further development of magnetic suspension and balance systems.

Elementary Magnetic Concepts

Only the most superficial description of a magnetic suspension or balance system does not rest upon ideas from the theory of magnetism. Consequently, some ideas from classical magnetostatics will be presented from a phenomenological viewpoint. (A viewpoint, incidentally, whose basis is justified to a greater or lesser extent by modern atomic theories.) It is convenient to present a discussion of magnetization first. Then the interaction of the magnetized body with different classes of applied fields leading to torques and forces will be discussed. Lastly, the problem of controlling the components independently is presented and several possible solutions discussed.

In a general sense, the state of any body is changed by immersing it in a magnetic field. After the body has been so immersed, it is said to be magnetized. The level of magnetization depends upon the shape of the body in comparison with the magnetic field lines, the history and strength of the magnetic field in which the body has been immersed, and the constitutive properties of the material from which the body was fabricated.

Since magnetization is a term used to describe the distribution of magnetic field inside the body, it is a vector. The magnetization vector is denoted by $\bar{M}$. $\bar{M}$ can be shown to be constant if the magnetic permeability, μ, is constant, the applied field is uniform, and the body is ellipsoidal in shape. If all these conditions are not satisfied $\bar{M}$ will be defined by a suitable integral over the body, or from measurements of phenomena in which these integrals appear. If a body is placed in a region where there is no initial magnetic field and a weak field is then applied to the body, the magnetization will be linearly related to the strength of the applied field. For strong applied fields the linear relationship ceases to be accurate. In fact the magnetization tends to become constant. In this case the body is said to be magnetically saturated. Some part of the magnetization may remain after the field is removed. Under these circumstances the body is said to be permanently magnetized. Throughout the remaining discussion, the magnetization for a permanently magnetized body will be assumed to be known. The behavior of a body in the linear range of applied field strength will be discussed.

If the inner field of a magnetized body is less than the applied field, the body is said to be diamagnetic. (That is, the diamagnetic body appears to exclude magnetic field lines.) If the inner field is proportional to the applied magnetic field and is greater than the applied field, the body is said to be paramagnetic. (In this case the applied field appears to be attracted into the body.) In the special case where the inner field depends non-linearly upon the applied field and is non-zero after the applied field is reduced to zero, the material is said to be ferro-magnetic.

The inner field tends to reduce the effectiveness of the applied field in changing the magnetization level. This effect is easy to imagine if one thinks of the body as being made up of many little magnets that are more or less free to rotate. These magnets line up N–S, N–S in response to a S–N applied field. Their very alignment tends to oppose the tendency to increase the inner field. This effect can be estimated for diamagnetic, paramagnetic and ferro-magnetic materials if they are weakly magnetized. That is, the magnetization is proportional to the applied field over some range of applied field strengths. The constant of proportionality is the magnetic permeability, μ. The relation between the magnetization and the applied field strength may be written, using the summation convention,

$$M_i = \mu H_i - D_{ij}(\mu - \mu_0) M_j. \tag{1}$$

That is to say the magnetization vector with components M_i is proportional to the applied field (μH_i) less an amount proportional to the level of demagnetization [$D_{ij}(\mu - \mu_0)M_j$]. The term ($\mu - \mu_0$) is necessary to differentiate between diamagnetic materials and non-diamagnetic materials. μ is less than the free space value of permeability, μ_0, in the diamagnetic case, and greater in the others. D_{ij} is the so-called demagnetization factor, which is purely dependent upon geometry, i.e. the shape of the magnetized body, and the slope and curvature of the field lines. D_{ij} represents the reduction in magnetization in the ith direction due to magnetization in the jth direction. In other words, the direction of magnetization can be different from the direction of the applied magnetic field, so the demagnetization factor is generally a tensor quantity. Equation (1) can be solved for M_j to give, if I is the idem tensor,

$$M_j = [I + (\mu - \mu_0) D_{ij}]^{-1} \mu H_i. \tag{2}$$

For a ferro-magnetic body $\mu \gg \mu_0$, $\mu D_{ij} \gg 1$, one finds that

$$M_j = (H_i/D_{ij}) + 0\left(\frac{\mu_0}{\mu} + \frac{1}{\mu D_{ij}}\right). \tag{3}$$

This approximation implies that the magnetization level in the body is essentially independent of μ in the linear range of magnetization. Thus to a first order in the magnetization level the body is not temperature sensitive. Strictly speaking, eqs. (1) to (3) are based upon the concept of a body immersed in a uniform field. They will be applied to more general situations, in which the demagnetization factors will be determined by measurement.

In the application of these ideas to wind tunnel problems, the principal magnetic axes will be initially aligned with the balance axes. x is against the wind, z is down and y completes a right-handed triad. The diagonal demagnetization factors will be denoted D_A, D_B and D_C in the x, y and z directions. Unlike the principal inertia factors the demagnetization factors are not independent, but rather are related by the requirement that

$$D_A + D_B + D_C = 1. \tag{4}$$

Thus for a body for which the x axis is the axis of rotational symmetry $D_B = D_C = (1 - D_A)/2$. For a body possessing spherical symmetry $D_A = D_B = D_C = 1/3$. The notation D_A and so forth will be used during the rest of this article. In the general case in which the magnetic field is not aligned with the principal magnetic axis, it can be transformed by the rotation matrix, $[R]$, to align its components with the magnetic axis. Then the magnetization will be computed using eq. (3). The resultant magnetization vector will then be rotated backward to the initial axis (usually the balance axis). Formally this results in

$$\begin{pmatrix} M_x \\ M_y \\ M_z \end{pmatrix} = [R]^{-1} \cdot \begin{pmatrix} \frac{1}{D_A} & 0 & 0 \\ 0 & \frac{1}{D_B} & 0 \\ 0 & 0 & \frac{1}{D_C} \end{pmatrix} \cdot [R] \cdot \begin{pmatrix} H_x \\ H_y \\ H_z \end{pmatrix} \tag{5}$$

The magnetization is dependent upon the volume of magnetic material. Thus in eq. (5) the magnetization is really dM_i and this infinitesimal is to be integrated over the volume filled with magnetic material. So the definition of D_A can be written

$$D_A = \frac{\mu \int H_x d\,\mathrm{Vol} - \int dM_x}{(\mu - \mu_0) \int dM_x} \tag{6}$$

which is a valid definition for a variety of general situations. Note unless otherwise stated that

$$M_x = \int dM_x, \text{ etc.} \tag{7}$$

Generation of Forces and Torques

The relation between the magnetization ($\bar{M}$) and the applied field ($\bar{B}$) and the forces and torques acting on the magnetized body may be deduced in a simple way from the concept of magnetic poles. That is, let the magnetized body be represented by a body of length L having a north magnetic pole of strength P at one end, and a south magnetic pole of equal strength at the other. Between these two poles the magnetization will be assumed to be zero.* Experimental evidence suggests like poles repel and unlike poles attract. Hence if this magnetized body is inserted into a uniform magnetic field, it will tend to rotate until its north pole points at the "south pole" of the applied field. Similarly, the south pole of the magnetized body points toward the "north pole" of the applied force. When the magnetized body is rotated through an angle θ each pole is displaced from the equilibrium position a distance $(L/2) \sin \theta$, one end being displaced in one direction and the other end in the opposite direction. These two poles no longer act along a line of force of $\bar{B}$ field line. This causes a torque but no net force, since the two forces ($\pm PB$) cancel each other out. If $\bar{M} = PL$, the torque is

$$T = M \cdot B \sin \theta. \tag{8}$$

This is the form of the vector cross-product. As long as $\bar{B}$ is constant and uniform the relation can be written

$$\bar{T} = \bar{M} \times \bar{B}. \tag{9}$$

When $\bar{B}$ is non-uniform, but steady, this result may be written (cf. eq. (7))

$$\bar{T} = \int (d\bar{M} \times \bar{B}). \tag{10}$$

The integration is over the volume filled with magnetized material.

The only way the elemental magnetized body can experience a net force of any kind is when the applied field $\bar{B}$ is non-uniform. The magnetized body is attracted in the direction of the stronger field, whether this part of the field is along the axis connecting the magnetic poles, or is transverse to this axis. In a cartesian coordinate system the relation for the force takes the form

$$F = \int (dM \cdot \nabla)\bar{B}. \tag{11}$$

* Some readers may prefer to derive the same results from the concept of magnetizing current. The results are expressed in terms of volume integrals of the Faraday force $\bar{J}_M \times \bar{B}$ and the torque resulting from this volume force.

As before, the integral is taken over the volume of magnetized material. The vector operator under the integral sign is made unique by taking the gradient of $\bar{B}$ first, then premultiplying by $d\bar{M}$. Note the divergence of $\bar{B}$ and $\bar{M}$ are each identically zero since no currents are flowing in the region of the model.* Equations (10) and (11) provide a means of estimating the magnetic forces and torques generated through the interaction of the magnetized body and the applied field. Thus a necessary condition for magnetic suspension has been presented. To demonstrate a sufficient condition it will be shown that the individual components of $\bar{T}$ and $\bar{F}$ can be developed arbitrarily. In other words, the magnitude and direction of both the force and torque can take arbitrary values. It is a matter of little importance whether $\bar{M}$ is due to a permanent magnet or is induced by a magnetizing field. The result is the same. Consider first the former. From eq. (9) in component form $T_x = M_y B_z - M_z B_y$; $T_y = M_z B_x - M_x B_z$; and $T_z = M_x B_y - M_y B_x$ (9a); if $\bar{M} = M_x i$, then T_x is identically zero, and $T_y = -M_x B_z$, $T_z = M_x B_y$. In other words if the magnetization vector lies along the x or wind axis, a pitching moment or torque is developed by a vertical or $z\bar{B}$ field and a yawing torque is developed by a lateral or $y\bar{B}$ field. These two moments can be developed independently as required. Consider next the case of induced magnetization. For an axially symmetric body magnetized in the linear range, $M_y = H_y/D_B$ and $M_z = H_z/D_B$. By substituting these values for the magnetization into the expression for T_x, it can be seen that T_x is again identically zero. Hence if the body or its magnetic core is axially symmetric, and if the body is magnetized along the axis of symmetry, the torque about that axis is always zero. Note $\bar{T} \equiv 0$ for a sphere if hysteresis effects are negligible. Methods of developing only T_x will be explained after the problem of developing force components is discussed.

By considering the uniform magnetization vector $\bar{M}$ one can compute the axial force

$$F_x = M_x \frac{\partial B_x}{\partial x} + M_y \frac{\partial B_x}{\partial y} + M_z \frac{\partial B_x}{\partial z} \tag{12a}$$

or from use of the fact curl $\bar{B} = 0$,

$$F_x = M_x \frac{\partial B_x}{\partial x} + M_y \frac{\partial B_y}{\partial x} + M_z \frac{\partial B_z}{\partial x}. \tag{12b}$$

By assuming the magnetization lies along the wind direction and similarly

$$F_x = M_x \frac{\partial B_x}{\partial x} \tag{13}$$

and similarly

$$F_y = M_x \frac{\partial B_x}{\partial y}; \quad F_z = M_x \frac{\partial B_x}{\partial z}.$$

Clearly independent control of the gradient of the magnetic field along and normal to the magnetization allows the force components to be controlled independently, no matter what the direction of $\bar{M}$. Parker[2] has proved that eq. (11) allows one to develop orthogonal force components in a variety of ways, and that a clear optimum exists. That is a maximum force can be produced for a minimum power consumption by well-defined coil configurations.

The integral forms for the force and torque eqs. (10) and (11) can be used to show that an

* But curl $\bar{M} = \bar{J}_M$. Here $\bar{J}_M$ is the so-called magnetizing current. It produces the same magnetic field outside the body as **M**.

axially symmetric magnetized body or core possessing fore and aft symmetry has no coupling between the force and torques. The applied field can be expanded by Taylor's theorem to give, for each component of $\bar{B}$,

$$B_x(x, \ldots) = B_x(0, \ldots) + \left.\frac{\partial B_x}{\partial x}\right|_{x=0} x + \left.\frac{\partial^2 B_x}{\partial x^2}\right|_{x=0} \frac{x^2}{2!} + \ldots \tag{14}$$

Substituting into eq. (10), and integrating, and taking account of the assumed symmetry gives the torque

$$\mathbf{T} = \int d\mathbf{M} x\, \hat{\imath} B_x(0, \ldots) + \int d\mathbf{M} x\, \hat{\imath} \left.\frac{\partial^2 B_x}{\partial x^2}\right|_{x=0} \frac{x^2}{2!} + \ldots \tag{15}$$

Similarly substituting into eq. (11) for the force, integrating, and accounting for the symmetry

$$\mathbf{F} = \int \left(d\mathbf{M} \cdot \hat{\imath} \left.\frac{\partial B_x}{\partial x}\right|_{x=0} \right) + \int \left(d\mathbf{M} \cdot \hat{\imath} \left.\frac{\partial^3 B_x}{\partial x^3}\right|_{x=0} \cdot \frac{x^2}{2!} \right) + \ldots \tag{16}$$

If fore and aft magnetic symmetry exists, the torque depends only on the zeroth and even order terms in the expansion of $\bar{B}$, while the force depends only on the odd ordered terms in the expansion of $\bar{B}$. The results given in eqs. (15) and (16) have a practical implication. Stephens[3] has shown that eq. (15) suggests an independent way to provide roll torque. Suppose that B_y and B_z can be expressed in quadratic form, i.e.

$$B_y = \mu H_y(0) \left\{ 1 + \kappa_y \frac{x^2}{2} + \ldots \right\} \tag{17}$$

where

$$\kappa_y = \frac{1}{H_y(0)} \left.\frac{\partial^2 H_y}{\partial x^2}\right|_{x=0}$$

by substituting into eq. (15) one obtains the expression for axial torque

$$T_x = \frac{\mu H_y(0)\, H_z(0)}{2 D_B} (\kappa_y - \kappa_z) \int x^2 d\,\text{Vol.} \tag{18}$$

The roll torque depends upon the difference in the curvature of B_y and B_z in the x direction. Stephens indicated that a double curved magnetic core in a non-uniform field may also be used to generate an independent roll torque. Goodyer[4] has generalized these results to obtain an independent roll torque with non-uniform three-dimensional magnetization.

The concept of non-uniform magnetization may be generalized in an alternate way. Suppose a non-steady field is arranged so that it does not interact with the steady state force and torque on the average. This concept is similar to that used by a magnesyn type of device. Suppose that

$$B_y = B_y(0) + \epsilon_y \cos \omega t \tag{19a}$$

and

$$B_z = B_z(0) + \epsilon_z \sin \omega t. \tag{19b}$$

The unsteady part of these fields interact with a loop of conducting material in the equatorial plane (say the y–z plane) of the core. The unsteady B_y induced a ring current that looks like an unsteady magnetization in the vertical direction. That is $M_z =$ (geom. factor) times $\omega \epsilon_z \sin \omega t$. This magnetization interacts with the unsteady lateral field B_y to give the roll torque. Thus from eq. (19a)

$$T_x = -\text{(geom. factor)}\ \epsilon_y \epsilon_z \sin^2 \omega t. \tag{20}$$

As long as ω is sufficiently large compared to the other frequencies in the system, the roll torque is essentially steady and independent of other force and moment components.

Independent means exist for applying arbitrary components of torque and force to a magnetized core or model. In the event that restrictions of symmetry are not satisfied, cross-couplings will exist. These cross-couplings are regular and well behaved. Thus cross-over networks can be built for the control system. Interactions can be deduced from calibration data (see Section III). In Section II the problem of finding coil configurations to generate these fields that produce the force and torque will be discussed.

System Analysis

Having demonstrated the possibility of independent control of the magnitude and direction of the force and torque vector developed by magnetic fields acting on a magnetized body, a next step might be to show how these forces and torques can be used in a magnetic suspension system. An alternate step is taken, however. It consists of studying a one-degree of freedom suspension in more detail. The simpler system is really a microcosm of a complete system, without all the attendant complications. Thus the central difficulties are clearly illustrated. The discussion will be further simplified in that real hardware problems will largely be omitted. Hopefully the basic elements remain. Once they are understood, complete systems can be synthesized more or less by repeated use of these elements.

It is important to know whether or not the system is stable. Suppose a sphere is suspended at an equilibrium position and then subjected to a small disturbance. Stability implies it will return to the initial equilibrium state. That is, positive work on the system is required to disturb a stable system from its equilibrium point. A generalization of Earnshaw's theorem* implies that any d.c. system is unstable.[5] In the absence of a perfect diamagnetic material no three-dimensional d.c. potential well exists. A perfect diamagnetic material or a material in the superconducting state can be used to hold a permanent magnet against gravity. For example, Arkadiev[6] floated a small bar magnet above a slightly hollowed disk of superconducting material. Alternately the induced current on the surface of a conducting body acts to exclude the magnetic field giving rise to so-called bulk diamagnetism. In 1939, Peer and Tonks proposed an alternating current suspension using bulk diamagnetism. In practice the I^2R losses heat the suspended object to above the melting temperature.† Conversely it is also possible to suspend superconductors in alternating current magnetic fields.

* Earnshaw published this theorem in connection with early studies of the atomic properties of matter. He proved that there is no array of bodies influencing each other by means of an inverse distance squared law, like gravity, possessing static stability. Thus he concluded that Mossoti's atomic theory of matter was incorrect. Earnshaw would have been correct if interatomic forces decayed as slowly as he hypothesized.

† Okress's[7] suspension was applied to the problem of vacuum melting of active metals, and for making exotic alloys. Ultimately he and his co-workers were awarded a series of patents on this process.

The instability in d.c. systems may be eliminated through use of position sensing and a feedback control for the power supply. Strictly speaking this alternative is not a true d.c. system. However, the frequencies are so low that the magnetic field acts as if it were a quasi-steady phenomena. This idea was first used by Holmes and Beams[8] in 1937. This automatic control element leads to the sort of suspension system to be studied here. The combination of all these ideas can be best illustrated by consideration of the problem of suspending a simple body against gravity by means of a one-dimensional suspension system. Many such systems have been built.*

The one-dimensional system to be discussed consists of a single N turn circular coil of radius a whose axis of revolution is in the vertical direction. For this argument let z be positive downwards and let $z = 0$ lie in the plane of the coil. The vertical field strength at $z = 0$ and $r = 0$ is $NI/z = H_0$. This field can be adjusted by varying the current I in the coil. The field in the z-direction is distributed approximately as

$$H_z = H_0/[1 + (z^2/a^2)]^{3/2} \tag{21}$$

so

$$\frac{\partial H_z}{\partial z} = -\frac{3z}{a^2}\frac{H_0}{[1 + (z/a)^2]^{5/2}}. \tag{22}$$

We assume the object to be suspended is a small sphere. It will either be uniformly magnetized to a level M, or made of ferromagnetic material in the linear range. Since the magnetic force is the product of the magnetization and the gradient of the magnetic field the force obtains, in the first case

$$F_M = -\frac{3\mu z}{a^2}\frac{M_z H_0}{(1 + z^2/a^2)^{5/2}}. \tag{23}$$

(Note the magnetized sphere will rotate until the magnetization vector is aligned with the z direction.) F_M acts in the negative z direction. In the case of induced magnetism

$$F_M = -(H_0)^2\left(\frac{3\mu}{aD_A}\right)\frac{z}{a}\frac{\frac{4}{3}\pi a^3}{(1 + z^2/a^2)^4}. \tag{24}$$

D_A is the demagnetization factor for a sphere and is 1/3 from the definition given above (eq. (6)). The magnetic force is zero at $z = 0$ in both cases. In both cases F_M has a maximum for a particular value of z/a. In the first case this maximum occurs at $z/a = 1/5$ and in the second case the maximum occurs at $z/a = 1/7$.

* A complete list of these one-dimensional systems can probably never be prepared for reasons that will become clear shortly. The list below includes all the authors are aware of.

E. L. Tilton III and L. A. Baron, The design and construction of an automatic control system for a wind tunnel magnetic suspension system. Thesis submitted for S.B., MIT Department of Aeronautics and Astronautics, 1960.

H. Schreiber, Photomagnetic toy is a true servo-mechanism, *Radio Electronics*, January 1962.

Anon., *Electronic Experimentor's Handbook of 1967*. We suspect the latter article is closely related to Schreiber's article. Mr. Frank Regan of the Aerodynamics Branch of the Naval Ordnance Laboratory, White Oak, Maryland, U.S.A., says the latter article is very complete, and the suspension he built works well.

R. H. Kilgore *et al.* (NASA Langley), cited in F. L. Daum, ed., *Summary of ARL Symposium on Magnetic Suspension*, USAF OAR Report 66-0135, July 1966.

K. R. Sivier, A one component magnetic support and balance system, *Journal of Aircraft*, vol. 6, no. 5, 1969, pp. 398–404.

A good easy bibliography of magnetic suspension for wind tunnel use is given by Clemens and Cortner.[9]

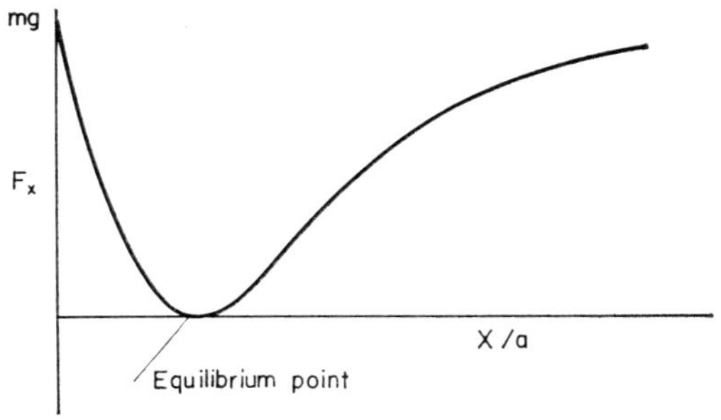

SKETCH I. Equilibrium under gravity and magnetic fields.

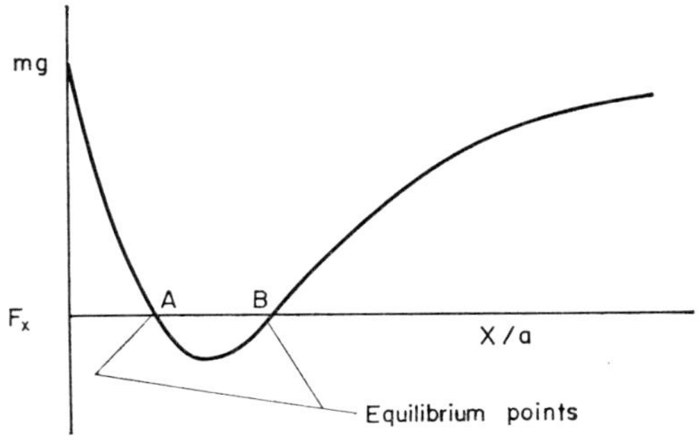

SKETCH II. Equilibrium under gravity and magnetic fields.

The weight of the sphere is mg. The net force is $F_z = mg + F_M$. If mg exceeds the value of the maximum, $F_z > 0$, no suspension is possible. If F_M just equals mg, then an equilibrium point exists (see sketch I). Examination of this sketch shows the equilibrium is unstable; any disturbance will cause the weight to exceed the magnetic force and the sphere will fall. If the magnetic force exceeds the weight there are two equilibrium points (see sketch II). Point B is an unstable equilibrium point. Depending upon the direction of a particular disturbance the sphere will either fall or move towards equilibrium point A. The latter point is stable to small disturbances. For a slight increase in z the force is negative; that is towards A. Similarly a slight decrease in z causes the magnetic force to be less than mg and the sphere tends to move in the positive direction. If the disturbance about A is too large the sphere will always fall. A detailed three-dimensional analysis shows that point A is an unstable point in the radial direction, while the point B is radially stable. This is a simple illustration of Earnshaw's theorem. The obvious step would be to suspend the sphere at point B and control the value of H_0 in such a way that the system is made stable.

To study this possibility it is necessary to study the motion of the sphere from Newton's Law

$$z = g + F_M/m. \tag{25}$$

The control law must require the H_0 to increase when the sphere falls and conversely. Such a law can be written

$$H_0 = H_0' + f(y, v). \tag{26}$$

A more specific relation is

$$H_0 = H_0' + K_1 y + K_2 v \tag{27}$$

where $y = z - z_B$ is the displacement from the equilibrium point,

$v = dy/dt$ is the velocity of the sphere,

$f(y, v)$ is the control law.

In eq. (25) z is the present position of the center of the sphere and z_B is the unstable equilibrium point (in z not r). The variable y and v are state variable of the system under discussion. This control law (eq. (27)) will be denoted I. Substituting from eq. (27) into eq. (25) gives the matrix equation

$$\frac{d}{dt}\begin{pmatrix} y \\ v \end{pmatrix} = \begin{pmatrix} 0 & 1 \\ \dfrac{L_2 - K_1 L_1}{m} & -\dfrac{K_2}{m} \end{pmatrix} \cdot \begin{pmatrix} y \\ v \end{pmatrix} \tag{28}$$

where

$$K_1 = \frac{3}{a}\frac{\mu M_z}{(1 + z/a_B^2)^{5/2}}$$

$$L_1 = K_1 (z_B/a)$$

$$L_2 = \frac{K_1 H_0'}{a}\left(\frac{\dfrac{5z_B}{a}}{1 + z^2/a^2} - 1\right).$$

The system dynamics are uniquely determined from the eigenvalues of the matrix on the right-hand side of eq. (28). The undamped natural frequency of this system is $\omega_n = \sqrt{\dfrac{L_1 K_1 - L^2}{m}}$ and the damping factor is $\zeta = \frac{1}{2}\dfrac{K^2}{\sqrt{m}\,(L_1 K_1 - L_2)}$. The reason for the correction term in f, Kv is now quite clear. This term is a source of damping. Denote the eigenvalue by ω. ω is a complex quantity. The locus of the eigenvalue in the ω plane for control law I is shown in sketch III as a function of K_1. When K_1 is zero the system is unstable as indicated by the positive real root. As K_1 is increased to the value L_2/L_1 the eigenvalue moves along the real axis to the point 1. Any further increase in K_1 causes the system to be statically and dynamically stable, although it is overdamped in the sense that no oscillation results from an impulsive input. As K_1 is increased to the value $(L_2/L_1) + (K_2^2/4m)$ the eigenvalues are equal (point 2), and take the value $-K_2/2$. As K_1 continues to increase the locus moves vertically, the frequency of oscillation increasing with K_1. The fastest response occurs at $\zeta = 0.7$, that is where $K_1 = (K_2^2/2mL_1) + L_2/L_1$, at point 3. For greater values of K_1 the system oscillates longer before the disturbances die out. Control law I is adequate for a pure suspension system. Suppose a drag load is applied to the sphere. According to the differential equation the new equilibrium position is

$$y = -\frac{D}{L_2 - K_1 L_1}. \tag{29}$$

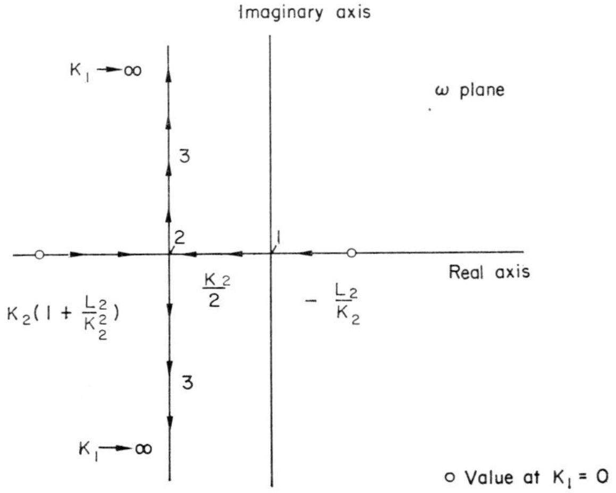

SKETCH III. Root locus diagram, control law I.

That is to say the control law admits a displacement under load. In aerodynamic testing this displacement is undesirable. To reduce the errors due to a non-uniform velocity field to a minimum, the sphere should be held at a fixed position. This can be achieved by introducing a new state variable u. This is defined $du/dt = y$. The control law is modified by adding the term K_3u, i.e. $f = K_1y + K_2v + K_3u$. This is called control law II. Upon adding the drag, the new equilibrium is found to be $u = D/K_3$. But for u to be constant the position y must be zero. It is possible therefore to modify the control system so that the system becomes a null position system. With this change the state equation becomes

$$\frac{d}{dt}\begin{pmatrix} u \\ y \\ v \end{pmatrix} = \begin{pmatrix} 0 & 1 & 0 \\ 0 & 0 & 1 \\ -\frac{K_3}{m} & \frac{L_2 - L_1K_1}{m} & -\frac{K_2}{m} \end{pmatrix} \cdot \begin{pmatrix} u \\ y \\ v \end{pmatrix} \tag{30}$$

The eigenvalues of the characteristic matrix may be found by finding the roots of

$$\omega^3 + \frac{K_2}{m}\omega^2 - \left(\frac{L_2}{m} - K_1\frac{L_1}{m}\right)\omega + \frac{K_3}{m} = 0. \tag{31}$$

Equation (31) can be used to describe the locus as shown in sketch IV. Note there is a locus for each value of K_3. The diagram then appears to be a contour plot. Control law II has the property that the asymptotes to its loci of eingenvalues are, for large K_1, essentially independent of K_3 but the behavior of the loci when K_1 is very small is strongly dependent upon K_3. For small K_1 two limiting cases appear at once. The first is for K_3 large and K_2 small, and the second is for K_3 small and K_2 large. In case 1 the eigenvalues consist of one real damped root and a complex pair that are sometimes damped (if K_1 is large enough) and sometimes unstable (for small values of K_1). In case 2, the eigenvalues are initially real, with two unstable and one stable. If K_1 then increases from zero the two unstable roots coalesce, then become complex and ultimately become damped. These are shown in sketch IV. Examination of sketch IV suggests the choice of a value for K_1 is more difficult here, principally because of the small negative eigenvalue. As K_1 becomes much larger than L_2/L_1 this eigenvalue becomes asymptotically equal to zero. Methods of selecting the "best" K_1 in terms of the other parameters in the system are known, and are easier to describe than carry out. For example, one can define J as the integral of the y squared and find values of K_1 for which J is minimized. Usually this J is a relative minimum and depends upon K_3. Hence by varying K_3 an optimum can be found. Note the optimum may lie on a saddle-like surface and in this case the optimum is at the so-called mini-max.*

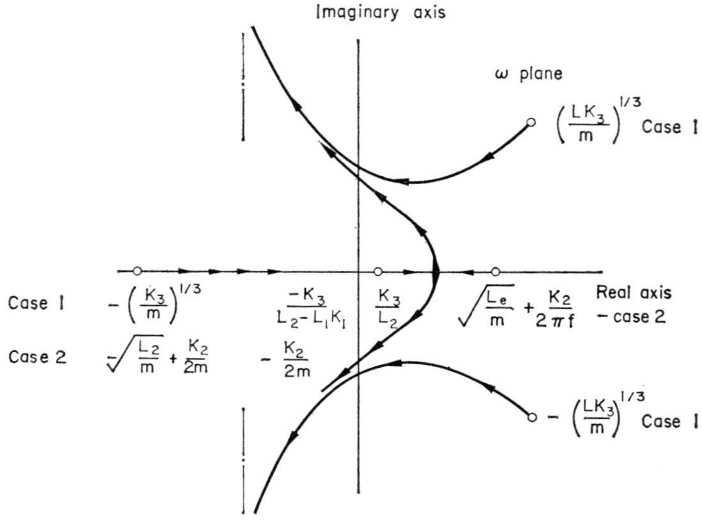

SKETCH IV. Root locus diagram, control law II.

Control law II holds the sphere in the same place by altering the field strength by D/K_3 by increasing the current in the coil. Thus the drag can be determined from current measurements. (This is also true for control law I in which y varies with the load, eq. (29).) If the current after loading is denoted I_a and the current before loading is denoted I_i, then it is easy to show that

$$D = \frac{I_i - I_i}{I_i} mg. \tag{33}$$

* Since this system may be slow, the alternative always exists of adding the term $K_4\dot{v}$ to the control law Ultimately this leads to, after some manipulation, the form

$$\frac{d}{dt}\begin{pmatrix} u \\ v \\ y \end{pmatrix} = \begin{pmatrix} 0 & 1 & 0 \\ 0 & 0 & 1 \\ -\frac{K_3}{m-K_4} & \frac{L_2 - L_1K_1}{m-K_4} & -\frac{K_2}{m-K_4} \end{pmatrix} \begin{pmatrix} u \\ y \\ v \end{pmatrix}. \tag{32}$$

K_4 has the effect of moving the vertical asymptote to the left which increases the damping, and tends to augment the value of K_1 as a parameter moving along the root locus. A more detailed discussion of the control (or compensation) system is given in Section II.

The stability analysis and the synthesis of the control system is similar for the case of induced magnetization. The problem is still linear since y is always small. Of course K_1 is different as are L_1 and L_2. The big difference is in the drag current relationship resulting from the introduction of u as a state variable. In the case of self-induced magnetization

$$D = \left[\left(\frac{I_a}{I_i}\right)^2 - 1\right] mg. \tag{34}$$

Thus a linear calibration can be expected only in the case for which the sphere is a permanent magnet or the magnetization current is independent of the gradient current that develops the force.

An actual system will be somewhat slower in its response than the simple one under discussion because additional lags will be introduced in the power supplies and the position sensing system. An actual system can be represented by a block diagram as shown in sketch V.

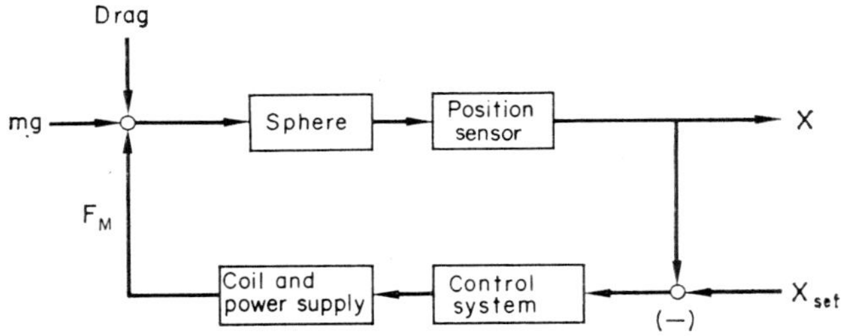

SKETCH V. System block diagram.

It should be clear that the one-dimensional system indeed is a microcosm of balance and suspension systems, lacking only the cross-coupling between the other five remaining degrees of freedom. As indicated in eqs. (11) to (13) these can be made small. An additional advantage of adding the state variable u, sometimes called reset or integral control, can be seen by applying a forcing function. This can be applied either at the position set point or at the force balance point. In the former the motion can be made to be sinusoidal and any non-linearities in the aerodynamics appear in the departure of the force from a simple image of the input. Alternately the forcing function can be applied to the force and the non-linearities will appear in the motion. This added alternative is useful in performing dynamic experiments on non-linear or even unstable aerodynamic configurations.

Closure

Having considered briefly the elements that enter into a magnetic suspension system and a magnetic balance system, it seems desirable to indicate briefly the contents of the remaining chapters. In section II a detailed study of magnetic suspension systems is undertaken. This study includes magnetic configuration including coil geometrics and materials that have been used. The description of power supplies, cooling systems and control systems will also be included in this section. Section II closes with a discussion of the several scaling laws needed to characterize magnetic suspension systems. Then the requirements and problems associated with wind tunnel applications are described. Section II concludes with a description of various uses to which these balance systems have been put.

The third section contains the generalization to magnetic balance systems, including calibration techniques, the need for absolute position information, the forms of data output and data handling systems. The elements are discussed within the framework of the evolution as contained in the practice in the various laboratories where such systems are in use.

Next a comparison between the aerodynamic data obtained with these instruments and the conventional instruments is presented. The final section contains a discussion of limitations, future uses and other applications of the ideas presented here.

References to Section I

Cited references

1. W. Gilbert, *De Magnete*, Dover Publications, New York, p. 3.
2. A. W. Jenkins and H. M. Parker, Electromagnetic support arrangement with three-dimensional control, *Journal of Applied Physics*, vol. 30, no. 4 (Supp.), 1959, pp. 2385–2395.
3. T. Stephens, Methods of controlling the roll degree of freedom in a wind tunnel magnetic balance, USAF OAR ARL, Report ARL 65-242.
4. M. Goodyer, Roll control of magnetically suspended wind tunnel models by transverse magnets, *The Aeronautical Quarterly*, vol. 18, February 1967, pp. 22–42.
5. S. Earnshaw, On the nature of the molecular forces, *Trans. Camb. Phil. Soc.*, vol. 7, 1842, pp. 97–112.
6. V. Arkadiev, A floating magnet, *Nature*, vol. 160, 1947, p. 3302.
7. E. C. Okress, D. M. Wroughton, G. Commenetz, P. H. Brace and J. C. R. Kelly, Melting without crucibles, *Journal of Applied Physics*, vol. 23, 1952, pp. 545–552.
8. F. T. Holmes, Axial magnetic suspension, *Review of Scientific Instruments*, vol. 8, 1937, p. 444.
9. P. L. Clemens and A. A. Cortner, Bibliography: The magnetic suspension of wind tunnel models, AEDC-TDR-63-20.

General references

R. M. Bozorth, *Ferromagnetism*, Bell Laboratories Series D, D. von Nostrand Company, Inc., New York, 1951.

P. S. Epstein, *Textbook of Thermodynamics*, J. Wiley & Sons, New York, 1937.

E. N. Lee and L. Markus, *Foundations of Optimal Control Theory*,* J. Wiley & Sons, New York, 1968.

F. W. Sears, *An Introduction to Thermodynamics, the Kinetic Theory of Gases, and Statistical Mechanics*, Addison-Wesley, Reading, Mass., 1953.

A. Sommerfeld, *Electrodynamics*, Academic Press, New York, 1952.

S. B. Starling, *Electricity and Magnetism*, Longmans, Green & Co., London, New York and Toronto, 1956.

J. A. Stratton, *Electromagnetic Theory*, McGraw-Hill Co., New York, 1938.

II. GENERAL SURVEY OF MAGNETIC SUSPENSION

A general survey of suspension has been given by Boerdijk.[1] This survey includes ideas contained in a number of more recent discussions.[2,3] Boerdijk considers

(a) Suspension of diamagnetics by steady magnetic fields (as first shown by Braunbeck, 1939).

(b) Suspension of permanent magnets by superconducting substances (as first shown by Arkadiev in 1946).

(c) Suspension of permanent magnets by a combination of permanent magnets and diamagnets or superconductors (as shown by Boerdijk in 1958).

* This field is changing rapidly. For example see:
T. R. Crossley, and B. Porter, Synthesis of helicopter stabilization using model control theory, *Journal of Aircraft*, vol. 9, no. 1, pp. 3–8, 1971.

(d) Suspension of electric conductors in alternating fields (as first shown by Bedford, Peer and Tonks in 1939).
(e) Suspension in a quasi-stationary electromagnetic field with a feedback system (as first shown by Holmes in 1937).

The suspension contained in paragraph (d) has been discussed at length by Hatch.[4] He generalizes the earlier results of Okress *et al.*, to include suspension of plasma,[4a] thin foils,[4b] and cryogenic gyroscopes.[4c,4d].

However, as indicated in the introduction it is the suspension of paragraph (e) that is of primary interest here. Here two alternatives exist in which the a.c. field is relatively slowly varying—the so-called "quasi steady" situation in which it is possible to suspend either a permanent magnet (Wilson and Luff, 1966),[5] or a piece of ferromagnetic material (Holmes, 1937)[6] or a ferrite (Zapata and Dukes, 1964).[7] H. Kemper reported a suspending efficiency of 1.3 W/kg.[8] Kemper applied his results to levitation of vehicles such as trains, or automobiles. In either of these alternatives the applied a.c. field could in principle be either earth fixed or suspended. In all the cases cited above, the applied field is earth fixed. As mentioned in the quasi steady cases an automatic control system is required, but some cases exist when

TABLE 1. DESCRIPTION OF MAGNETIC CORE FOR MODEL

	a	b	c	d
1. Magnetic core	Solid	Laminated	Powdered	
2. Magnetic core shape	Spherical	Ellipse of revolution	Conical	Aerodynamic model
3. Source of roll asymmetry	Permanent magnets	Non-axially symmetric	Bent core	Imbedded hysteresis coil

the a.c.–d.c. system is self-stable.[4] The self-stable systems lack general utility in wind tunnel applications because of the problem of high power consumption and of removing the heat due to eddy current heating. The oft-cited limitation due to control of direction can be overcome using the scheme first suggested by Parker (1959).[9]

The description of the suspension must include the nature of the object being suspended (Table 1) and several features such as source of magnetization, source of gradient, and a control system to overcome the basic instability in the six degrees of freedom (Table 2). The control system depends in turn upon sensing the model positions. These features can be arranged in a form that also includes the alternatives for each feature. Thus Table 2 defines a suspension system by selecting one element from each row. For example 4c, 5c, 6a, 7c, 8b, 9c, 10a, 11a, 12a, 13a and 14a represents the original Laurenceau and Tournier[10] suspension. Hence Table 2 represents an enormous number of suspension systems, over 10^6 in fact.* With the exception of the E system (Fig. 4) and the orthogonal B (Fig. 7) only the systems outlined in Figs. 1–8 have been used to successfully suspend models.

Naturally all the remaining systems have not been built and are not likely to be built. Several geometric configurations shown in Figs. 1 through 8, will be discussed in detail below.

There are three common lineages for the early systems as shown in Table 3.

* It can be argued that entry 5b implies 8a in the classification scheme. While this is true, there is a distinct difference in the treatment of the control system. It is for this reason that the two types are entered.

TABLE 2. DESCRIPTION OF MAGNETIC SUSPENSION CHARACTERISTICS

	a	b	c	d	e	f	g	h	i
4. Source of M	Separate external coils	Separate internal coils	Combination with force coils	Permanent magnet					
5. Direction of M	Flow direction	Transverse direction	Skew direction						
6. Magnetizing coil winding material	Normal	Super-cooled	Super-conducting						
7. Source of gradient B	Separate air core per gradient	Separate iron core per gradient	Separate iron core per force	Separate air core per force	Combination with M coils				
8. Source of pitch and yaw torque	Tilt M	Difference in force	None						
9. Source of roll	D.c. (quasi-steady)	A.c.	None						
10. Gradient coil winding material	Normal	Super-cooled	Super-conducting						
11. Geometric arrangement	L system	Double-L system	V system	E system	Single-axis system	Ortho-gonal A	Ortho-gonal B	Ortho-gonal C	Ortho-gonal D
12. Position sensor	Optical[8] small beam	Optical image or large beam	Optical reflected	Magnetic induction					
13. Cooling	Free convection	Forced convection layered	Forced convection hollow core	Solid conductivity					
14. Control	Without integral and cross-over	Without integral with cross-over	With integral no cross-over	With integral and cross-over					

Notes: (1) Model 3d must accompany 9b.
(2) Entries 4c and 7e must occur together.
(3) Includes EM waves from radar to light to X-rays.

TABLE 3. DESCRIPTION OF EARLY SYSTEMS

ONERA Ref. 10	MIT-ARL 11	Univ. of Southampton 12	RAE Farnborough 6	AEDC 13	Univ. of Virginia 9	Princeton University 7	Univ. of Michigan 14	NASA[1] 15	NOL 16
4c	4c	4c	4c	4c	4a	4a	4a	4c	4c
5a	5c	5c	5c	5c	5b	5a	5a	5a	5a
6a	6a	6a	6a	6a	6a	6a	6a	6a	6a
7c	7c	7c	7c	7c	7a	7a	7a	7e	7e
8b	8b	8b	8b	8b	8c	8c	8c	8c	8c
9c	9c[3]	9c[4]	9c	9c	9c	9c	9c	9c	9c
10a	10a	10a	10a	10a	10a	10a	10a	10a	10a
11a[2]	11a	11b	11a	11c[5]	11f	11f	11e	11e	11e
12a	12b	12b	12a	12a	12a	12b	12a	12a	12a
13a	13c	13c	13c	13a	13d	13c	13c	13c	13c
14a	14c	14c	14c	14c	14c	14c	14a	14a	14a

[1] Demonstration only.
[2] An 11b was also built and is in use (ref. 17).
[3] Was modified to 9b (ref. 18).
[4] Was modified to 9a (ref. 19).
[5] Changed from visible light to X-rays (ref. 20).

As Table 3 shows, there was a heavy dependence upon the first ONERA suspension of Laurenceau and Tournier[10] in England and the United States in the early systems. A second system was that developed at University of Virginia by Parker[9] and his co-workers. The NOL demonstration model[16] and the NASA[15] demonstration are similar in concept and precede the one-degree of freedom Michigan[14] system, but there may be no special relationship here.

The discussion that follows is arranged in the order that the component parts are listed in Table 2.

Magnetizing System

The model to be suspended can be magnetized in a variety of ways. In the original work the fields from the drag coil, the lift coils and side force coils were used to magnetize a soft iron model. All the subsequent French systems used this scheme, as did some other systems. The one-degree of freedom systems also had this feature. Thus the magnetization of the core was dependent upon the applied loads leading to strong interactions between components. Parker[9] (1959) offered a more logical solution to this problem, a separate set of coils (actually a Helmholtz pair) provide a uniform field whose sole purpose is to magnetize the model. Thus there is the potential to reduce the coupling between the diverse components. It is likely that all subsequent systems[7,14,21] used this idea. The magnetizing coils are usually wet wound from hollow-cored conductors[22] or superconducting Cu–Ni clad litz wire.[23,24] Such a coil is compact and easy to manage. The field is usually sufficient to magnetize a sphere, that is 8000 to 10,000 G (0.8 to 1.0 Wb/m^2). The direction of the magnetizing field is perpendicular to the flow[9] or along the direction of the model longitudinal reference axes[7] which may be at an angle of attack or yaw.[21,22,25] In principle the magnetization could be supplied to the model by an a.c. coupled system with coils in the model for power pick-up. The model also needs a rectifier in the circuit to generate the internal field.*

Permanent magnets could be used in two ways, either to generate the magnetizing field (which is presently impractical) or as the model.[5] The latter use of permanently magnetized material, particularly with the platinum cobalt and samarium cobalt alloys, is eminently practical.

Gradient Coil System

As explained in the introduction, the force is a result of the change in the magnetic field over the length of the model. The gradient can be produced simply. A single coil produces a field that decreases with distance; this is well known.

In the L system[10] the gradients are produced by five coils (Fig. 1). These coils are functionally named for the force they produce, i.e. drag, forward lift, aft lift, forward lateral and aft lateral. Note the lateral coils may also contain bias windings coils to provide a side force if the lateral power supply is not two-sided. Note in this system the total lift is essentially the sum of the two lift forces and the pitching moment is the difference between the lift forces plus a component due to the drag coil field acting on a model at angle of attack. Similar statements hold for side force and yawing moments. As shown in the sketch the L system is a five degree of freedom system. By adding an a.c. coil opposite the lift and lateral coils an a.c. roll system can be added.[18] These coils may be air-cooled.[10] They may be hollow

* This was suggested to one of us (E.C.) by Alan Wilson of RAE, but was not put into practice as far as we know.

conductor water-cooled.[11] An L system of this kind has also been used by Wilson and Luff.[5] In principle the coils could be super-cooled or superconducting, but no such system has yet been built. It should be noted this kind of a system has been used for tunnels from 7.5 cm to 30 cm[10] (ONERA) and up to 7 in.[11] The nature of this suspension system can be illustrated best by sketches of the field configuration. These sketches are based upon the center vertical plane. The solid lines are lines of constant B from the drag coil, the dashed lines are from the lift system. The model of soft iron will be attracted to regions of higher flux density. The configuration in sketch VI will cause a forward and an upward force. Note the field lines have a zero slope region at the midpoint between the lift coils. Hence a model in the axis has zero moment if its center of magnetism is located there. Alternately one can say the lift is equal on both the forward and aft pole of the magnetized model. If, as in sketch VII the forward lift is stronger the model will tend to rotate as to increase its angle of attack. If the

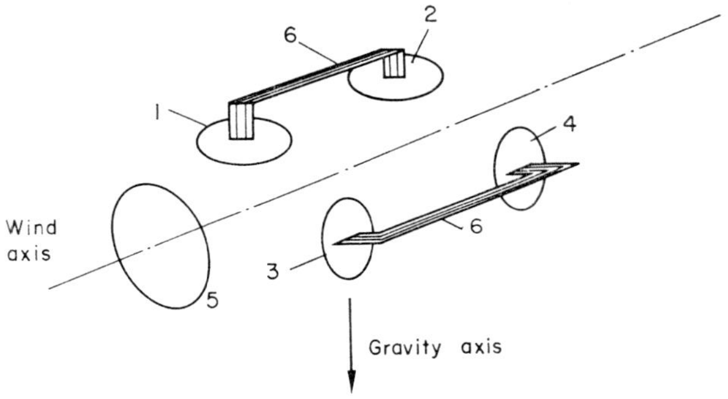

FIG. 1. L system schematic diagram.

aft lift is reduced as much as the forward lift is increased the model is still in force equilibrium. The model whose center of magnetism is at the center is between the lift coils where the field is inclined upward. Thus the model will want to rotate to reduce the moment and thus increase its angle of attack. Naturally the ultimate stability will be fixed by the control system. The lateral field configuration is similar to the lift field except that the absence of gravity causes the field to be much smaller. Finally note that a bent model would also possess roll stability.[26]

The performance of the MIT-ARL system (for a 5-in. tunnel) can be expressed in terms of the forward lift current I_{L_1}, the aft lift current I_{L_2}, and the drag current as

$$H_x = 5.5\, I_{L_1} + 5.5\, I_{L_2} + 3\, I_D \qquad \text{(gauss)} \tag{35}$$

$$H_z = -6.5\, I_{L_1} + 6.5\, I_{L_2} \qquad \text{(gauss)} \tag{36}$$

and

$$\frac{\partial H_z}{\partial z} = 1.7\, I_{L_1} - 1.7\, I_{L_2} - 0.37\, I_D \qquad \text{(gauss/inch)} \tag{37}$$

$$\frac{\partial H_z}{\partial x} = 0.2\, I_{L_1} + 0.2\, I_{L_2} + 0.75\, I_D \qquad \text{(gauss/inch)} \tag{38}$$

$$\frac{\partial H_x}{\partial z} = \frac{\partial H_z}{\partial x} = 3.9\, I_{L_1} + 3.9\, I_{L_2} \qquad \text{(gauss/inch)} \tag{39}$$

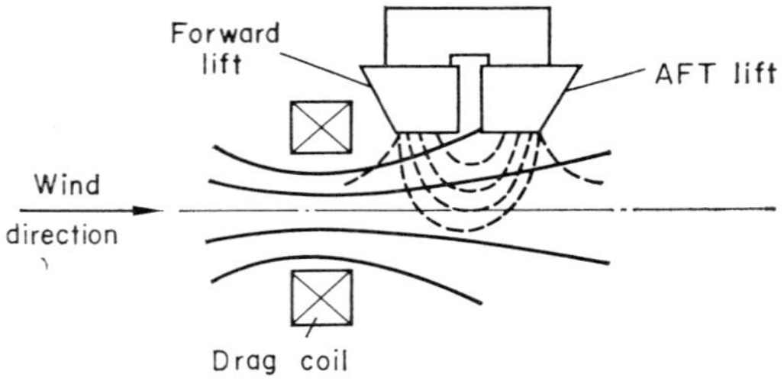

SKETCH VI. Lift and drag fields.

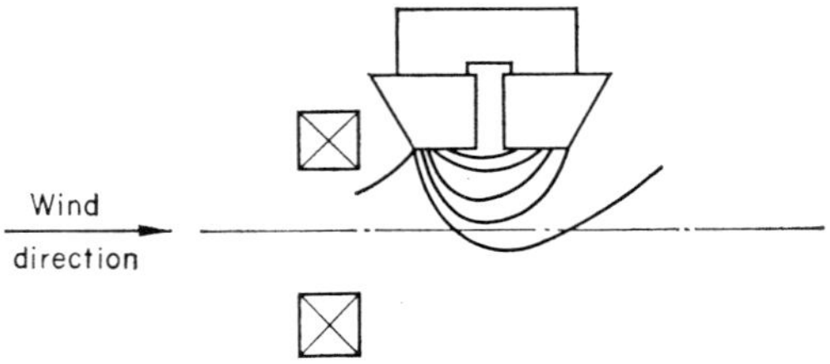

SKETCH VII. Moment field.

where the coefficients are determined from measurements of the field on the axis below the midpoint between the coils. The magnetization takes the form (Vol is the volume of magnetic material in cubic inches)

$$\mathbf{M} = \frac{\text{Vol}}{D_A}(5.5\, I_{L_1} + 5.5\, I_{L_2} + 3.0\, I_D)\,\hat{\imath} + \frac{\text{Vol}}{D_B}(-\, 6.5\, I_{L_1} + 6.5\, I_{L_2})\,\hat{k} \quad \text{gauss} \tag{40}$$

Thus in this system the model tends to be magnetized by the lift currents acting against the model weight. The forces and torques are clearly quadratic functions of the currents. For example consider F_z and T_y

$$F_z = \text{Vol}\left(\frac{21.45\,(I_{L_1} + I_{L_2})^2 + 11.7\, I_D\,(I_{L_1} + I_{L_2})}{D_A} - \frac{11.05\,(I_{L_1} - I_{L_2})^2 - 2.4\, I_D\,(I_{L_1} - I_{L_2})}{D_B}\right) \tag{41}$$

and

$$T_y = \text{Vol}\left(\frac{1}{D_A} - \frac{1}{D_B}\right)]35.75\,(I_{L_2}{}^2 - I_{L_1}{}^2) + 19.5\, I_D\,(I_{L_2} - I_{L_1})]. \tag{42}$$

This sort of relation is common to all "L systems" unless the model is permanently magnetized in which case

$$F_z = M_x \, (3.9 \, I_{L_1} + 3.9 \, I_{L_2}) \tag{43}$$

$$F_x = M_x \, (0.21 \, I_{L_1} + 0.2 \, I_{L_2} + 0.75 \, I_D) \tag{44}$$

$$T_y = M_x \, (6.5) \, (I_{L_2} - I_{L_1}) \tag{45}$$

which is a linear system.

Double-L System

The double-L system offers some flexibility in its use for dynamic stability testing and for d.c. or permanent magnet roll control.[12,19] This balance is used in a 6-in. tunnel at the University of Southampton.

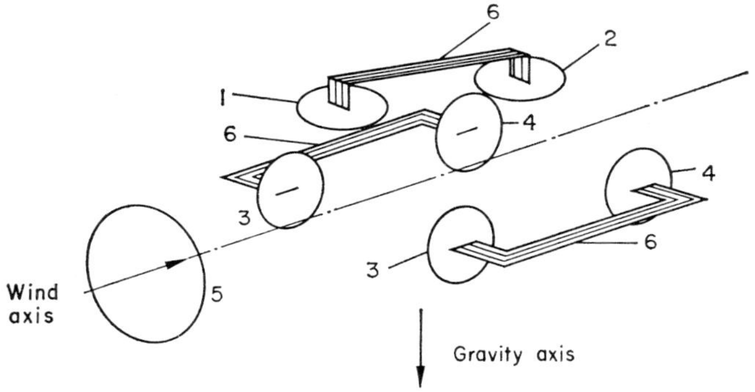

FIG. 2. Double-L system schematic.

Goodyer[19] has shown the double-L system has an additional advantage of offering positive roll control. By use of a rectangular core, or a double-finned model, roll torques can be produced without coupling into either lift or pitch. Roll torque is generated by an extra coil in each of the double-L cores making four in all. These coils are connected so that a relatively uniform lateral field is set up across the space between the pole faces. The rectangular core or finned model, like the horizontal tail, lines itself up such that the transverse magnetic moment lines up with the field. The roll torque is therefore proportional to twice the roll angle, just as in the classical $\bar{M} \times \bar{B}$ system. The roll torques so generated are shown to be more than large enough to make the procedure practical. Roll position sensing, which will be discussed below, is accomplished by modulating an optical beam by a slot in the tail. Damping is introduced in the control system.

In the same article Goodyer uses currents induced in a loop around the tail to produce a d.c. magnetic field that can be used to produce roll torque. Again the data shows an adequate roll torque can be developed. This latter system can be made to work with any system whose power supply has some noise that can be picked up and rectified.

V System

The V system was conceived as an improvement of the L system to eliminate the need for lateral bias currents or a two-sided (plus or minus current) lateral power supply. It also provides a clear observation path. At least two such systems were built for 12-in. tunnels.[12,17]

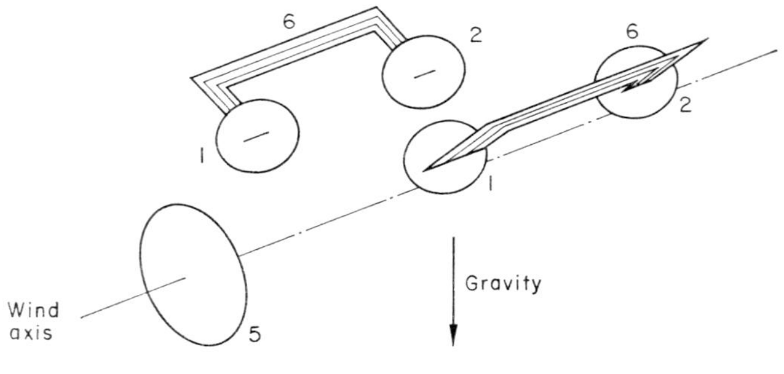

FIG. 3. V-coil system schematic.

They easily satisfied the goals set for them. The calibration of these systems is similar to the L system rotated 45 deg except for the resolution of error signal in the control system. The V system represents the first step towards greater symmetry.

E System

The E system represents a generalization of the L system that includes a d.c. roll system[21] (Fig. 4). That feature will be discussed briefly, since in other respects the E system is included in the discussion of the L system. A nonuniform core produces a roll torque in a uniform

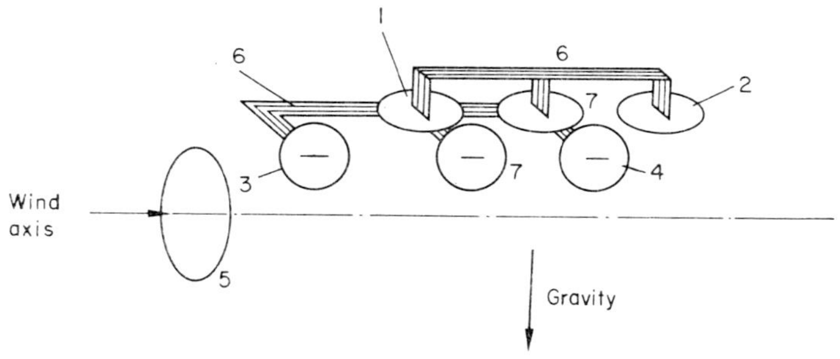

FIG. 4. E system schematic.

field (sketch VIII). Alternately the roll torque could be produced by a nonuniform field as shown in sketch IX. Thus the M vector changes direction along the model roughly as the magnetized core direction changes. By having similar currents in the forward and aft lateral coils and an opposing current in the mid-coil a field configuration like that in sketch IX is found. For further data see appendix of ref. 21.

Single-axis System

This system is shown in Fig. 5. It is laterally stable requiring an automatic control and stabilization only in the axial direction. From the discussion of the L system it is clear how this coil exerts an upward force. A number of systems based on this principle have been

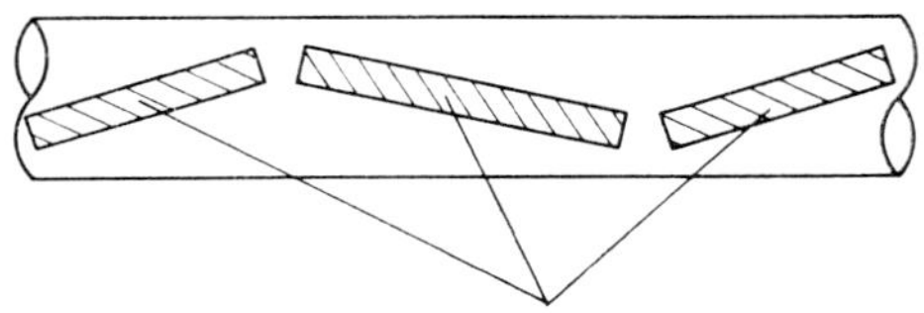

SKETCH VIII. Typical nonuniform core.

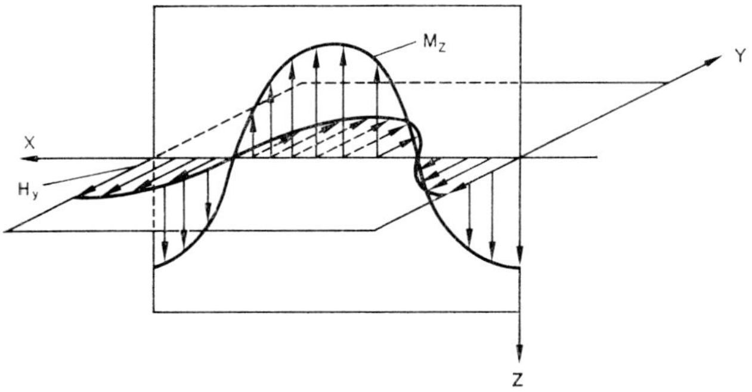

SKETCH IX. Nonuniform field and magnetization.

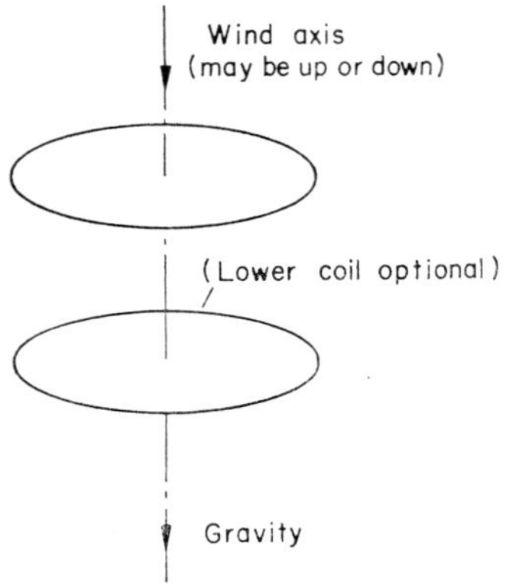

FIG. 5. Single-axis system.

built. The one-coil is an excellent demonstration system[15,16] because it is cheap and reliable. The two-coil system is a practical instrument.[14,27] They may in fact have separate magnetizing and gradient coils.

Orthogonal A System

The remaining configurations to be discussed are approached from the idea of magnetic symmetry.[9] Parker's analysis is set forth in a cylindrical coordinate in which the direction of the magnetization vector is in the principle direction. The gradient fields result in z and radial components, the azimuthal components being zero. Let us set (as usual) $z = z$, $x = \rho \cos \theta$, $y = \rho \sin \theta$ and substituting into the relation $d\bar{F} = d\,\text{Vol}\,(\mathbf{M}_s \cdot \nabla)B$. Note ϕ is the angle between M_s and the line connecting the gradient coil axis of symmetry.

$$dF_x = M_s d\,\text{Vol}\left[-\sin\phi\left(\cos^2\theta\,\frac{\partial B_\rho}{\partial \rho} + \frac{1}{\rho}\sin^2\theta\, B_\rho\right) + \cos\phi\cos\theta\,\frac{\partial B_\rho}{\partial z}\right] \quad (46)$$

$$dF_y = M_s d\,\text{Vol}\left[-\sin\phi\sin\theta\cos\theta\left(\frac{\partial B_\rho}{\partial \rho} - \frac{B_\rho}{\rho}\right) + \cos\phi\sin\theta\,\frac{\partial B_\rho}{\partial z}\right] \quad (47)$$

$$dF_z = M_s d\,\text{Vol}\left[-\sin\phi\cos\theta\,\frac{\partial B_z}{\partial \rho} + \cos\phi\,\frac{\partial B_z}{\partial z}\right]; \quad (48)$$

recalling $\nabla \cdot \mathbf{B} = 0$ $\qquad \dfrac{\partial B_\rho}{\partial \rho} + \dfrac{B_\rho}{\rho} = -\dfrac{\partial B_z}{\partial z}.$

Assuming a spherical core and integrating over the volume filled with magnetic material:

$$F_x = \pi M_s \sin\phi \iint \frac{\partial B_z}{\partial z}\,\rho\,d\rho\,dz \quad (49)$$

$$F_y = 0 \quad (50)$$

$$F_z = 2\pi M_s \cos\phi \iint \frac{\partial B_z}{\partial z}\,\rho\,d\rho\,dz. \quad (51)$$

The direction of the force in the x–z plane is determined by selecting ϕ. By choosing three independent sets of coils one obtains ϕ_1, ϕ_2, ϕ_3.

Thus

$$F_x = {}_1F_x + {}_2F_x + {}_3F_x \quad (52)$$

$$F_z = {}_1F_z + {}_2F_z. \quad (53)$$

By rotating the third set 90 degrees about **M** one finds $F_y = {}_3F_y$ which is perpendicular to ${}_1\mathbf{F}$ and ${}_2\mathbf{F}$.

When the direction angles of the three axes are chosen properly one can have three orthogonal non-interacting forces. One rather feels Parker favors the solution for which $\phi = \tan^{-1}\sqrt{8}$ (as opposed to $\phi_1 = 0$, $\phi_2 = \phi_3 = \tan^{-1}\sqrt{2}$, giving direction cosines

coil axis 1	$\cos\alpha = \frac{2}{3}\sqrt{2}$	$\cos\beta = 0$	$\cos\gamma = \frac{1}{3}$
coil axis 2	$\cos\alpha = -\frac{1}{3}\sqrt{2}$	$\cos\beta = \sqrt{2/3}$	$\cos\gamma = \frac{1}{3}$
coil axis 3	$\cos\alpha = -\frac{1}{3}\sqrt{2}$	$\cos\beta = -\sqrt{2/3}$	$\cos\gamma = \frac{1}{3}$.

Here α is the direction angle between x and the coil axis,
β is the direction angle between y and the coil axis,
γ is the direction angle between z and the coil axis.
This configuration is shown in Fig. 6.

A variation of this system with the wind axis corresponding to M then $\phi = 0$, etc., coil set was built and operated at Princeton in a 6-in. $M = 16$, He tunnel.

A family of balances based on Parker's solution[9] has been built and operated at the University of Virginia to provide drag of sphere in rarified gas flow.[28–32]

Further it is sufficient at this point to point out that coils for a similar family of balances have been designed, developed and constructed of superconducting material.[33–37] Such a system is very economical from a power consumption standpoint.[33] The start-up time is long and presently uses considerable liquid helium. No satisfactory lifetime economic analysis comparing costs of alternatives has been reported at this time.

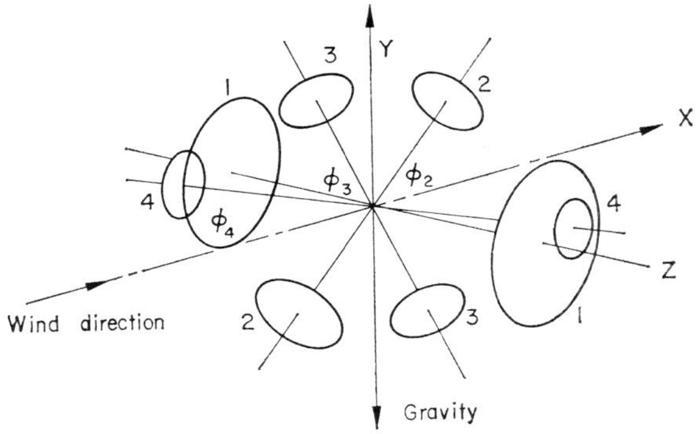

FIG. 6. Orthogonal A system.

Orthogonal B and C Systems

These two systems also possess symmetry and orthogonality. These configurations evolved from the MIT L system. A study of the configurations[21,22,25,38] shows the evolution quite clearly. Again the first step was separate magnetizing coils (also a Helmholtz pair). The axis of these coils was along the wind axis. Then the second step was to install transverse Helmholtz coil pairs to incline the uniform field with respect to the wind axis in both pitch and yaw. These pitch and yaw coils could all be excited by a low frequency voltage (400 to 1200 Hz) at a 90 deg phase to apply roll torque. The drag gradient was a second set of coils about the wind axis (actually wound into the magnetizing coils). The problem of the lift and side force was solved in two ways—in the orthogonal A, two double-L iron core coils were used, one set each for lift and side force. The result (also described in ref. 39) is shown in Fig. 7.

A different system, based upon the same principle is the so-called MIT-NASA prototype[22,40] or orthogonal C system which has been used up to a 7-in. tunnel. This system is straightforward for scale up as shown in the design study.[38] There are two primary differences between the orthogonal C and the orthogonal B systems: the location of the pitch and yaw coils and the arrangement of the lift and sine force coils. The former have been moved from outside the magnetizing and drag coils to inside adjacent to the test section. This change reduces the voltage for unsteady measurements, particularly in roll. The forces are produced more or less in the same way, except the iron path is outside the magnetizing and drag coils to reduce saturation. The result is shown in Fig. 8. Each ring of iron has four pole faces, two for lift and two for side force. For pure lift the coils at the left end are excited so that a vertical field (say downwards) exists between pole faces numbered four. The coils on the right end are excited so the field between the pole faces is upwards. Hence $\partial H_z/\partial x$ exists. The vertical force is M_x $(\partial M_z/\partial x)$. Similarly a side force can be made to exist.

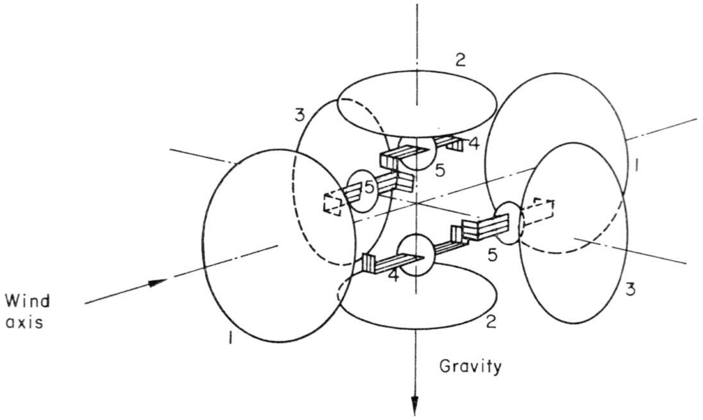

FIG. 7. Orthogonal B schematic.

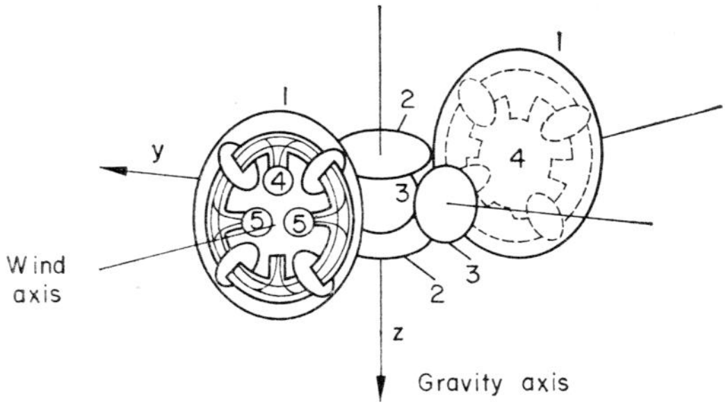

FIG. 8. Orthogonal C schematic.

Near the center of the balance (and in fact over a sphere of nearly 6 in. radius) the measured performance of the magnetizing coils are[22]

$$B_x = 20\ I_x \text{ gauss} \quad (I_x \text{ amperes}) \tag{54}$$

and the drag gradient is

$$B_{xx} = 1.70\ I_{xx} \text{ gauss/inch.} \tag{55}$$

The constants are more than 40% above the air core predictions due to the iron ring. However

$$B'_y = 4.7\ I'_y \text{ and } B'_z = 4.7\ I'_z \text{ gauss.} \tag{56}$$

These values average 25% below the air core estimates due to the iron rings. Similarly for the side force and lift system provide measured gradients

$$B_{xy} = 1.4\ I_{xy} \text{ gauss/inch} \tag{57a}$$

$$= B_{xz}. \tag{57b}$$

These fields and gradients appear to be orthogonal to better than 0.1 deg.

There exists obviously a fourth orthogonal system that is all air core. It consists of coils 1, 2, and 3 from the orthogonal C system plus the side force and lift coils from the Dukes–Zapata[7] or Dancy–Towler systems.[32] This would be an all air core system and would be easy to scale up as well as being a promising candidate for superconducting coils. It is not clear if the roll should be a.c. or d.c. at the time of writing.

The new orthogonal systems are characterized in Table 4. The principle difference between the orthogonal A and the orthogonal C and D is that the former produces forces only whole, the latter produces forces and moments. The former requires spherical cores while the non-magnetic material can assume a variety of shapes.

TABLE 4. DESCRIPTION OF RECENT ORTHOGONAL SYSTEMS

Orthogonal A Univ. of Virginia–NASA Cold suspension	Orthogonal C MIT–NASA Prototype balance	Orthogonal D Possible balance
4a	4a	4a
5b	5c	5c
6c	6a	6c
7a	7b	7a
8c	8b	8b
9c	9b	9 (a or b)
10c	10a	10c
11f	11h	11i

Cooling System

The power-handling characteristics of the coils are relatively small. Bitter[41] has defined structural and cooling limits (Fig. 9). It is clear from this figure that the fields (less than 50 kG or 5 Wb/m^2) used in magnetic balance systems are sufficiently small that neither cooling nor stress are limited in present technology. For normal (as opposed to superconducting) coils three alternative schemes have been used. The first and subsequent balances at ONERA[10,17,44,49] and the V system at AEDC[13] are all free convection air-cooled pancake coils although fans have been recently added to the latter. The orthogonal A system of Dancy and Towler[32] is conduction cooled, with ribbon windings being attached at the top and bottom to a cooled plate (see sketch X).

The third alternative[11] consists of square hollow-cored triple varnished tubing wet wound with epoxy cement. In addition to being strong and compact this type of coil has enormous cooling reserve if each layer of this coil is brought out to a manifold that allows it to be hydraulically in parallel and electrically in series. In this way the pressure drop is sufficiently small that the cooling can be connected to ordinary water mains. This winding arrangement, shown in sketch XI, has become common. Wilson and Luff[5] use this type of winding in a closed system, the heat ultimately being removed through an ordinary automobile radiator. They report no problem in eight years of operation because the closed system eliminates all need for filtering, chlorination and other water treatment. The closed system has many advantages and would seem to be desirable for new systems.

The superconducting coils[37] can also be cooled several ways. However, since they are wet wound from a Cu-clad niobium tin "litz" wire it seems likely in the long run that the litz wire can be wet wound on a thin nylon tube. After the coil is layed up, again with pigtails

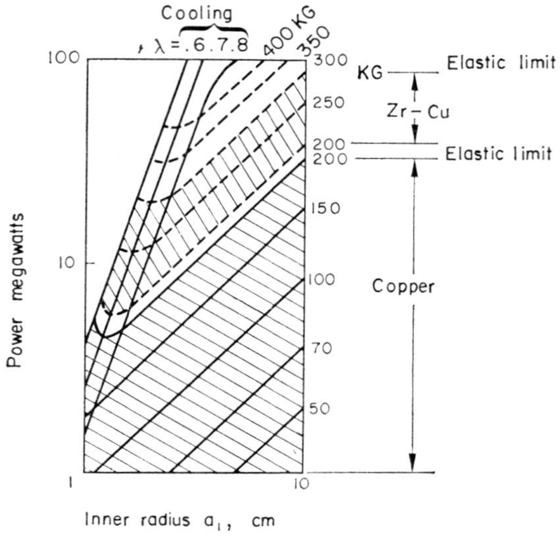

FIG. 9. Coil stress and cooling limitations.

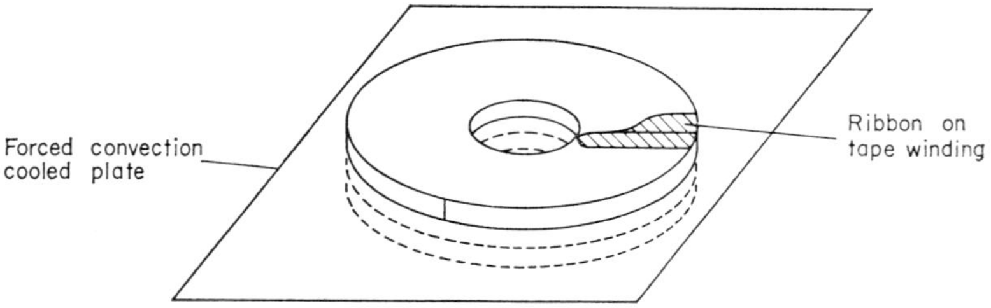

SKETCH X. Conduction cooled coil.

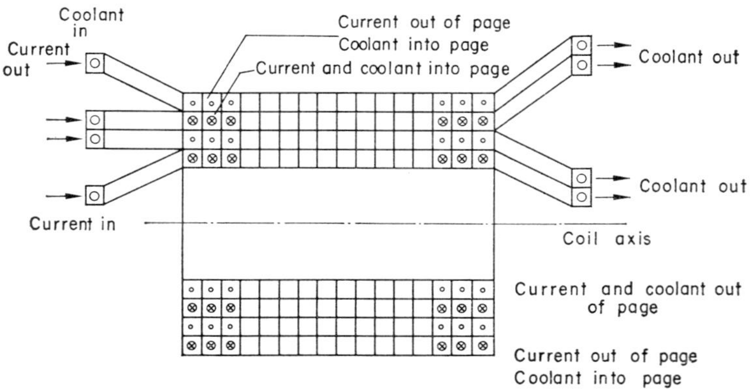

SKETCH XI. Schematic winding diagram illustrating series electrical path and parallel coolant path.

brought out at opposite ends of the winding, the nylon can be dissolved by pumping a solvent through the coil; a high-"Q" compact coil exists.[42]

Without appearing to dismiss this aspect of the design situation as superficial, we have not found any reports of cooling problems with magnetic suspension balance systems, particularly in the hollow-cored forced convection system. The thermal insulation problem for superconducting coils also seems to be well within the state of the art.[34]

Control and Position Sensing

Since all these systems are statically and dynamically unstable some sort of automatic control system is needed. This in turn requires some sort of position sensing equipment. In this section we will discuss position sensing first, then the control systems.

Position Sensing

At the outset it would appear that magnetic position sensing would be ideal. By sensing the change in the flux linkage between the model and the coil, a built-in sensing element would seem to be available. Detail studies, however, indicate the large air gap and relatively small model size combine to reduce the sensitivity to the point that this "ideal" scheme

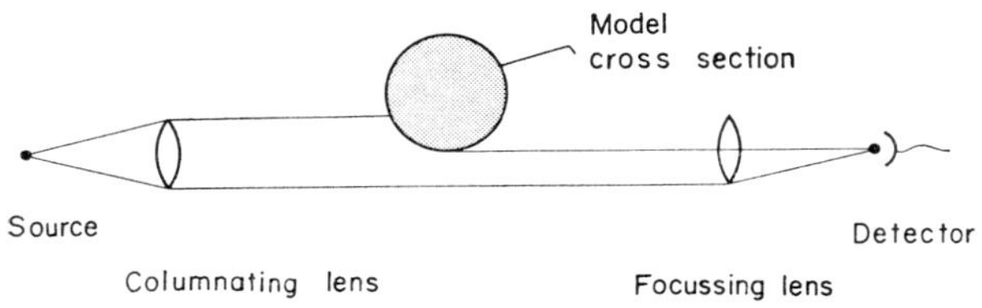

FIG. 10. Narrow light beam position sensor.

is impractical. The desired characteristics for a position sensing system that have been established are the following:

(a) No mechanical contact with the body.

(b) Translational displacements of the model parallel to the orthogonal wind tunnel axes will be transformed into three independent high level electrical signals proportional to the translational displacements over a suitable range.

(c) Angular displacements of the model about the wind tunnel axes will be transformed into three or more independent, high level electrical signals proportional to or strong functions of the angular displacements, over a suitable range.

(d) Resolution, repeatability, and frequency response must be compatible with desired accuracy of static and dynamic wind tunnel results.[43]

(e) The system must be easily adaptable to a broad range of model geometry.

(f) The system must be compact, and must not interfere with access to the test section or prevent visual contact with the model.

(g) The system must operate without interference from the magnetic fields used to support the model.

Laurenceau and Tournier[10] resorted to an optical system that is more or less in use up to the present time. The original form of each sensor is shown in Fig. 10. At ONERA five

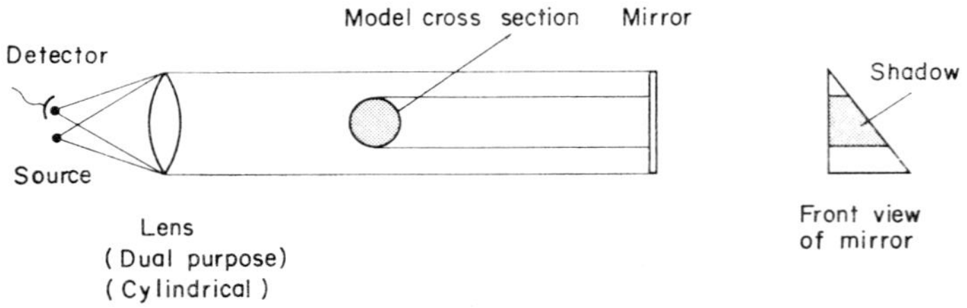

FIG. 11. Large beam position sensor.

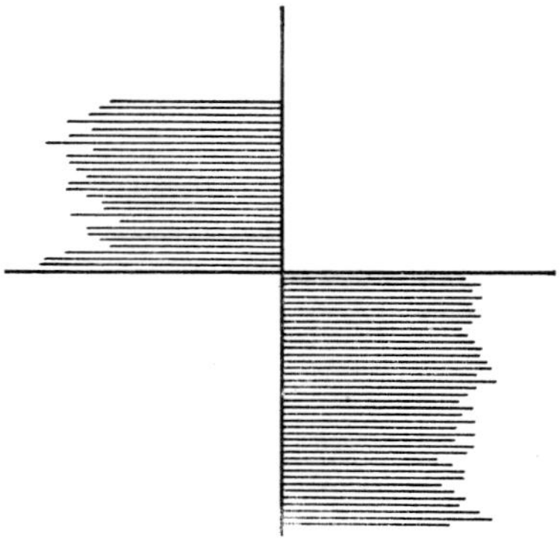

FIG. 12. Flying spot scanner target.

such assemblies were used, forward lift, aft lift, forward lateral, aft lateral and drag. In this arrangement the signal output from the detector is proportional to the position of that part of the model cutting off the part of the light beam. As shown, the lower the model, the less light, which implies more current in the coil. The large beam system[11] permits the model to be controlled over a larger distance. This system is shown in Fig. 11. Here, it is the triangular mirror that controls the amount of light reflected to the detector. As the model falls, less light is reflected and more current is needed in the coils. Again one of these assemblies is needed for each degree of freedom controlled. The use of light presents some problems and at AEDC[20] a low energy X-ray system has been put in to use and as expected works nicely. Any of these systems, when properly tuned, will measure to ± 0.001 to ± 0.0015 in. Adaptation of optical method to measure roll is straightforward.[18,19,26]

Three additional schemes warrant some discussion: a flying spot scanner technique,[44] and edge-looker[18] and the electro-magnetic position sensing system.[21] (Incidentally, there is a good summary of a large class optical position sensing system in ref. 18 including shadow scanning and image tracking as well.) Moreau's system is based upon the idea of scanning across a high contrast target on the model, resembling that shown in Fig. 12. A rectangular scan compares the time of light and dark area with a reference in the iconoscope. The system will resolve 8.4×10^{-2} mm or 0.003 in. on an arbitrarily shaped model for the worst possible case. The model angle is measured to 0.1 deg up to 45 deg.

The edge looker is based upon the light passing through a vibrating slit. This light falls on a photodetector whose output drives a servo that moves the mean position of the slit to a point where the slit is illuminated half the time. That is, the slit is centered in the edge of the

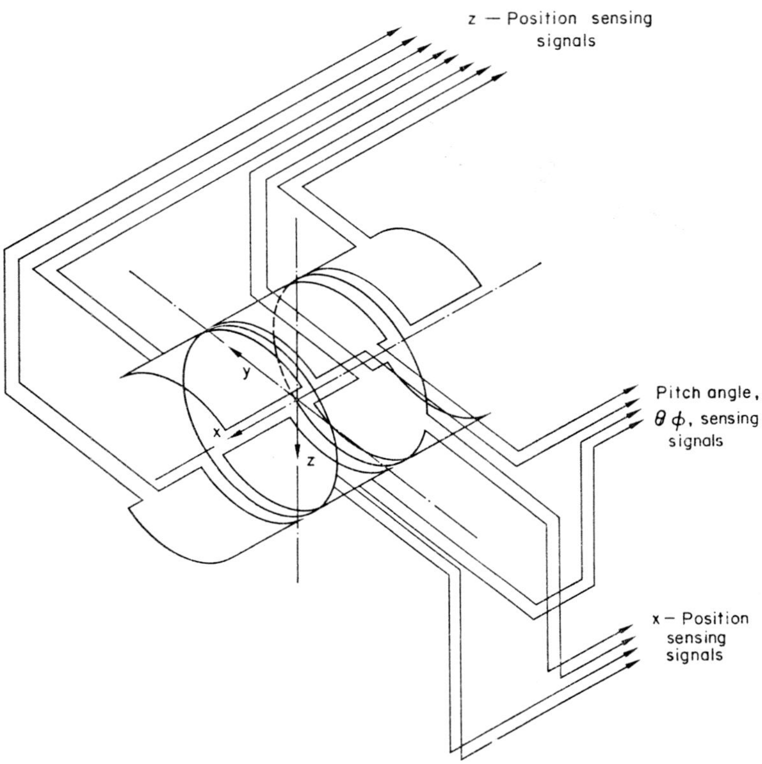

FIG. 13. Electromagnetic position sensor—pitch sensing plane.

model. The system actually measured to within $\pm$ 0.0005 in. over a frequency range of zero to 60 Hz over a total range of 1.14 in. A bench model of this system measured up to 30 deg angle of attack to $\pm$ 0.03 deg.

Although promising, the edge looking system was not fully independent of model geometry and its development was deferred in favor of the electromagnetic position sensing system.[21] This system consists of a Helmholtz coil pair located around the flow (i.e. the coil axis coincides with the wind axis) excited at 20 kHz. There are four saddle coils sequentially located on the inner cylindrical boundary of the Helmholtz pair and between them. Four saddle coils are arranged upstream of the forward coil of the Helmholtz pair (like a mirror image of these saddle coils between the Helmholtz pair). Four more are located downstream of the aft coil of the Helmholtz pair. These coils operate differentially (like the coils in a Schaevitz transformer) to produce the appropriate signals. Figures 13 and 14 show how different combinations give different output signals. Roll can be detected by measuring difference between the top and bottom and side coils of the center four, if the model has a degree of asymmetry. The asymmetry may be either in the core shape[42] or in a loop of copper inclined at 20 degrees or so to its longitudinal axis of symmetry.

The operation of the system is briefly as follows:

(a) Excitation: The power required to drive the excitation windings of the E.P.S. trans-

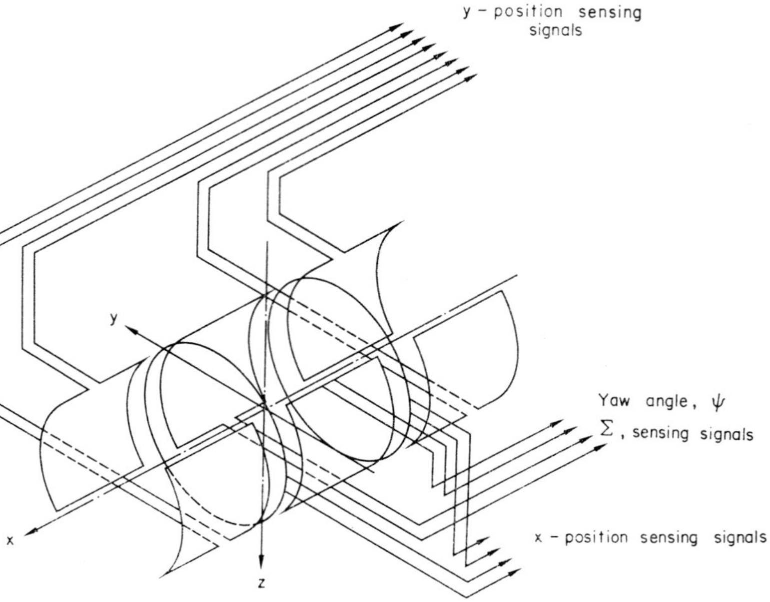

FIG. 14. Electromagnetic position sensor—lateral plane sensing.

ducer coils is provided by an audiofrequency power amplifier driven by a crystal-controlled master oscillator.

(b) Preamplification: The output a.c. signals from the transducer coil (seven channels) are filtered to remove noise outside the desired bandwidth, and residual excitation-frequency voltages are nulled at this stage.

(c) Signal combination: In order to provide separate a.c. voltages proportional to lateral displacement, yaw angle, roll angle, vertical displacement and pitch angle, the preamplified signal voltages from the saddle-coil pairs must be combined. A further stage of amplification is provided for these functions.

The axial channel signal, which is not combined with any other signal, is also amplified further in a stage following the preamplifier.

(d) Phase-sensitive demodulation: The a.c. voltages from the second amplifier stages are rectified in full-wave phase-sensitive demodulators to produce d.c. voltages proportional to the a.c., and with polarity corresponding to the phase of the a.c. signal before demodulation.

This system measures to ± 0.0008 in. and 0.1 deg in all components. The calibration of this system must be shifted with model geometries. Moreau[44] considers this a handicap, but since all automatic control systems must be retuned for each model, and since the adjustment is simple this is not a serious drawback. Moreau[44] and Zapata[37] in their use of this system have reported a more serious shortcoming due to electromagnetic interference between the noise pick-up from the power signals and the sensing coils. Filtering the power supply output can reduce this problem to a manageable level,[39] but the success depends upon system time constants. The electromagnetic compatibility between all these systems is a problem whose solution requires careful attention.

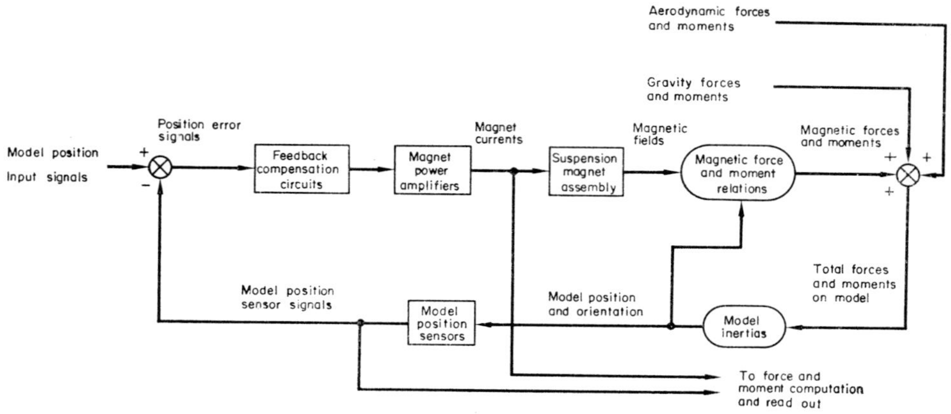

FIG. 15a. Functional block diagram of suspension.

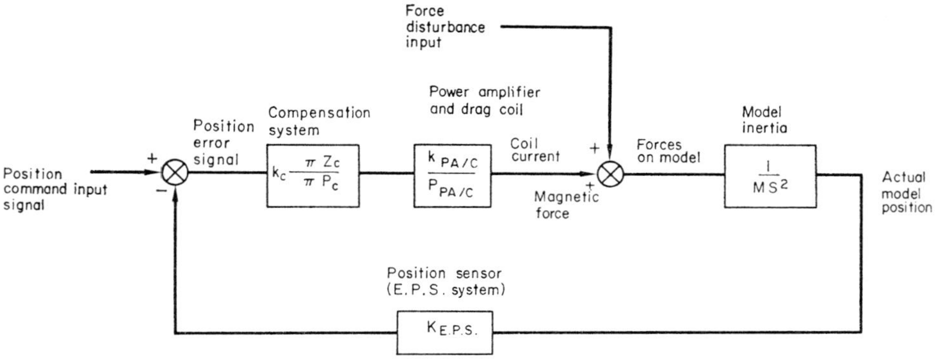

FIG. 15b. Servo-loop diagram of suspension system.

Compensation System

This section describes the design of the electronic compensation system which provides control of the stability and the response characteristics of the magnetic suspension system. This system is located in the control loop between the model position sensor system and the magnet power amplifiers, as shown in Fig. 15a. The compensation system can be quite simple and effective for a wide range of models and angles of attack; refinements may be added to extend the range of the system.

The most simple compensation scheme is based upon the assumption that each degree of freedom is uncoupled from the others. The compensation then takes the form of electrically uncoupled channels which may be designed and analyzed independently. This assumption is approximately valid for all magnet system designs described in this review, at least for small angles of attack. Consider first the translational components. One channel is shown in simplified form in Fig. 15b.

The magnetic force is assumed to be proportional to and in phase with the magnet coil current. The magnet coil current-power amplifier input signal transfer function is assumed

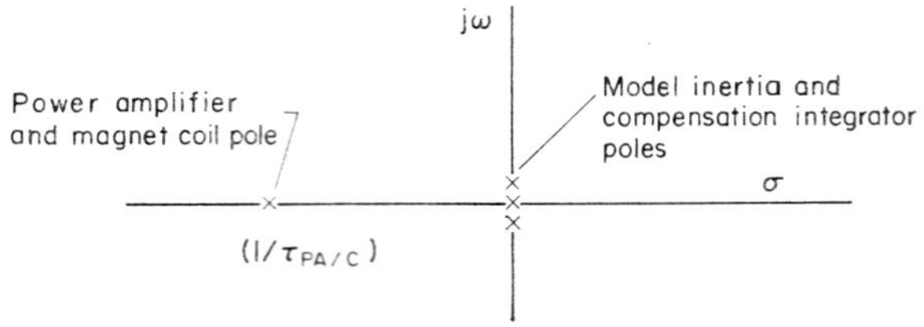

FIG. 16. Pole-zero plot of magnetic suspension control loop with integral compensation only.

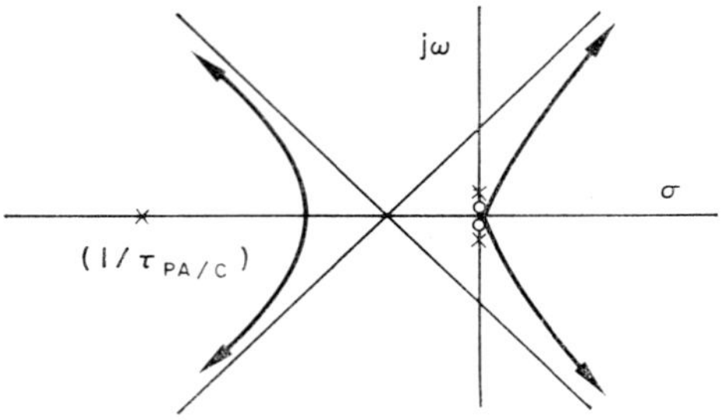

FIG. 17. Root locus of integrally compensated suspension loop.

to be characterized as a first-order lag. The measured model position is assumed to be proportional to and in phase with the actual model position. The effect of model inertia, defined as the displacement force transfer function, is characterized by a gain inversely proportional to the model mass and a double integration. The compensation system is assumed to be a ratio of two polynomial complex functions, factored into poles and zeros.

The design problem is resolved into three basic parts. These are:

(a) Stabilization of the loop.

(b) Minimization of position error for given force disturbance inputs.

(c) Minimization of position error for given position command inputs.

The design proceeds as follows: for steady disturbance inputs, it is desirable to maintain zero position error. This requires an integration in the compensation system. The pole-zero plot of the control loop will then appear as in Fig. 16. The loci of the closed-loop poles of this system are sketched in Fig. 17.

Two closed-loop poles immediately branch into the right half of the *s*-plane, and the loop is unstable for all values of loop gain.

In order to stabilize the system, lead compensation is required. That is, zeros must be introduced into the left half of the *s*-plane in order to bring all of the root loci into the left half, or stable region of the *s*-plane. Satisfactory stabilization of the system is not practical with a single zero. Two zeros can provide the loci shown in Fig. 18. The closed-loop response of the system shown in Fig. 18 is oscillatory, with small damping ratio, and is not satisfactory. Some improvement may be made by placing the compensation zero closer to the origin, as shown in Fig. 19.

In this case, the closed-loop poles can be critically damped (non-oscillatory). However, the speed of response is decreased, since the closed-loop poles are closer to the origin.

In practice, due to noise considerations, it is usually not practical to use a pure zero

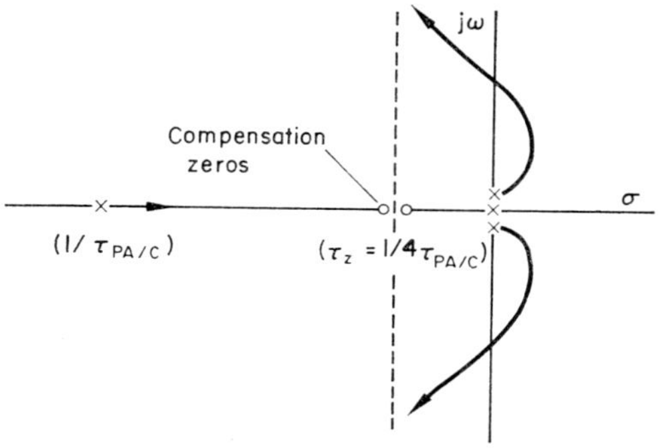

FIG. 18. Effect of two negative real zeros on root loci ($\tau_z = 4\tau_{PA/C}$).

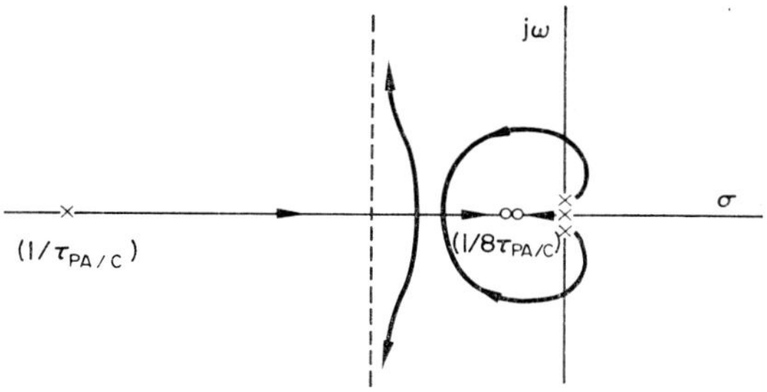

FIG. 19. Effect of increasing compensation zero time constants ($\tau_z = 8\tau_{PA/C}$).

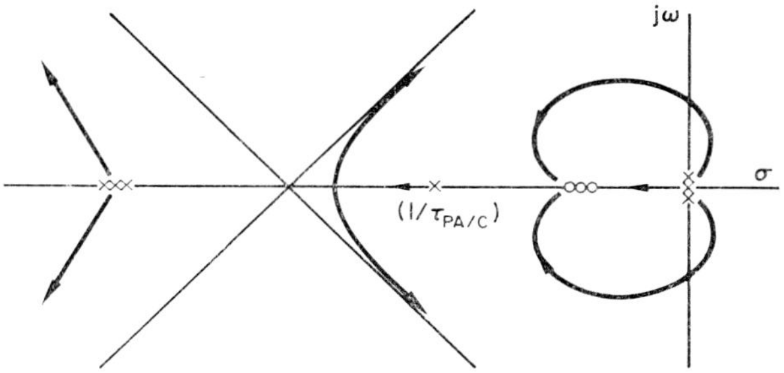

FIG. 20. Root locus of three-zero and integral compensated translation component loop.

without introducing a smaller time-constant pole in association with it. Usually the associated pole is no more than twenty times as far from the origin as the zero. In the case of the system of Fig. 19, for example, two additional poles would be required at approximately two and one-half times the distance of the power amplifier and coil time constant from the origin. These additional poles would have only a minor effect on the dominant closed-loop poles, and the response of the system.

Further improvements are possible by addition of a third compensation zero. Three zeros allow the closed-loop poles to be moved farther from the origin, and hence speed up the transient response. An example of a three-zero locus plot is shown in Fig. 20.

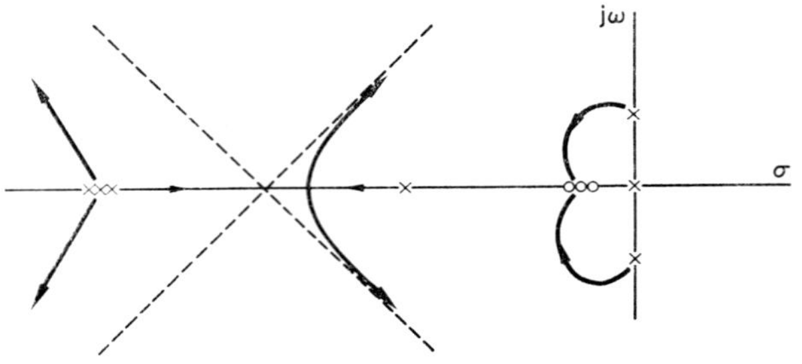

FIG. 21. Root locus of three-zero and integral compensated angular component loop.

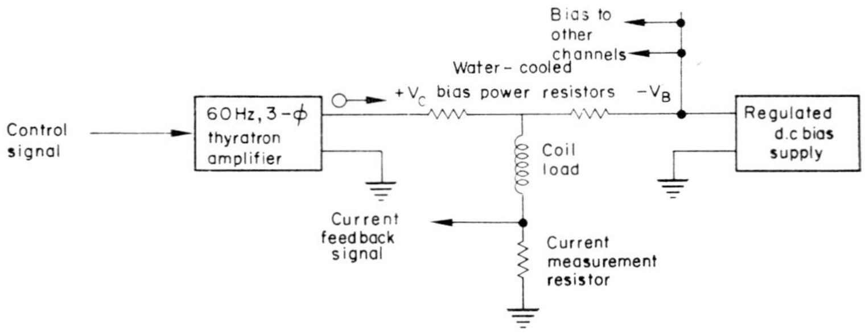

FIG. 22. Schematic of prototype power amplifiers (ref. 11).

Pitch, Yaw and Roll Component Compensation

The pitch, yaw and roll loop dynamics are approximately the same as the translational components. The main difference comes in the angular response of the model to coil current or moment inputs. The model behaves as if held by a torsional spring, and possesses natural frequencies in pitch, yaw and roll corresponding to the magnetic and aerodynamic restoring moments and the moments of inertia of the model about the pitch, yaw and roll axes. This has the effect of separating the model dynamics poles symmetrically from the origin, along the imaginary axis. A typical three-zero plus integrator root locus is shown in Fig. 21. This arrangement has proved satisfactory.[39]

Power Supplies

The initial power amplifiers consist of 60 Hz,* three-phase full-wave thyratron controlled unipolar output amplifiers d.c. biased through resistor networks with opposite polarity. The general arrangement is shown in Fig. 22. Similar circuits have been built up using power diodes.

With this arrangement, reasonable response can be expected up to 30 Hz; above this, the response breaks up into modulation products or side lobes of 60 Hz, and overall system response is relatively poor. This presents problems at the highest pitch and yaw natural frequencies in the prototype, and in fact imposes the most important limit on the response and stability of the system. Power supplies of this kind perform much better with current

* The 60 Hz is the line frequency. 50 Hz is equally useful.

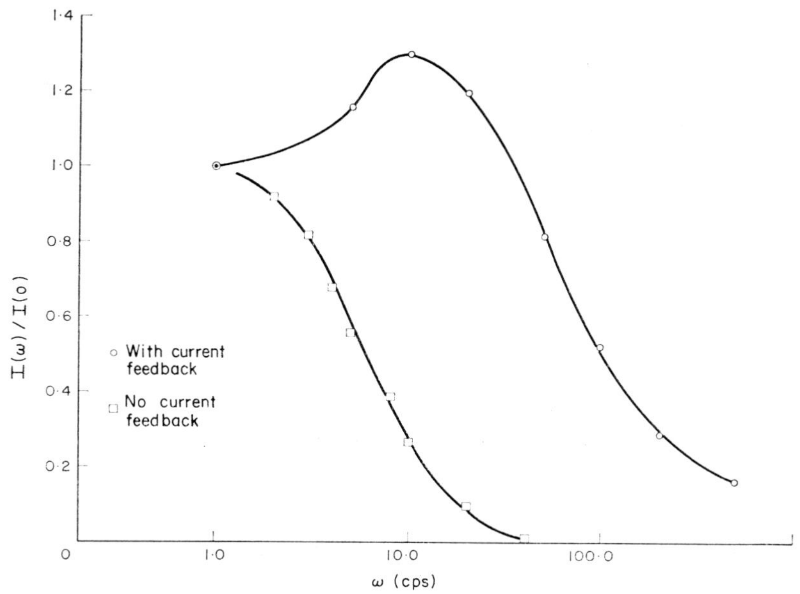

FIG. 23. Frequency response magnetic balance power supply.

feedback (Fig. 23). The current feedback also has the effect of linearizing this class of power amplifiers.

Since the pitch and yaw natural frequencies decrease with increasing system size, and since it appears appropriate to accordingly reduce system bandwith, the 30 Hz limit inherent in this basic power amplifier design is satisfactory for systems larger than the prototype. This presents important advantages, since reversing, three-phase, full-wave SCR controlled amplifiers are readily available as off-the-shelf systems and have all the corresponding cost and reliability advantages. Furthermore, the market for such systems (primarily variable speed mill motor drives) is quite large and competitive.

It is necessary to minimize high-frequency noise at the output of the power amplifiers in order to prevent interference with the electromagnetic position sensing system. If the power amplifier is of the phase-controlled SCR or thyratron type, then relatively high noise levels are encountered. The most straightforward method of attenuating this noise is by passive filtering of the output. Two filter configurations are shown in Figs. 24a and 24b.

In configuration 24a, the shunt capacitance C1 is made to tune the load inductance at approximately 150 Hz, and the series inductance L1 is made no larger than 0.1 times the load inductance. At the electromagnetic position sensor carrier frequency (20 kHz, for prototype), the attenuation contributed by the filter is therefore approximately 60 db, and insertion loss is approximately 1 db, at low frequency. The series filter inductor L1 must be designed to carry the maximum rated current and must have negligible parallel capacitive reactance at the position sensor carrier frequency.

The band-reject filter scheme of 24b offers better performance; the attenuation is proportional to the product of the quality-factor Q ($= \omega L/R$) of each of the two filter inductors, and the ratio of the series inductance to the shunt inductance. The series filter inductance must carry the full load current, and since it is tuned it must be air-cored (inductance independent of current). The shunt inductor carries no d.c. and very little a.c., and thus can be relatively easily designed with a high Q. If $L_2/L_3 = 10$, $Q_2 = 30$, $Q_3 = 300$ (at 20 kHz) and stray coupling effects are negligible, the attenuation is thus approximately 96 db, or almost a factor of 100 better than the low-pass filter design shown in Fig. 24a.

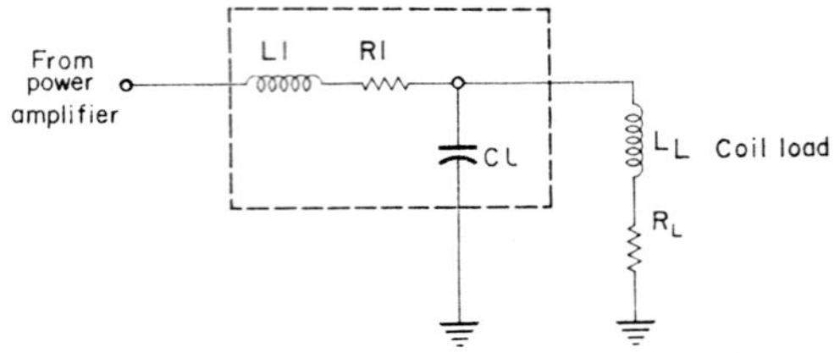

FIG. 24a. Low-pass single-stage output filter.

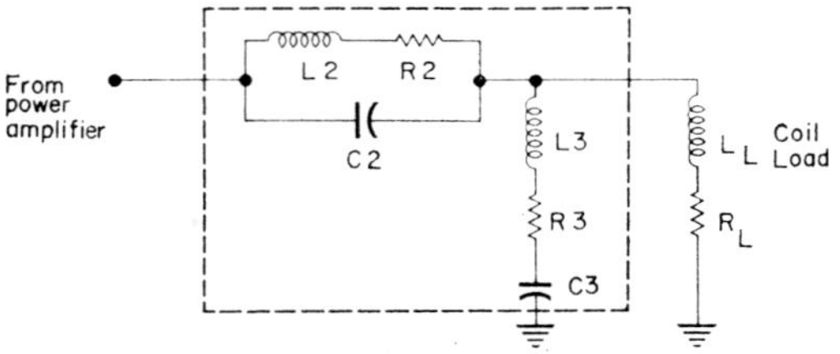

FIG. 24b. Band-reject single-stage output filter.

Roll-control Power Amplifiers and Coupling Networks

The power amplifiers required for roll control consist essentially of a pair of amplitude-modulated fixed center-frequency power supplies. For a given model configuration and size relative to the suspension system, the minimum required center frequency varies inversely with the square of the size. For the prototype system[22] ($D = 7.5$ in.), a roll-control center frequency of 400 Hz appears to be acceptable. For systems twice the size of the prototype or larger, the line frequency (50 or 60 Hz) would be acceptable and has the advantage that it is readily available and easily controlled.

The roll-control amplifiers must be coupled to the saddle coils in such a way as to ensure the maximum power transfer to the coils and a minimum loss through the pitch and yaw power amplifiers. The prototype arrangement is shown in Fig. 25. Parallel-tuned trap filters present high impedance to the roll-control power and prevent significant losses through the pitch and yaw control amplifiers. The saddle coils are series-tuned with the roll-control amplifiers, and optimum coupling is obtained by adjustment of the impedance-matching output autotransformers.

Performance of the a.c. roll system is dependent to a large degree upon the model geometry. Therefore, only general limit specifications can be given. The maximum steady rolling moment for a particular model, system size and carrier frequency is proportional to the roll power supplied to the saddle coils. The maximum power is related to the saturation limits. This power limit corresponds to roll control fields of the order of 2500 G and roll power levels at the saddle coils of 120 kVA for the prototype. From the limited experience with the prototype, usable rolling moments have been obtained with 1 kVA.

The saturation-limited power increases in proportion to the system size, and the maximum rolling moment for a given model configuration varies as the cube of the size.

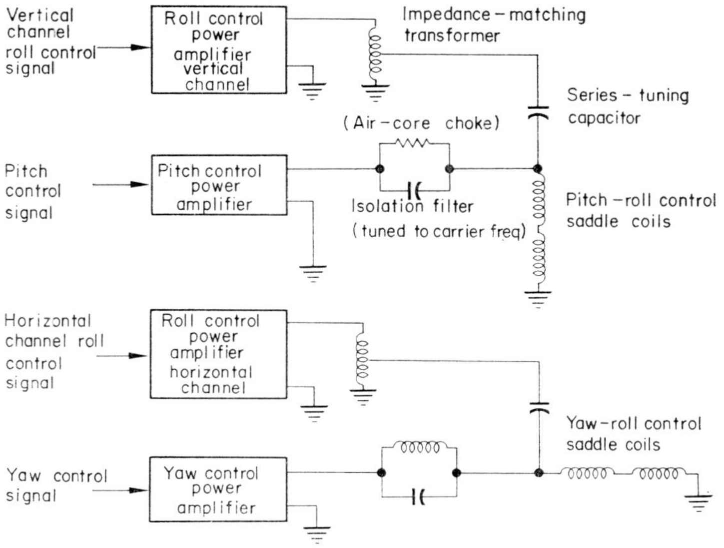

FIG. 25. Pitch-yaw-roll power amplifier coupling scheme using separate roll power amplifiers.

Amplifiers for Superconducting Coils

The results of Hamlett[24] and those of Frank Moss[45] indicate copper-clad niobium litz wire superconducting coils behave dynamically in the same way as normal coils. These findings are important. They imply suspension and balance systems made with superconducting coil systems are practical. The power supplies for these systems require certain important differences from the phase modulated "rectifiers" discussed above. For the superconducting system the current rises linearly in response to a constant supply voltage. Hence filtered and smoothed fluctuations are imperative. Typical requirements are of the order of 10^{-6} V per volt. D.c. output at $\pm$ 210 V is supplied by a rectifier with sufficient internal impedence to limit the current to safe levels.[46] The power supply acts in an off–on mode with three positions: + on, off (neutral) and — on. The power supply can supply 0 to 350 A in 16 msec. The current level in turn acts to build up the magnetic field, linearly in time, and so too the acceleration of the magnetic model. Thus the model position obeys the law

$$\mathbf{r} = \frac{\mathbf{F}_0}{2m} \int_0^t \epsilon(\tau)\,(t - \tau)^B\, d\tau \tag{58}$$

where $\epsilon(\tau)$ is the output from the control system $\epsilon(\tau) = 1, 0$ or -1 times a signal duration. Incidentally, this power supply is well suited to a sampled data digital control system. There seems to be no particularly unsafe aspects of this use of superconducting coils because the actual energies are not extreme.

Scaling Laws

The problem of setting forth a set of scaling laws is complicated by deciding *a priori* what will be held fixed in the process. Individual processes are well understood and can be discussed simply.[21,22,36,40,47] It is also possible to even refer to standard texts on electromagnetic theory for the coil scaling law. For purposes of discussion consider separation into electric scaling, magnetic scaling, inertial scaling and aerodynamic scaling.

Electrical scaling laws: magnet coils

(i) The low-frequency resistance $R_{\text{d.c.}}$ of a magnet coil of a given configuration, having a characteristic linear dimension d and composed of n turns of uniform size, of which the conductive portion accounts for a proportion P_f (packing factor) of the total cross-section of the coil, and of resistivity ρ, will be proportional to ρ, n squared, and inversely proportional to the linear dimension d, and the packing factor P_f, i.e.

$$R_{\text{d.c.}} \propto \frac{\rho n^2}{dP_f}. \tag{59}$$

(ii) The self-inductance L of a magnet coil of a given configuration, having a characteristic linear dimension d and composed of n turns of uniform conductor, is proportional to the number of turns squared, and the linear dimension, i.e.

$$L \propto n^2 d. \tag{60}$$

(iii) The voltage/current time-constant, $L/R_{\text{d.c.}}$, of a magnet coil of given configuration is proportional to the linear dimension d squared and the packing factor, and is inversely proportional to the resistivity (i and ii), i.e.

$$\frac{L}{R_{\text{d.c.}}} \propto \frac{d^2 P_f}{\rho}. \tag{61}$$

Magnetic scaling laws: fields and gradients

(i) The components of the static or low-frequency magnetic field at any point of interest can be expressed as the sum of components due to the ampere turns of the yth coil in the system. After normalizing with respect to a characteristic dimension D

$$\mathbf{B}\left(\frac{x_i}{D}\right) = \frac{1}{D}\sum_{j=1}^{N} \mathbf{f}_{ji}\left(\frac{x_i}{D}\right)(NI)_{ji}. \tag{62}$$

(ii) Similarly the gradient $\mathbf{B}$, which is a tensor field of rank 2, can be written

$$\nabla\mathbf{B}\left(\frac{x_i}{D}\right) = \frac{1}{D^2}\sum_{j=1}^{N} \nabla f_{j2}\left(\frac{x_i}{D}\right)(NI)_{j2}. \tag{63}$$

These five laws allow definite statements to be made if the magnetic configuration is held fixed. However, this is not sufficient. Suppose we postulate geometric similarity for the coils and the model, and hold the magnetization per unit volume fixed,

$$\mathbf{M}_1 \simeq \frac{\mathbf{H}_1}{D_A} = \frac{NI)_{1M}}{R_1 D_A} \cdot \mathbf{d} \tag{64}$$

for a second system

$$NI)_{2M} = NI)_{1M} \frac{R_2}{R_1} = NI)_{1M} \bar{\bar{S}}. \tag{65}$$

So if $R_2/R_1 = \bar{\bar{S}}$ is the scale factor $\bar{\bar{S}}$ is greater than one if the second system is larger than the first. Thus the ampere turns for magnetization of the model (to the same level) increases linearly with the scale factor.

Inertial Scaling

(i) The linear momentum involves mass times acceleration, which can be written as

$$m\omega^2 \text{ or } F = \rho_M \text{Vol } \omega^2 \tag{66}$$

if the mass is expressed in terms of the density of the model and its volume.

(ii) Similarly the angular momentum takes the form (D is a characteristic length)

$$T = \rho_M \text{ Vol } D^2 \omega^2. \tag{67}$$

Aerodynamic Scaling

The well-known relation for forces and torques here is

(i) $$F = qS\, C_F\, (\alpha, Re, M, k, \ldots) \tag{68}$$

(ii) $$T = q\, S\, D\, C_T\, (\alpha, Re, M, k, \ldots) \tag{69}$$

(q = dynamic pressure)

The latter two relations show a generalized force coefficient (C_F) and the torque coefficient (C_T) that depend upon airframe attitude with respect to the relative wind (α). Reynolds number, Mach number and reduced frequency $k = \dfrac{wD}{U_\infty}$.

To carry the scaling further, one must relate the aerodynamic, inertial and magnetic forces and torques.[47] Two clear ideas become apparent, static requirements ($\omega = 0$) and dynamic testing requirements.

Static Scaling

To carry out static scaling the ratio of magnetic forces and torques will be held constant. Thus, for example, the lift system scales up in the following way, since Mach number and C_L = constant,

$$\frac{qS}{\text{Vol } M_x \dfrac{\partial B_z}{\partial x}} = \text{constant.} \tag{70}$$

Thus

$$\left.\frac{\partial B_z}{\partial x}\right|_2 = \left.\frac{\partial B_z}{\partial x}\right|_1 \frac{q_2}{q_1} \cdot \frac{S_2}{S_1} \frac{(\mathrm{Vol}_1\ M_{x_1})}{(\mathrm{Vol}_2\ M_{x_2})} \tag{71}$$

$$= \frac{q_2}{q_1} \cdot \frac{1}{\overline{\overline{S}}} \left.\frac{\partial B_z}{\partial x}\right|_1 \tag{72}$$

where, as before, the magnetization is constant.

In terms of ampere turns the gradient scaling (sub G)

$$NI)_{G2} = \frac{q_2}{q_1} NI)_{G1}\ \overline{\overline{S}}^{-1}. \tag{73}$$

Note the increased model volume acts to reduce the amount of gradient ampere turns from 1 to $\overline{\overline{S}}$. Note the gradient ampere turns increase linearly with dynamic pressure. Similar relations may be written for the ampere turns required for torques (sub T):

$$NI)_{T2} = \frac{q_2}{q_1} NI)_{T1}\ \overline{\overline{S}}. \tag{74}$$

In this case the volume effects just cancel out and the increase in ampere turns is due only to scaling of the coil geometry.

Dynamic Scaling

Dynamic scaling is more difficult because additional parameters, the reduced frequency and the model density are now important. Assuming the model density is the same, scaling angular acceleration requires

$$NI)_{T,d_2} = NI)_{G,d_1} \cdot \left(\frac{k_2}{k_1}\right)^2 \cdot \overline{\overline{S}}^3 \cdot \left(\frac{U_2}{U_1}\right)^2 \tag{75}$$

and for linear acceleration the gradient scaling law becomes

$$NI)_{G,d_2} = NI)_{G,d_1} \left(\frac{k_2}{k_1}\right)^2 \left(\frac{U_2}{U_1}\right)^2 \tag{76}$$

$$= \left(\frac{\omega_2}{\omega_1} \overline{\overline{S}}\right)^2 NI)_{G,d_1}. \tag{77}$$

Usually one desires to have ω as large as possible which implies a large increase in NI.

Finally one can scale in terms of acceleration.

(a) Axial force per pound of iron in model, with model at zero angle of attack and zero pitching and yawing moments:

$$F_x/W = 4.7 \times 10^{-6} \frac{B_x}{D_A} B_{xx} \quad (g\text{'s}) \tag{78}$$

where D_A = demagnetizing factor of iron part of test model, in axial direction.

(b) Vertical force per pound of iron in model, with model at zero angle of attack and zero pitching and yawing moments:

$$F_z/W = 4.7 \times 10^{-6} \frac{B_x}{D_A} B_{xz} \quad (g\text{'s}). \tag{79}$$

Restraints for Use with Wind Tunnels

The practical wind tunnel magnetic suspension or balance system must satisfy several conditions. A clear quasi-cylindrical volume in the center of the system is needed for the wind tunnel test section. For a closed jet, the tunnel walls must be made of low electrical conductivity, non-magnetic material. Non-magnetic stainless steel and fiberglass and resin nozzles have been used successfully for this purpose. In the case of an open-jet nozzle there must be sufficient volume so the jet clears the position sensors, the pole faces, and the inside coil layer. Usually the length of the balance in the suspension system will exceed 4–6 test section diameters. Hence either the tunnel will have a long test section of the effusor nozzle and the diffusor inlet will be located within the confines of the balance. Both alternatives have been adopted and work equally well. For a fixed balance size, a transonic test section will be smaller than either a subsonic or supersonic test section because of the volume that must be included in the plenum chamber.

Provision must be made for model insertion. This is simple for the L, double-L, and V systems, indeed it may be manual. Insertion is increasingly difficult and complicated for the more orthogonal systems. Stephens[22] took advantage of the Eiffel tunnel configuration and mounted the inlet bell on rollers to permit ready access to the tunnel test and thus allows manual insertion of model sections. More commonly, the lateral and pitch coils have air cores, and model access is possible using this region. In fact the air core is also used as a viewing port.[7,22,25,32,38]

A variety of clever mechanical and pneumatic instruments have been built for model insertion and removal.* Dukes and Zapata,[7] Wilson and Luff,[5] and Crane[48] have described a system that approaches the model downstream and grabs the base with a three-prong clamp. This seems superior to the center line spike inserted in the model base that Mirande[49] describes. A pneumatic three-clamp external system was built and operated by Dubois and Rogue[17] at ONERA at $M = 7.5$. In this system considerable care is taken to insure that the shock waves interact with the model symmetrically. The simplest system[32] consists of a small hollow cone or funnel attached to a suction system. The model, a small sphere, is held by air pressure in the funnel, and can be released at will. Bouseman[18] constructed a two-position lateral system that worked well at $M = 4.26$ and 4.8. This system consists of two annular clamps, forward and aft, with four adjustable tabs in the interior. The inner radius of the annular is sufficiently large that the shock interaction is weak. In operation the clamps open, then the arm retracts, and the model is free. For model retrieval the arm is inserted and the clamps close, turning off the integrator at the same time.

The necessity for model insertion retraction equipment is somewhat dependent upon system time constants. Each case must be treated on an individual basis, as shown below. In many cases manual insertion before the wind is turned on and manual retrieval after the wind is turned off seem to be the most practical approaches.

Power-amplifier-limited Starting Loads[21]

The chain of events that occurs during the transient period in which the wind tunnel flow is established is as follows. The test model is held in the magnetic suspension system in the attitude corresponding to minimum drag. The magnetization of the model is adjusted to the maximum level. The drag coil current and resultant drag field gradient $\partial H_x/\partial_x$ are zero at

* Note that the integral control must be turned off during this kind of model insertion.

this time. The tunnel air flow is then started. It is assumed that the drag force on the suspended model can be characterized by a step change equal to the steady aerodynamic drag load corresponding to the dynamic pressure at equilibrium, and in addition, an impulse corresponding to the passage of the starting shock downstream past the model. (There is considerable debate about the magnitude of this impulse; a reasonable assumption might be that the peak force is three times the steady drag force, and the effective duration of this force is of the order of 10 to 100 μsec.) At the instant the aerodynamic drag force begins, it is assumed that the model position signal, as modified by the control networks, calls for full control power from the drag power amplifier. That is, the power amplifier immediately "saturates". This means that the full amplifier supply voltage is applied to the drag coils. The response of these coils is characterized by a single lag, of time-constant equal to the ratio of the inductance to resistance. The consequence of this is that the drag gradient, and hence the drag force, increases with time, and this force asymptotically approaches the maximum magnetic drag force available for the particular model. If it is assumed that the amplifier remains "saturated", it is possible to predict the maximum downstream displacement of the model due to the starting transient. This displacement is the crucial factor in the consideration of the starting problem. For instance, the model must not travel very far downstream, since the position sensor has a total axial measurement range of no more than one-half the tunnel radius. A quantitative analysis of the maximum displacements for the typical test models is given in ref. 21. The results from ref. 21 are given in the following examples.

Consider a balance system whose coil time-constant τ is 0.20 sec, and the maximum drag gradient $(\partial H_x/\partial_x)_{max}$ is tentatively set at 0.380 kG/in. For typical test models with base diameters of one-eighth of tunnel diameter, the corresponding values of α and $[x(0) - x_{min}]$ are readily found, for a maximum value of dynamic pressure of 2.5 psia. For a given value of $[x(0) - x_{min}]$ there is a corresponding value of F_A/M. This in turn is uniquely related to the semivertex angle Γ of the standard model and the base diameter d. If $[x(0) - x_{min}]$ is set at the maximum value then the minimum diameter of each standard model can be computed (in fractions of a tunnel diameter). These are as follows:

Model Γ	d_{min} $[x\ (0) - x_{min}] = \frac{1}{2}$ diameter	d_{min} $[x\ (0) - x_{min}] =$ one diameter
10°	0.0214	0.0160
10°	0.1250	0.0953
30°	0.4390	0.3325
40°	0.9250	0.700

These constraints can be used in a description of the overall operating ranges of the balance system. In particular, these contours define part of a region in the d, Γ-plane in which it is necessary to use an injection mechanism in order to start the model. Within the band defined by these curves and contingent upon the suspension performance limits, the bodies may be tested, but must be injected by some mechanical device. A further boundary in the k, Γ-plane consists of the envelope of minimum diameters of the standard bodies that may be used in static testing over the standard range of angle of incidence, as limited by the maximum field and gradient properties.

The overall operating range of the balance system for standard bodies and standard incidence range, including the effects of the starting transient can be displayed in the d, Γ-plane. Three distinct regions are shown: one in which the model may be started without an injection system, one in which the model may be tested but must be started using an injection device, and a third in which testing is not practical because the system lacks power. A chart showing these regions and boundaries in the d, Γ-plane is given in Fig. 26. An analysis of this kind is useful in deciding whether or not the system response time is adequate or whether auxiliary equipment is needed.

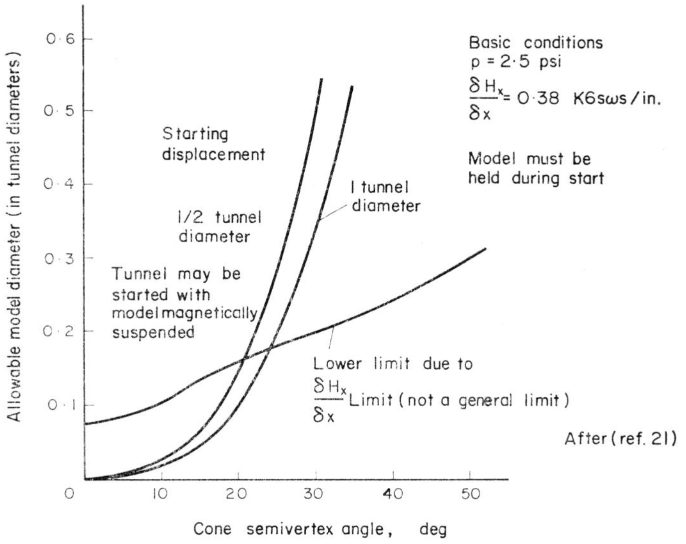

FIG. 26. General operating regimes of magnetic balance system, for standard bodies, 7.5 in. tunnel diameter.

Flow Field Measurements

The remainder of this section deal with types of aerodynamic measurements that can be made with the aid of magnetic suspension or balance systems. The discussion will be kept general; particular cases will be treated in the next section.

The most obvious use for a magnetic suspension system is for flow field measurements in the wake of the model since there is no support interference. In making wake measurements care must be taken to avoid interference between probes and the position sensing system, and to insure electromagnetic compatibility with transducers, hot-wire anemometers and other instrumentation. Care must be taken to insure the probe drive mechanism is compatible with the environment in which it is to be used.

Static Force and Moment Measurements

These measurements have been made in one of two ways. If the control system contains no correction signal proportional to the integral of the position error, the model deflections are directly proportional to the aerodynamic force acting on the model. Otherwise aerodynamic forces cause a change in the currents in the several coils. Through a calibration procedure the change in current can be related to the aerodynamic forces. In measuring aerodynamic forces and moments the primary sources of error are due to calibration, model position and unsteady airloads.

Dynamic Measurements

Dynamic characteristics may be deduced by one of two classes of experiment. In the first class the external shape of the model is built about a spherical core made of low magnetic hysteresis material. This core is suspended rigidly against forces but is allowed to rotate. The model is put into motion and its position recorded as a function of time. The time-position data is reduced using known procedures adapted from standard ballistic range data reduction procedures.[50] Alternately the force suspension can be soft and then the data involves 6 degrees of freedom in the data reduction process.

The second class of procedure involves forced or controlled oscillations. Two distinct processes can be used to handle the data here. The first involves the difference in phase shift between wind-on and wind-off.[51] Here the primary problem is that the data involves the difference between two large numbers, which results, inherently, in data of reduced accuracy. The second involves varying the magnetization field so a "resonance" exists. Care is taken to insure the "resonant" frequency lies well away from system frequencies. The data is reduced by curve-fitting the resonance in the Bode (amplitude vs. frequency), for the case of wind-on and wind-off.[52,53] One can find by iteration the aerodynamic damping, for these cases where the aerodynamic damping is much larger than the magnetic damping for all values of the aerodynamic spring. The aerodynamic spring may be found when it is large compared to the magnetic spring.

By applying suitable motions to the model it seems likely that cross-coupled parameters like C_{lr} and C_{np} may be deduced.

Finally note that both damping and roll due to roll rate and magnus effects can be measured.

References to Section II

1. A. H. Boerdijk, Technical aspects of levitation, *Philips Research Report* 11, pp. 45–56, 1959.
2. R. J. Duffin, Free suspension and Earnshaw's theorem, *Archive for RATNL Mech. and Anal.*, vol. 14, no. 4, pp. 261–263, 1963.
3. V. M. Ponizoeskii, Free suspension of a diamagnetic body in a constant magnetic field, *Progress in Physical Science*, vol. 100, no. 3, pp. 511–512.
4. A. J. Hatch, Potential well description of electromagnetic levitation, *Journal of Applied Physics*, vol. 36, no. 1, pp. 44–52, 1965. (a) A. J. Hatch, *Physical Review Letters*, vol. 6, p. 53, 1961; (b) R. Redlick, *Journal of Applied Physics*, vol. 33, p. 321, 1965; (c) I. Simon, *Journal of Applied Physics*, vol. 24, p. 19, 1953; (d) J. R. Harding and R. H. Tuffies, *Space/Aeronautics*, vol. 36, p. 133, 1961.
5. A. Wilson and B. Luff, The development, design and construction of a magnetic suspension system for the RAE 7 inch by 7 inch hypersonic tunnel, RAE TN 66248, August 1966.
6. F. T. Holmes, Axial magnetic suspensions, *Review of Scientific Instruments*, vol. 8, pp. 444–447, 1937.
7. T. A. Dukes and R. N. Zapata, Magnetic suspension with minimum coupling effects for wind tunnel models, *Transactions IEEE* (*AES*-1), vol. 1, pp. 20–28, August 1965.
8. H. Kemper, Schwevende Aufhangung duich Electromagnetische Krafte, *Electrotech. Z.*, vol. 59, pp. 391–395, 1938.
9. A. W. Jenkins and H. M. Parker, Electromagnetic support arrangement with three-dimensional control, *Journal of Applied Physics*, Suppl. to vol. 30, p. 2385, April 1959.
10. M. Tournier and P. Laurenceau, Suspension magnétique d'une maquette en soufflerie, *La Recherche Aeronautique*, no. 59, July/Aug., pp. 21–27, 1957.

11. J. E. CHRISINGER *et al.*, Magnetic suspension and balance system for wind tunnel application, *Journal of the Royal Aeronautical Society*, vol. 67, pp. 717–724, 1963.
12. M. JUDD and M. GOODYER, Magnetic suspension system, *Annual Report for the Year 1965*, Department of Aeronautics and Astronautics, University of Southampton.
13. C. D. CRAIN, M. D. BROWN and A. H. CORTNER, Design and initial calibration of a magnetic suspension system for wind tunnel models, AEDC-TR-65-187, September 1965.
14. K. R. SIVIER and M. HENDERSON, One component, magnetic, support and balance system for wind tunnel models, *Journal of Aircraft*, vol. 6, no. 5, p. 398, 1969.
15. R. A. KILGORE and I. L. HAMLET, Some aspects of an air-core single-coil magnetic suspension system, in *Summary of ARL Symposium on Magnetic Wind Tunnel Model Suspension and Balance Systems*, ed. F. L. Daum, ARL 66-0135, July 1966.
16. F. REGAN, Private communication, NOL Letter 312:FJR:tmb, 13 November 1967.
17. R. DUBOIS and C. ROGUE, Sur une méthode de mesure de la pression de culot. Mesure et visualisation sur une maquette cylindro-conique suspendue magnétiquement *à* $M_0 \simeq 7{,}6$, *La Recherche Aeronautique*, Bulletin Bimestriel ONERA No. 79, Nov./Dec. 1960, pp. 35–44.
18. W. BOUSEMAN *et al.*, Recent advances in magnetic balance systems IV, ARL 66-0200, October 1966, also MIT-AL-TR 121.
19. M. GOODYER, Roll control of magnetically suspended wind tunnel models by transverse magnets, *Aeronautical Quarterly*, vol. 18, pp. 22–42, February 1967.
20. R. K. MATTHEWS, M. D. BROWN and J. M. LANGFORD, Description and initial operation of the AEDC magnetic model and suspension facility: Hypersonic wind tunnel (E), AEDC TR-70-80, May 1970.
21. V. V. BASMAJIAN, A. B. COPELAND and T. STEPHENS, Studies related to the design of a magnetic suspension system, NASA CR-66233, February 1966, also MIT TR 128.
22. T. STEPHENS, Design, construction and evaluation of a magnetic suspension and balance system for wind tunnels, NASA CR-66903, Novemebr 1969, also MIT TR 136.
23. R. N. ZAPATA, The University of Virginia superconducting magnetic balance, *Proceedings of the 2nd International Symposium on Electro-Magnetic Suspension*, Southampton, England, July 1971.
24. I. L. HAMLET and R. A. KILGORE, Superconducting magnets for AC and modulated DC operation, *Proceedings of the 2nd International Symposium on Electro-magnetic Suspension*, Southampton, England, July 1971.
25. T. STEPHENS, Design study of a magnetic balance system for a twenty inch $M = 14$ hypersonic wind tunnel, MIT TR 101, November 1964.
26. T. STEPHENS, Methods of controlling the roll degree of freedom in a wind tunnel magnetic balance. Part I: Production of rolling moments, ARL-65-242, December 1965, also MIT AL-TR 78.
27. H. ALTMANN, An optical scanning detection system and its use with a magnetic suspension system for low density sphere drag measurements, *Proceedings of the 2nd International Symposium on Electro-magnetic Suspension*, Southampton, England, July 1971.
28. W. M. PHILLIPS, The measurement of low density sphere drag with a University of Virginia magnetic balance, *Summary of ARL Symposium on Magnetic Wind Tunnel Model Suspension and Balance Systems*, ARL 66-0135, July 1966, ed. F. L. Daum.
29. R. N. ZAPATA, The correction of drag measurements made in free expansion jets, Snippet Session, *Proceedings of the 2nd International Symposium on Electro-magnetic Suspension*, Southampton, England, July 1971.
30. H. S. FOSQUE and G. MILLER, Electromagnetic support arrangement with three-dimensional control. II. Experimental, *Journal of Applied Physics*, Suppl. to vol. 30, No. 4, pp. 240S–241S, April 1959.
31. H. M. PARKER and A. R. KUHLTHAU, A magnetic wind tunnel balance, AFOSR-64-0567, February 1964.
32. W. H. DANCY and W. R. TOWLER, Three dimensional magnetically supported wind tunnel balance, *Review of Scientific Instruments*, vol. 37, no. 12, pp. 1643–1648, December 1966.
33. H. M. PARKER, Principles, typical configurations and characteristics of the University of Virginia magnetic balance, *Summary of ARL Symposium on Magnetic Wind Tunnel Model Suspension and Balance Systems*, ARL 66-0135, July 1966, ed. F. L. Daum.
34. H. M. PARKER, Theoretical and experimental investigation of a three-dimensional magnetic-suspension for dynamic-stability research in wind tunnels, AST-4040-103-67U, April 1967, also NASA CR-66344.
35. H. M. PARKER, R. N. ZAPATA and G. B. MATTHEWS, Theoretical experimental investigation of a three-dimensional magnetic-suspension balance for dynamic stability research in wind tunnels, Report AST-4030-105-68U, March 1968.
36. H. M. PARKER and R. N. ZAPATA, The University of Virginia cold magnetic wind tunnel balance, 1969.
37. R. N. ZAPATA, A. R. KUHLTHAU and S. S. FISHER, Research in rarefied gas dynamics using an electro-magnetic wind tunnel balance, *Proceedings of the 2nd International Symposium on Electro-magnetic Suspension*, Southampton, England, July 1971.
38. T. STEPHENS, Summary of the design of a magnetic suspension and balance system, ARL 69-0019, January 1969.
39. T. STEPHENS, The general features of a six-component magnetic suspension and balance system, *Summary*

of ARL Symposium on Magnetic Wind Tunnel Model Suspension and Balance Systems, ed. F. L. Daum, ARL 66-0135, July 1966.
40. T. STEPHENS *et al.*, Recent developments in a wind tunnel magnetic balance, AIAA paper no. 72-164, January 1972.
41. F. BITTER, Water cooled magnets, *The Review of Scientific Instruments*, vol. 33, no. 3, pp. 342–349, March 1962.
42. C. W. HALDEMAN, J. SULLIVAN and E. E. COVERT, A variable phase velocity travelling wave pump, *AIAA Journal*, vol. 9, pp. 1389–1395, 1971.
43. A. B. COPELAND *et al.*, Recent advances in the development of a magnetic suspension and balance system for wind tunnels (Part III), ARL 65-114, June 1965.
44. R. MOREAU, J. BESSON and R. HOARAU, Detecteur opto-électronique pour suspension magnétique et étude du mouvement de maquette en chute libre (Electro-optical detectors for magnetic suspension and the study of the free motion of models), *Proceedings of the 2nd International Symposium on Electro-magnetic Suspension*, Southampton, England, July 1971.
45. F. MOSS, Use of superconductivity in magnetic balance design, *Proceedings of the 2nd International Symposium on Electro-magnetic Suspension*, Southampton, England, July 1971.
46. N. D. JAYSINGHANI, Power supply for a magnetic suspension system, *Proceedings of the 2nd International Symposium on Electro-magnetic Suspension*, Southampton, England, July 1971.
47. E. E. COVERT, Remarks on the design of magnetic balance and suspension systems with particular reference to the ARL 20-inch hypersonic tunnel, MIT Aerophysics Laboratory TN 113, 1965.
48. J. F. W. CRANE, Performance aspects of the R.A.E. magnetic suspension system for wind tunnel models, RAE Tech Memo Aero 1056, March 1968.
49. J. MIRANDE, Mesure de la résistance d'un corps de révolution à $M_0 = 2{,}4$ au moyen de la suspension magnétique ONERA, *La Recherche Aeronautique* Bulletin Bimestriel, ONERA No. 70, May/June 1959, p. 24.
50. I. D. JACOBSON, J. L. JUNKINS and J. R. JANCAITIS, Data acquisition and reduction for the University of Virginia superconducting magnetic suspension and balance facility, *Proceedings of the 2nd International Symposium on Electro-magnetic Suspension*, Southampton, England, July 1971.
51. E. L. TILTON, Dynamic stability testing with a wind-tunnel magnetic model suspension system, MIT S.M. Thesis, January 1963. Also contained in: E. E. COVERT and E. L. TILTON, Further evaluation of a magnetic suspension and balance system for application to wind tunnels, ARL 63-226, December 1963.
52. M. JUDD, Methods of the wind tunnel measurements of unsteady nonlinear aerodynamic forces, *AIAA Journal*, vol. 9, no. 7, p. 1302, 1971.
53. G. D. GILLIAM, Data reduction techniques for use with a wind tunnel magnetic suspension and balance system, MIT Aerophysics Laboratory Report 167, June 1970, also NASA CR-111844.

APPENDIX TO SECTION II

Measured Performance of Orthogonal C Magnet System[22]

The performance of the magnetic suspension system can be judged by analysis of the characteristics of the individual magnetic subsystem. These characteristics are estimated in the design phase but may be measured precisely once the prototype is assembled.

The following is an analysis of the performance of a prototype suspension system based on the measured characteristics of the completely assembled prototype. These characteristics are summarized below. All electrical quantities are referred to the terminals of the individual coil systems, and apply to operation of the system within the range of linear relationship between currents and fields; that is, below saturation of the iron cores.

(i) B_x. Axial magnetizing bias field ("magnetization")
Helmholtz coils
Total turns $N_x = 800$ (400 turns/coil)
Total resistance $R_x = 2.0\ \Omega$
Total inductance $L_x = 0.36$ H
Magnetic performance $B_x = 20\ I_x$ G
or: $B_x = 0.025\ N_x I_x$ G
(I_x = "magnetizing" current).

(ii) B_y, B_z. Traverse fields ("pitch, yaw, roll")
(a) Inner saddle coils — B_y'
Total turns $N_y' = 176$ (88 turns/coil)
Total resistance (d.c.) $R_y'_{\text{d.c.}} = 0.24\ \Omega$
(a.c.) $R_y'_{\text{a.c.}} = 0.5\ \Omega$
(400 Hz)
Total inductance $L_y' = 8$ mH
Magnetic performance $B_y' = 3.8\ I_y'$ G
or: $B_y' = 0.0216\ N_y' I_y'$ G
(I_y' = inner saddle — coil current).
(b) Outer saddle coils — B_z'
Total turns $N_z' = 266$ (133 turns/coil)
Total resistance (d.c.) $R_z'_{\text{d.c.}} = 0.44\ \Omega$
(a.c.) $R_z'_{\text{a.c.}} = 0.9\ \Omega$
(400 Hz)
Total inductance $L_z' = 17$ mH
Magnetic performance $B_z' = 3.5\ I_z'$ G
or: $B_z' = 0.0135\ N_z' I_z'$ G
(I_z' = outer saddle — coil current).

(iii) B_{xx}. Axial gradient of axial field ("drag")
Total turns $N_{xx} = 800$ (400 turns/coil)
Total resistance $R_{xx} = 2.0\ \Omega$
Total inductance $L_{xx} = 0.16$ H
Magnetic performance $\bar{B}_{xx} = 1.7\ I_{xx}$ G/in.
or: $\bar{B}_{xx} = 4.3 \times 10^{-3}\ N_{xx} I_{xx}$ G/in.
(I_{xx} = "drag" current).

(iv) B_{yx}. Axial gradient of lateral field ("side force")
Side and lift magnet assemblies
(Four coils in vertical plane)
Total turns $N_{yx} = 1160$ turns (290 turns/coil)
Total resistance $R_{yx} = 1.0\ \Omega$
Total inductance $L_{yx} = 0.40$ H
Magnetic performance $\bar{B}_{yx} = 1.4\ I_{yx}$ G/in.
or: $\bar{B}_{yx} = 1.2 \times 10^{-3}\ N_{yx} I_{yx}$ G/in.
(I_{yx} = "side-force" current).

(v) B_{zx}. (Same as above, change y to z.)

Performance Limits Due to Saturation of the Iron Magnet Cores

The performance parameters listed above do not imply any bounds on the magnetic performance of the system. However, due to the saturation characteristics of the steel magnet cores, limits do exist. The steel cores are magnetized to a greater or lesser degree by all coil systems, and therefore a combination of magnet currents at some level is capable of saturating the cores at some point.

The most likely point at which saturation would occur in this geometry is at the pole tips. The flux density at the surface of the pole tips is made up of the applied field from the coils, and the magnetization of the cores. In general, the magnetization component will predominate, and hence can be assumed (conservatively) to be approximately equal to the measured flux density at the surface of the pole tips. The pole-tip flux density is linearly related to all the magnet currents, at low current levels, and the relation can be easily found by successively energizing each coil system with low d.c. currents, and measuring the pole-tip flux density with a gaussmeter. It is desirable that the magnetization level of the pole tips remain below the saturation level. For the material used in the cores, this level is approximately 15 kG, at the lowest. This will consequently provide the constraint on the performance of the magnet system. The pole-tip flux density, $B_{\text{p.t.}}$, was found to be a function of the coil currents as follows:

$$B_{\text{p.t.}} = 33I_x \pm 4I_y' \pm 6I_z' \pm 25I_{xx} \pm 39I_{xy} \pm 39I_{xz} < 15{,}000. \tag{80}$$

This equation applies to the worst-case pole tip when all terms are of the same sign.

It will be useful to relate the actual field variables to the pole-tip flux density. Using the magnetic performance parameters this relation follows:

$$B_{\text{p.t.}} = \pm 1.6B_x \pm 1.05B_y' \pm 1.7B_z' \pm 14.7B_{xx} \pm 28B_{xy} \pm 28B_{xz} < 15{,}000. \tag{81}$$

III. MAGNETIC BALANCE SYSTEMS

Previous Usage

The suspension of a ferromagnetic model by means of its interaction with magnetic fields is only a part of the functional operation of a magnetic balance system. The suspension system produces the forces and moments necessary to automatically maintain the model at a desired position or force the model to perform a prescribed motion. The "balance" aspect of the magnetic balance suspension system involves the determination of the forces and moments on a suspended model from the measurements of variables related to the position and orientation of the model, and the magnetic fields.

The first application of a magnetic suspension system as a force balance for wind tunnel testing was performed at ONERA.[1] Though only drag measurements on a wind tunnel model were obtained with this balance, it was clear that the technique of magnetic suspension could be applied for multicomponent force and moment measurements. The magnetic balance system was shown to be particularly attractive for wind tunnel testing since it eliminated the need for mechanical model supports. This led to the development of magnetic balances at several laboratories.[2–6] The ultimate goal of these efforts was to produce a reliable and versatile wind tunnel instrument applicable to a broad class of aerodynamic simulation studies.

Model Position Sensing

In sections I and II it was pointed out that a model position sensing scheme was a necessary element for the closed loop operation of a magnetic suspension and balance system. Since no mechanical contact exists with a magnetically suspended model, its position sensing serves two purposes in the operation of such a system; first, it is necessary for the stabilization of a suspended model, second, it provides a measure of the model position and orientation relative to an arbitrary co-ordinate system (e.g., wind tunnel axes or balance axes).

It has been found that precise aerodynamic measurement requires precise positioning.[7] To determine a fixed absolute position three schemes are presently in use. First one can always use a surveyor's transit—clumsy and slow but accurate. At MIT photodetectors (silicon cells) were arranged in staggering stacks, with their edges accurately located by means of a transit. Vlajinac, using a laser, a mirror, and a photodetector covered with a slotted diaphragm mounted on a lead screw in place of the stacked array, has increased the accuracy of measuring angles.[8]

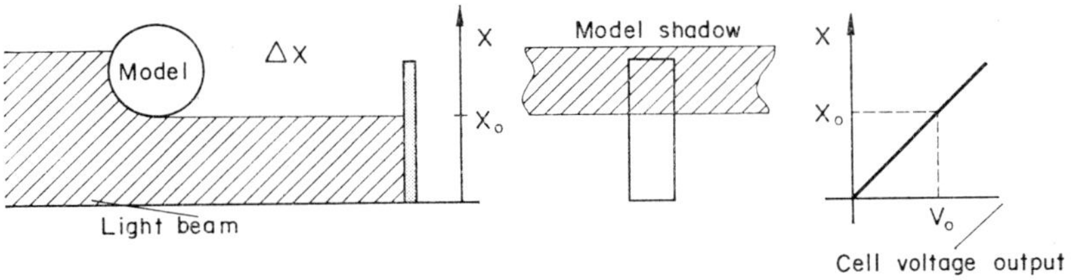

FIG. 27. Single degree of freedom optical position sensor.

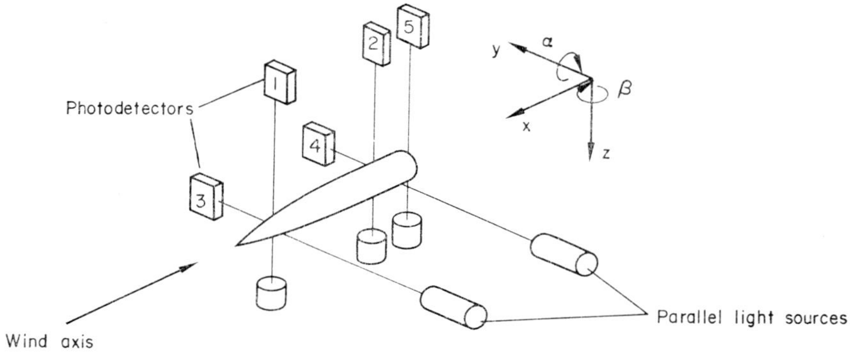

FIG. 28. Five-component optical sensing scheme.

Model Position Monitoring

The model position sensors described in Section II have proven adequate in providing model position information necessary for stabilization of the magnetic model suspension (Figs. 27 and 28). However, these sensors are usually subject to long-term drift which, though it does not affect the model suspension stability, will produce changes in the model's absolute position in the wind tunnel. In addition, the need for a means of indexing the model position and attitude (e.g. changing angle of attack) and verifying the absolute position of the model with respect to the wind axes during the course of a test has led to the development of absolute model position monitors which are used in addition to the regular position sensors previously described. A practical absolute model position indicator that is readily adaptable to the optical position sensors consists of an overlap of two photodetectors. This system is described in ref. 8. The sensor arrangement is shown in Fig. 29. By connecting the two photodetectors to read differentially, the output reaches a minimum when the model edge shadow is the photodetector interface. This concept can be extended to monitor the model position in 6 degrees of freedom by suitable arrangement of a number of such detectors. If a new model position setting is desired, the absolute position sensors are moved accordingly or other cells are piled on the stack.

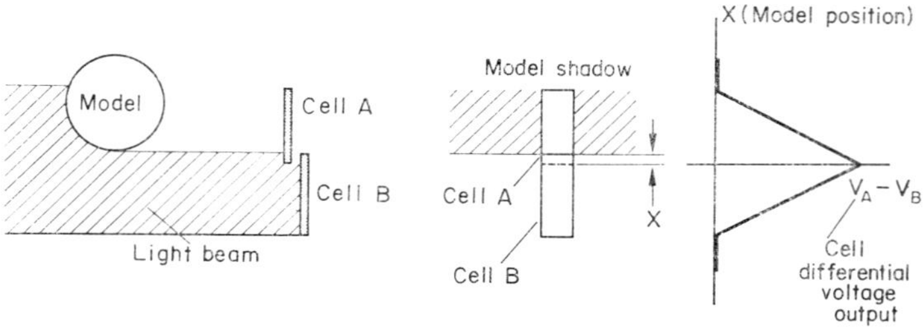

FIG. 29. Optical position monitoring scheme.

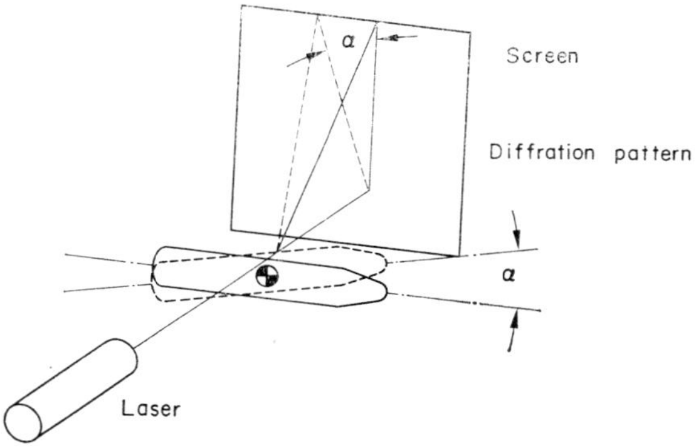

FIG. 30. Laser absolute positioning scheme.

Another technique for absolute model positioning (used with the electromagnetic position sensor)[10] is the use of surveying transits. This method represents a visual monitoring of the model position and is performed by lining up the model edge or other convenient model surface features with the transit cross-hairs. In addition the model's angular position could be indexed to a desired value by rotation of the transit's cross-hair reticule. The use of two transits, one for the wind tunnel pitch plane and the other for the yaw plane was required to monitor all 6 degrees of freedom.[9]

The absolute position monitoring methods described above are generally accurate to ± 0.001 in model translation and angles of ± 0.1 deg. While this accuracy is acceptable for most wind tunnel applications it is not sufficient for accurate measurement of cross-coupled parameters such as Magnus forces and moments.* This problem has been ameliorated sub-

* The Magnus force or lateral force due to the spin rate at angle of attack in general is an order of magnitude less than the lift due to angle of attack. Thus an uncertainty of $\pm$ 0.1 deg in yaw angle will produce errors in the measured Magnus force comparable to an error in measured lift due to a 1 deg angle of attack uncertainty.

stantially with the use of laser light diffraction emanating from the model's edge.[11] The scheme is shown in Fig. 30. A low power (3 mW) laser striking the edge of a model generates a diffraction pattern perpendicular to the model surface. The diffracted light striking a screen appears as a narrow line of light, the length being determined by the distance from the model to the screen. In practice this provides a sufficiently large lever arm to measure easily the angle to ± 0.02 deg by measuring the movement of the diffraction pattern.

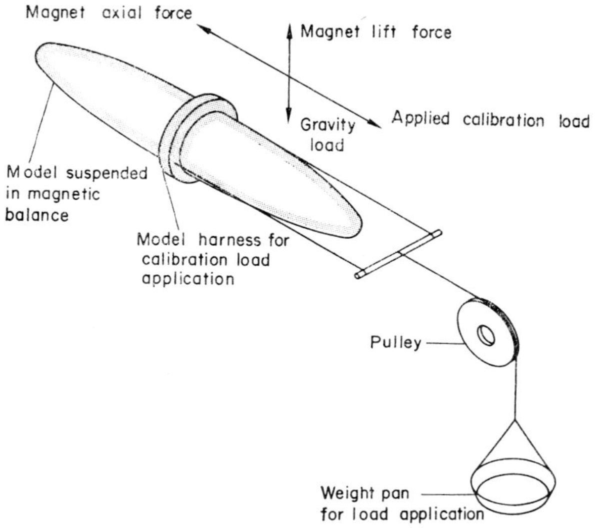

FIG. 31. Sketch showing calibration method. In wind tunnel: magnetic axis force is equal and opposite to aerodynamic drag on model. During calibration: magnetic axial force is equal to applied calibration load on model.

Calibration

The forces and moments acting on a magnetically suspended model (e.g. gravity and aerodynamic loads) can be obtained from the electric currents in the magnets used to suspend the model. The relationship between the magnet currents and the magnetic loads they produce must be determined experimentally by calibration. In general the technique used to obtain this relationship has been to apply known (calibrated) loads on a suspended model.[1–7,9] This is done with a model harness that has strings attached to it and allows the calibration forces and moments to be applied. A typical arrangement is shown in Fig. 31. Due to the spatial variation of the magnetic field produced by the various magnet coils, the proper combination of external forces and moments must be applied to the model at each position (e.g. angle of attack) and for each model configuration. For practical reasons, this approach requires that the external loads be kept constant instead of the magnet currents. Therefore, to find the loads on a given model for a given set of currents the calibration equations have to be inverted.[6,9] This is generally done with a computer iteration scheme.

An alternative approach to the calibration technique described involves the use of a pneumatic calibration rig reported in ref. 13. The rig is essentially a force resolver that operates with an arrangement of air bearings. The rig is capable of resolving the magnetic forces and moments in 6 degrees of freedom on a model held captive in the rig along the principal axes of the rig. This allows the loads on the model to be determined for an arbitrary set of magnet currents. Another practical use of the rig is to obtain directly the aerodynamic loads on a model from the measured wind-on currents. This is done by attaching the model to the rig and locating the model in the same location as during the wind on tests. The same magnet currents as measured with the wind on are applied, thus exerting the same loads required to hold the model against the flow. The forces and moments are then obtained from the calibration rig's output.[12]

The two techniques for magnetic balance calibration described above, though effective, do not take advantage of the full knowledge of the field structure. This in part led to the development of a procedure related to the physics of the magnetization and thus simplifying the calibration and data reduction equations. This will be discussed in the next sections.

Data Acquisition

The variable quantities that are measured with a magnetic suspension and balance system are the magnet coil currents and model position and orientation relative to the magnet coils. The measurement of model position has been discussed above. The measurement of the magnet currents is fairly straightforward. They are generally measured with current shunts in series with the magnet winding. The coil current measurement may be processed in a variety of schemes depending on whether an "average" force on the model is desired, in which case the currents are sampled over some time interval, or if time varying loads are being measured, in which case a continuous plot of magnet current versus time is required. In general the acquisition of magnet current data and model position information can readily be used with high-speed multichannel data processing systems. The wind tunnel parameters used in reducing the forces and moments on a model to a coefficient form are taken simultaneously with the magnetic balance variables.

Data Reduction

The relative ease with which the data from a magnetic balance can be reduced from the magnet current and model position information to forces and moments on the models depends to a large extent on the magnet arrangement. In general the data reduction schemes are similar to conventional multi-component strain gauge balances. As with strain gauge balances it is desirable to minimize interactions between orthogonal components of forces and moments. As was shown in Section II the production of magnetic field components by a suitable coil arrangement can minimize the interaction between the various coils. If such an arrangement is chosen the data reduction equations become relatively simple. The magnet system described in ref. 9 was designed to satisfy this requirement and the data reduction technique and equations have been extensively developed for an arbitrary model geometry.

The equations governing the forces and moments on an arbitrary ferromagnetic body in a magnetic field have been derived above. They can be used to illustrate a general approach to the reduction of these forces and moments to the measured magnet currents producing the fields. Recall the magnetic force vector **F** is approximately

$$\mathbf{F} \sim K_t \text{ Vol } (\bar{\mathbf{m}} \cdot \nabla)\, \mathbf{H} \tag{82}$$

and the magnetic torque is

$$\mathbf{T} \simeq K_t \text{ Vol } (\bar{\mathbf{m}}\, x\, \mathbf{H}) \tag{83}$$

where Vol = ferromagnetic model volume,
$\bar{\mathbf{m}}$ = average model magnetization vector,
$\mathbf{H}$ = magnetic field flux vector,
∇ = gradient operator,
x = cross-product operator,
K_t = magnetic moment constant.

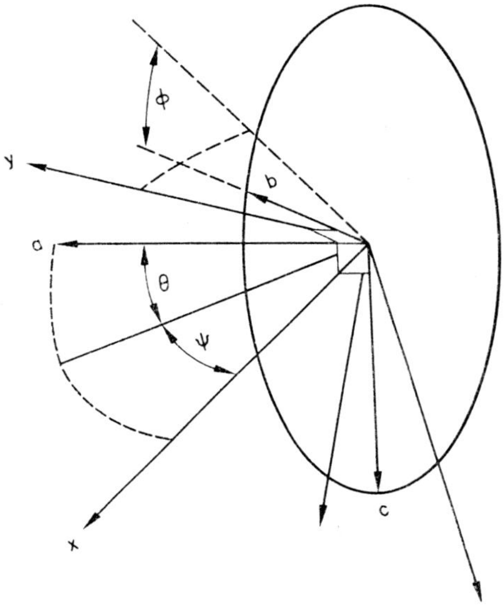

FIG. 32. Model and wing tunnel axis system.

The wind tunnel model can be considered in general to be a three-dimensional, non-axisymmetric shape, with principal axis a, b, c and set at angles θ, ψ and ϕ with the x, y and z axes of the wind tunnel. This is shown in Fig. 32. The average magnetization of the model along its principal axes was shown in section I to be

$$\bar{m}_a = \frac{H_a}{D_a} \tag{84}$$

$$\bar{m}_b = \frac{H_b}{D_b} \tag{85}$$

$$\bar{m}_c = \frac{H_c}{D_c} \tag{86}$$

where as before $D_{a,b,c}$ are the model demagnetizing factors along its principal and $H_{a,b,c}$ is the applied field component along these axes.

The model magnetization along the wind tunnel (or balance) axes can thus be expressed as

$$\begin{pmatrix}\bar{m}_x\\ \bar{m}_y\\ \bar{m}_z\end{pmatrix} = \begin{pmatrix} R^{-1} \end{pmatrix} \begin{pmatrix} \frac{1}{D_a} & 0 & 0 \\ 0 & \frac{1}{D_b} & 0 \\ 0 & 0 & \frac{1}{D_c} \end{pmatrix} \begin{pmatrix} R \end{pmatrix} \begin{pmatrix} C_x I_x \\ C_y I_y \\ C_z I_z \end{pmatrix} \tag{87}$$

where I_x is the current in the magnetization coil, I_y is the current in the pitch coil and I_z is the current in the yaw coil.

The moment equation becomes

$$\begin{pmatrix} T_x \\ T_y \\ Z_z \end{pmatrix} = K_t \text{ Vol} \begin{pmatrix} 0 & -\bar{m}_z & \bar{m}_y \\ \bar{m}_z & 0 & -\bar{m}_x \\ -\bar{m}_y & \bar{m}_x & 0 \end{pmatrix} \begin{pmatrix} C_x I_x \\ C_y I_y \\ C_z I_z \end{pmatrix}. \tag{88}$$

This equation contains at most three products of six constants, C_xC_z, C_xC_y, C_yC_z, so equation (88) contains three unknowns to be determined experimentally, from the data and the three equations. These constants, when determined can be cross-checked by angulating the model.

The force equation may be written in the same general way

$$\begin{pmatrix} F_x \\ F_y \\ F_z \end{pmatrix} = K_t \text{ (Vol)} \begin{pmatrix} C_{xx} I_{xx} \\ C_{xy} I_{xy} \\ C_{xz} I_{xz} \end{pmatrix} \cdot (\bar{m}_x \; \bar{m}_y \; \bar{m}_z) \tag{89}$$

The additional constants, C_{xx} etc., are gradient constants. Equation (89) nominally contains nine constants so that the constants may be determined by taking three independent sets of data, for different values of I_x, I_y, I_z, and I_{xx}, I_{xy} and I_{xz}. Usually the calibrations are made with the body axis lined up with the tunnel axis systems. The effect of pitch, roll and yaw rotations or of displacements is cross-checked afterwards.

Details, which are straightforward, for data reduction equations and computer program for the MIT-NASA prototype balance are given in ref. 14. A typical procedure used with the double-L configuration is given in ref. 15.

The measurement of "static" aerodynamic loads on a model with a magnetic balance represents only a portion of its capability as an instrument for wind tunnel testing. Dynamic testing involves both different data acquisition methods and data reduction equations. In this case the fields are generally time varying and must be related to the kinematics of the moving body. A description of one technique for extracting dynamic parameters from the magnetic balance is also found in refs. 14 and 15.

References to Section III

1. M. Tournier and P. Laurenceau, Suspension magnétique d'une maquette en soufflerie, *La Recherche Aeronautique*, no. 59, July/Aug. 1967, pp. 21–27.
2. J. E. Chrisinger *et al.*, Magnetic suspension and balance system for wind tunnel application, *Journal of the Royal Aeronautical Society*, vol. 67, pp. 717–724, 1963.

3. H. M. Parker and A. R. Kuhlthau, A magnetic wind tunnel balance, University of Virginia Report No. AST 3420-105-64U, Feb. 1964.
4. A. Wilson and B. Luff, The development, design and construction of a magnetic suspension system for the RAE hypersonic wind tunnel, RAE Technical Report 66248, 1966.
5. T. A. Dukes and R. N. Zapata, Magnetic suspension with minimum coupling effects, *IEEE Transactions* (*AES*-1), p. 1, August 1965. Also Report 682, Princeton University, Department of Aerospace and Mechanical Sciences, Princeton, New Jersey, July 1964.
6. C. D. Crain, M. D. Brown and A. H. Cortner, Design and initial calibration of a magnetic suspension system for wind tunnel models, AEDC TR-65-187, September 1965.
7. R. N. Zapata, The University of Virginia superconducting magnetic balance, *Proceedings of the 2nd International Symposium on Electro-magnetic Suspension*, July 1971.
8. A. B. Copeland *et al.*, Recent advances in the development of a magnetic suspension and balance system for wind tunnels, Part III, ARL 65-114, June 1965.
9. T. Stephens, Design, construction and initial evaluation of a magnetic suspension system for wind tunnels, NASA CR-66903, November 1969.
10. E. E. Covert and E. L. Tilton, Further evaluation of a magnetic suspension and balance system for application to wind tunnels, Wright-Patterson Air Force Base ARL Report 63-226, December 1963.
11. M. Vlajinac, Report on laser sensing scheme (to be published).
12. M. Vlajinac and G. D. Gilliam, Aerodynamic testing on conical configurations using a magnetic suspension system, ARL report 70-0067, April 1970.
13. M. Vlajinac, A pneumatic calibration rig for use with a magnetic suspension and balance system, ARL Report 70-0016, January 1970.
14. G. D. Gilliam, Data reduction techniques for use with a wind tunnel magnetic suspension and balance system, NASA CR-11184, June 1970.
15. M. Judd *et al.*, Aerodynamic data acquisition with the University of Southampton magnetic balance, *Proceedings of the 2nd International Symposium on Electro-magnetic Suspension*, July 1971, Southampton, England.

IV. AERODYNAMIC DATA OBTAINED WITH MAGNETIC SUSPENSION SYSTEMS

Introduction

The application of the magnetic suspension techniques to wind tunnels has been viewed with interest for a number of years. Except for Reynolds number and power-on effects, magnetic suspension of a model in a wind tunnel offers a very close approach to true aerodynamic simulation of a variety of configurations in flight. This is due to the complete absence of the aerodynamic interference from the mechanical model supports required in conventional balance systems. These interferences or "sting effects" can have a strong influence on the pressure distribution near the base of the body, generally distort the near and far wake flow field and in some cases influence the flow and pressure distribution over the entire body. While in some instances corrections may be applied to the measurements in order to reduce the error incurred by the presence of the supports, the corrections are usually difficult to estimate accurately and for some configurations the interference effects are not fully understood. This immediately suggests several areas that could be studied with a magnetic suspension system. Some of the experimental results obtained with magnetic balances are presented in this section thus giving an up-to-date summary of their uses in aerodynamic testing while at the same time giving the reader insight to potential applications and areas of interest in the future.

Static Aerodynamic Test Results

The most straightforward application of a magnetic suspension system is its use as a force balance to determine the static aerodynamic forces and moments on a model in a wind tunnel. In fact, the first wind tunnel data obtained with a magnetic suspension system was

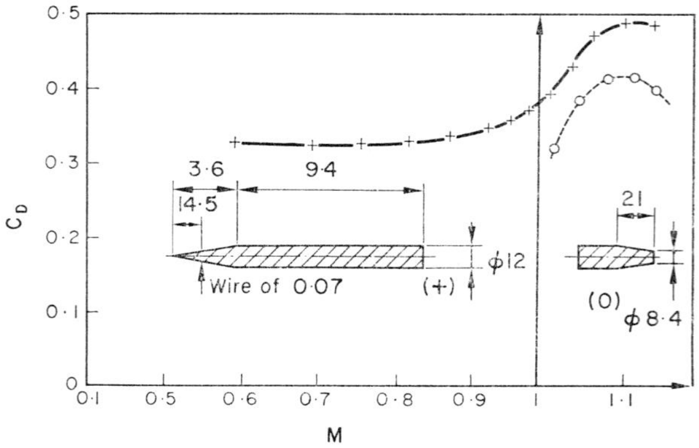

FIG. 33. Drag coefficient vs. Mach number.

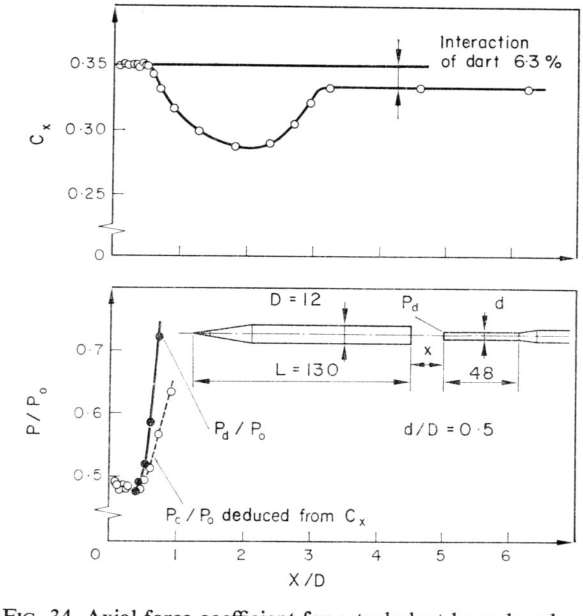

FIG. 34. Axial force coefficient for a turbulent boundary layer.

the measurement of the axial force (drag) on a model and is reported in ref. 1. The measurements were obtained over a Mach number range from 0.1 to 1.1 on slender-winged and non-winged models. These results proved possible the application of magnetic suspension systems for force measurement over a range of Mach numbers from subsonic to transonic and are shown in Fig. 33. These results led to magnetic balance development at several laboratories.

The Mach number range of magnetic balance operation was extended to the supersonic region ($M = 2.4$) and reported in ref. 2. While again only the drag component was obtained, the experiments showed the effects of mechanical model supports on the measured drag of the model. This was done by placing dummy stings of various sizes near the model base. The drag on the model was measured as a function of sting position from the model base and is shown in Fig. 34. From Fig. 34 one can see that as the sting approaches the model base the measured drag reaches a value corresponding to a sting-supported model. As the sting is moved sufficiently far away from the model base the measured drag becomes constant, simulating a free-flight body. The discrepancy in the model drag due to the presence of the sting is seen to be 6.3% when compared to the interference-free results.

The development of magnetic balance technology which allowed testing models at angle of attack to the flow led to the measurement of forces and moments in the aerodynamic pitch plane (lift, drag and pitching moment) described in refs. 3 and 4. These results showed that multicomponent force and moment data could be obtained with a magnetic suspension system. In addition the tests reported in refs. 3 and 4 were conducted at a Mach number of 4.8 thereby extending the Mach number range of magnetic suspension application to the hypersonic flow regime. The results reported in refs. 1–4 represent the initial experimental static aerodynamic data that was measured with magnetic suspension and balance systems.

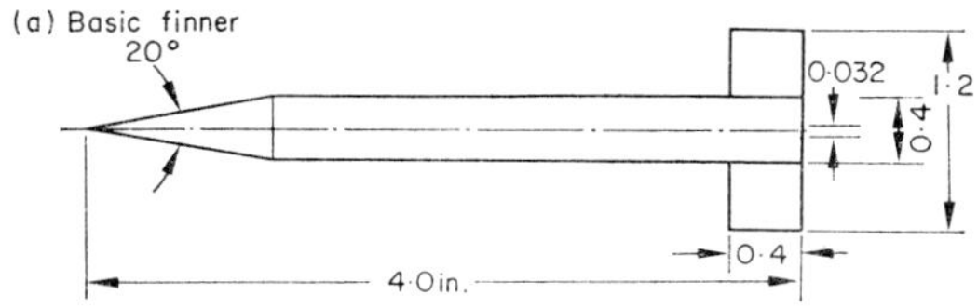

FIG. 35. Model tested. *Note*: All linear dimensions are in inches; tolerance on linear dimensions ± 0.0005 in.; tolerance on angular dimensions ± 0.05 deg.

Subsequent tests to determine static model characteristics were conducted at various laboratories on a variety of model configurations. These results have been reported in refs. 4–18. The data was taken over a Mach number range from low subsonic ($M \simeq 0$) to hypersonic ($M = 16$). A small sample of data taken with magnetic balances in refs. 4–18 is presented here. The data chosen for presentation was selected for those model configurations where comparison with either conventional wind tunnel balance or free flight results can be made. Additional data is included where interesting aerodynamic observations have been made with magnetic balances.

Finned Missiles

The basic finned model shown in Fig. 35 is a standard wind tunnel model used for comparison of various wind tunnels and techniques. Results obtained in ref. 10 with a magnetic suspension system are compared with data obtained in ref. 39. The magnetic suspension data shows a maximum scatter of 1% (based upon a force coefficient of 0.5) and is lower than the scatter obtained in ref. 39.

20° *Cone*

The drag coefficient of a 20 deg total angle sharp cone was measured with a magnetic suspension system at a Mach number of 8.5. These results reported in ref. 14 are compared with free-flight data from ref. 28 in Fig. 36.

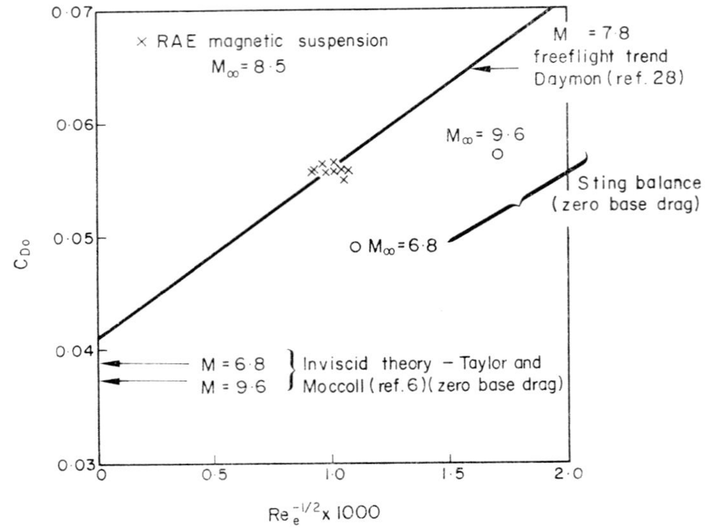

FIG. 36. Comparison of zero incidence drag measurements for a 20 deg cone.

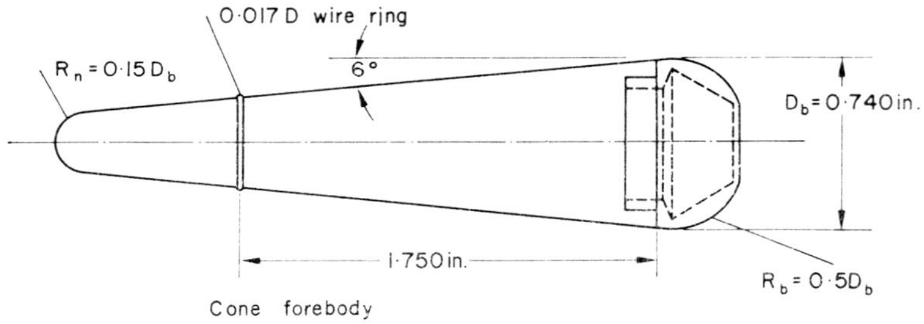

FIG. 37. Bulbous base cone model.

Bulbous Base Cones

A family of blunt 6 deg half-angle cones with varying base radii were tested with magnetic balance and the results are reported in ref. 17. One of these cones shown in Fig. 37 was tested both with and without the presence of a dummy sting since it was believed that static characteristics are affected by flow separation on the body base. The dummy sting shown in Fig. 38 was positioned inside the model base cavity without making physical contact with the model.

The present sting-free data together with sting interference results obtained with the magnetic balance are compared in Fig. 39 with results obtained in ref. 29. Though the Mach number in the two tests is approximately the same ($M \simeq 0.26$), the Reynolds number is an order of magnitude different. However, the agreement is good particularly in the nonlinear behavior near zero angle of attack. The magnetic balance test results indicate that sting interference is sufficient to account for the disagreement between the two and is in accord with the conclusions regarding bulbous base cones in both refs. 29 and 30.

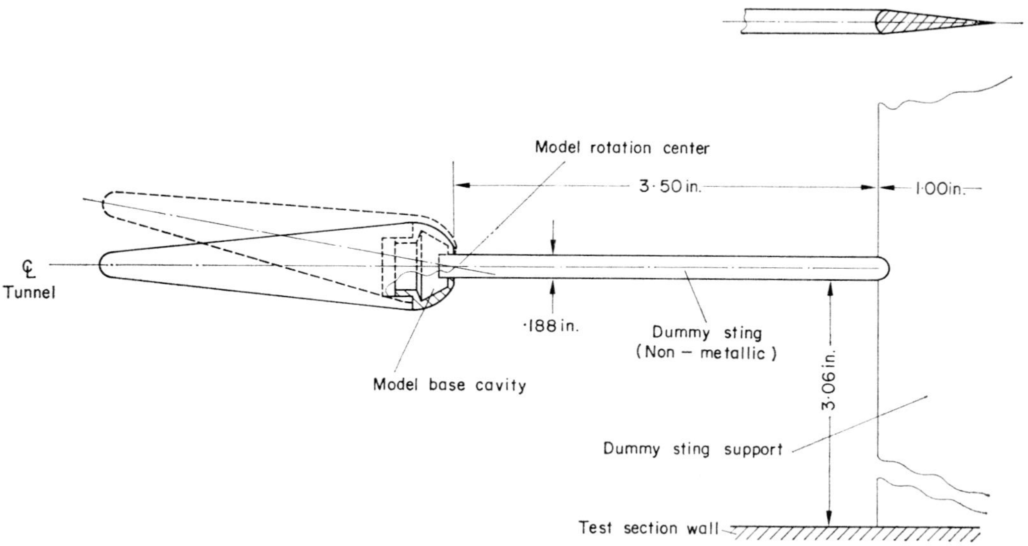

FIG. 38. Bulbous base cone with dummy sting.

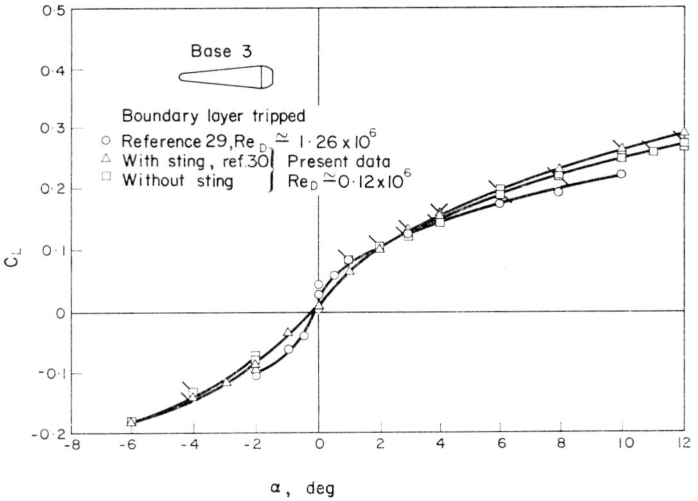

FIG. 39. Subsonic lift coefficient vs. angle of attack.

Slender-wing Results

The static lift, drag and pitching moment on the three 74 deg sweepback planforms shown in Fig. 40 were measured and a complete presentation of the data is available in ref. 15.

Comparison of lift coefficient data versus angle of attack for the delta wing planform with both theoretical and experimental results obtained is shown in Fig. 41. The lift coefficient is plotted versus angle of attack for a delta wing with 75 deg leading-edge sweep. The

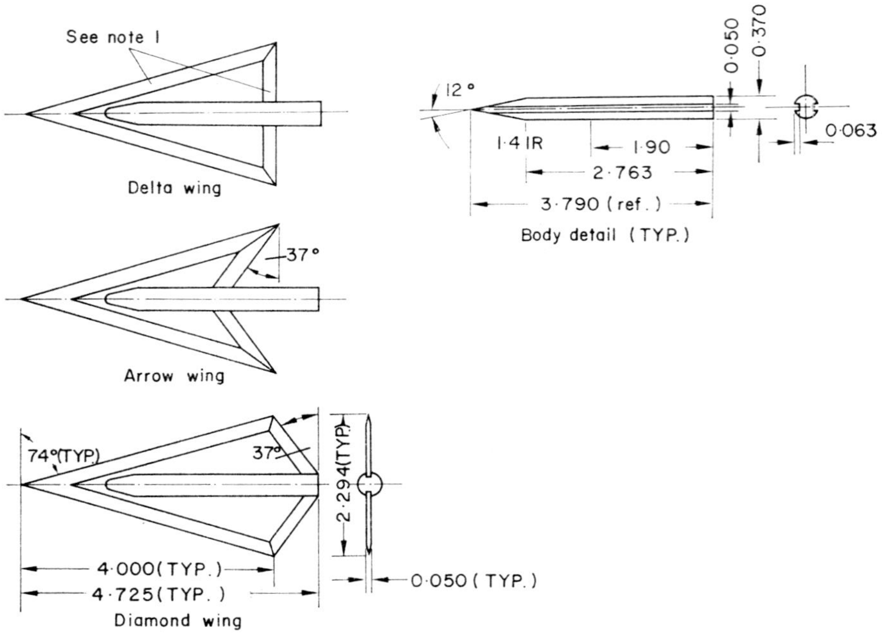

FIG. 40. Wing model configurations.

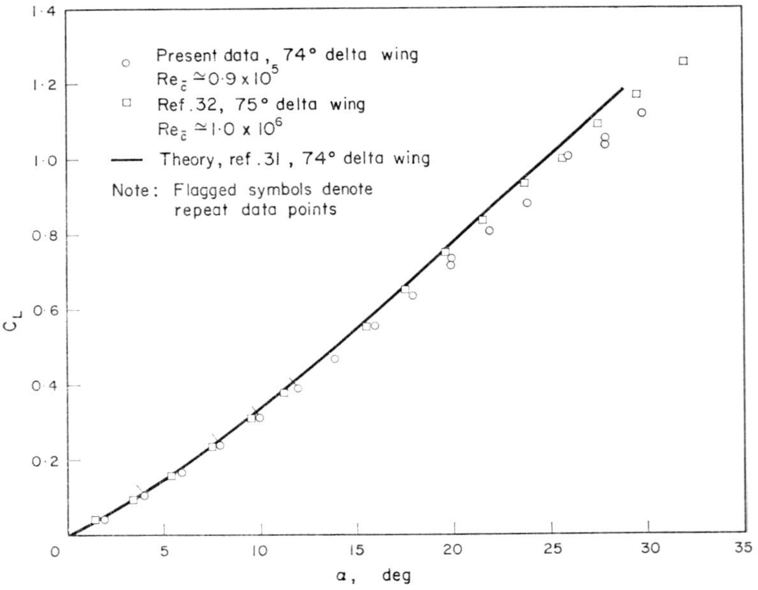

FIG. 41. Lift coefficient vs. angle of attack.

magnetic suspension data shows a lower value of lift coefficient than predicted by both Polhamus's theory[31] and the experimental results of ref. 32. One possible cause for the lower values of the lift coefficient in the study could be due to the model body extending aft of the wing trailing edge (see Fig. 40), thereby being in the down wash from the wing. This explanation is consistent with the behavior of the pitching moment curve in Fig. 42. The agreement of the present lift-coefficient data with that obtained in ref. 32 is within 2%, in spite of an order of magnitude difference between the two test Reynolds numbers.

The delta-wing drag-coefficient data plotted as drag due to lift ($C_D - C_{D0}$) versus angle of attack is compared in Fig. 43 with results obtained in ref. 32 for a 72.5 and 75 deg sweep delta wing. The magnetic suspension results (74 deg sweep wing) are seen to fall between the ref. 32 data.

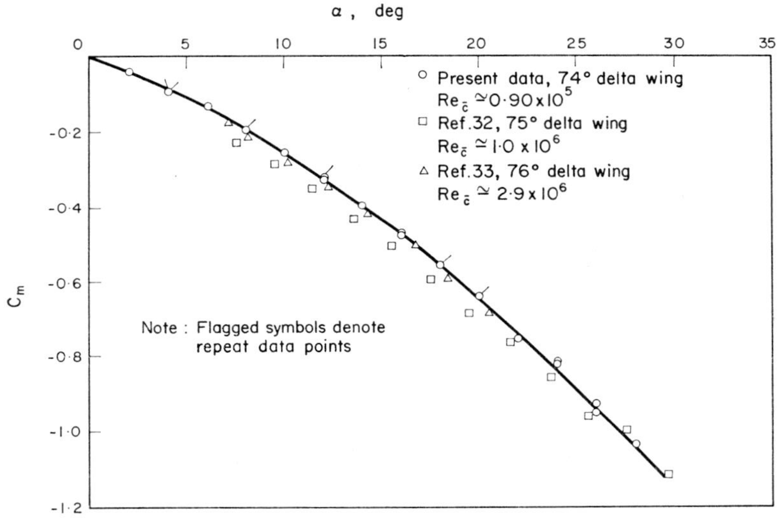

FIG. 42. Pitching moment coefficient vs. angle of attack.

Ellipsoidal Cylinders

The drag coefficient of a family of ellipsoidal cylinders was measured with a magnetic balance and is reported in ref. 13. The results of geometrically similar ellipsoids of fineness ratio 8 are shown in Fig. 44. The data is corrected using classical blockage interference corrections. In these tests there was no model support interference and the effect of the tunnel walls on the measured drag coefficient can be deduced. The present data indicates that the classical wind tunnel correction does not completely account for the effect of model size and wall interference, since the collapse of the data is not complete in the region of Reynolds number overlap.

Spheres

The drag coefficient of spheres of various sizes was obtained with a magnetic balance and is reported in ref. 26. The present data is shown in Fig. 45 which was taken from ref. 26.

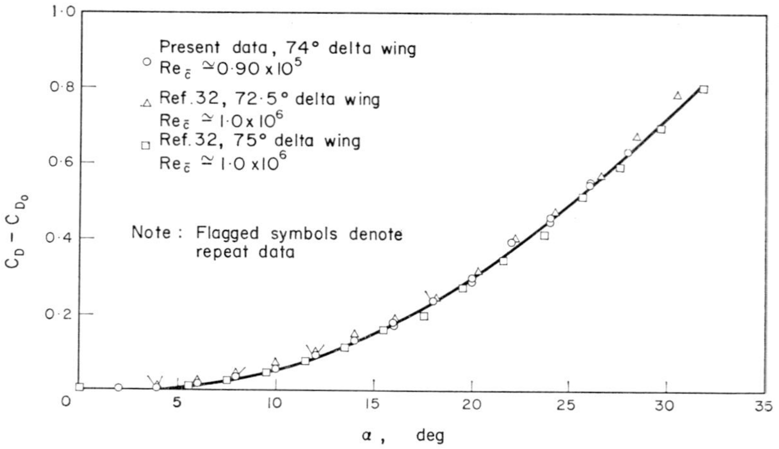

FIG. 43. Induced drag coefficient vs. angle of attack.

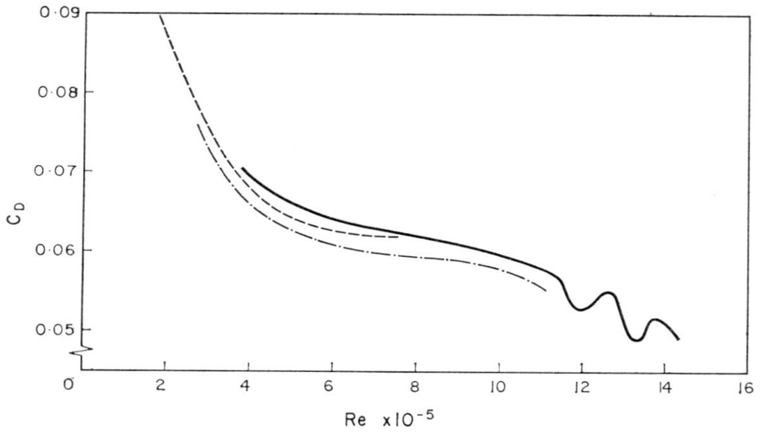

FIG. 44. C_D versus *Re* corrected for blockage. 8:1 ellipsoid.

The magnetic suspension data which is consistently lower agrees within 1.5% with ballistic range data. The case of the 1.5 in. sphere is in excellent agreement with free-flight (ref. 38) data at higher Reynolds number. One should also note that, as in the case of the ellipsoids (Fig. 44), the collapse of the measured drag for different size spheres tested is not complete when using classical wind tunnel corrections.

Dynamic Test Results

The theory for measuring dynamic coefficients with a magnetic suspension system have been outlined in Section II. However, the application of magnetic suspension systems thus far, in measuring dynamic coefficients, has been to a considerably lesser extent than its use in measuring static parameters in a wind tunnel. Consequently the amount of available dyn-

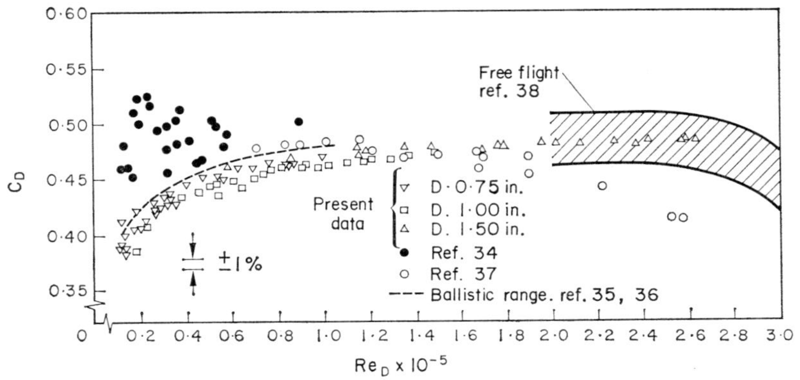

FIG. 45. Drag coefficient vs. Reynolds number.

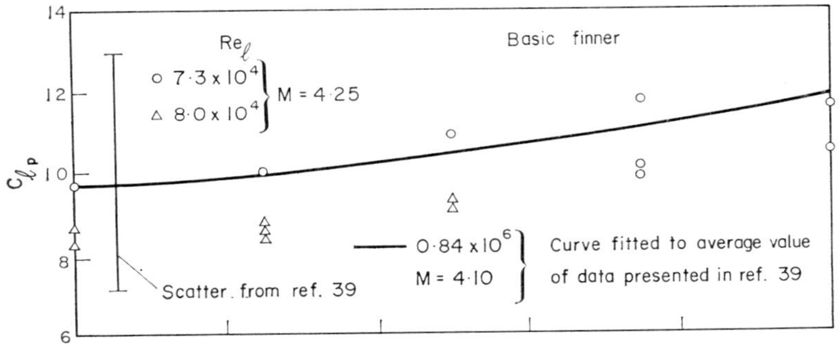

FIG. 46. Roll-damping moment coefficient derivatives vs. angle of attack.

amic data is limited. Dynamic tests with magnetic suspension systems are reported in refs. 3, 6, 7 and 19. The principal dynamic parameters in these investigations have been measurements of pitch damping due to pitch rate and roll damping due to roll rate. While no comparison data of pitch-damping results with those obtained in the above references is available, the measurement of roll-damping coefficients with magnetic balances has shown a substantial improvement in data accuracy over conventional methods.

The roll-damping coefficient results for the basic finner model (Fig. 35) obtained in ref. 39 with a magnetic suspension system are shown in Fig. 46.

Surface Pressure and Wake Surveys

In the introduction to this section, the effect of conventional mechanical model supports on the pressure surrounding the body and the near and far wake flow field was mentioned. The fact that the magnetic model suspension technique permits a model to be held against wind tunnel airstream without the interference from mechanical supports has led to a number of experimental studies investigating the wake flow field on a variety of model geometries.[2,20–25] Browand and Finston found that the laminar wake of $M \simeq 4.29$ was sensitive to angle of attack—assuming a "bifurcated" form at low angle of attack. In this state the

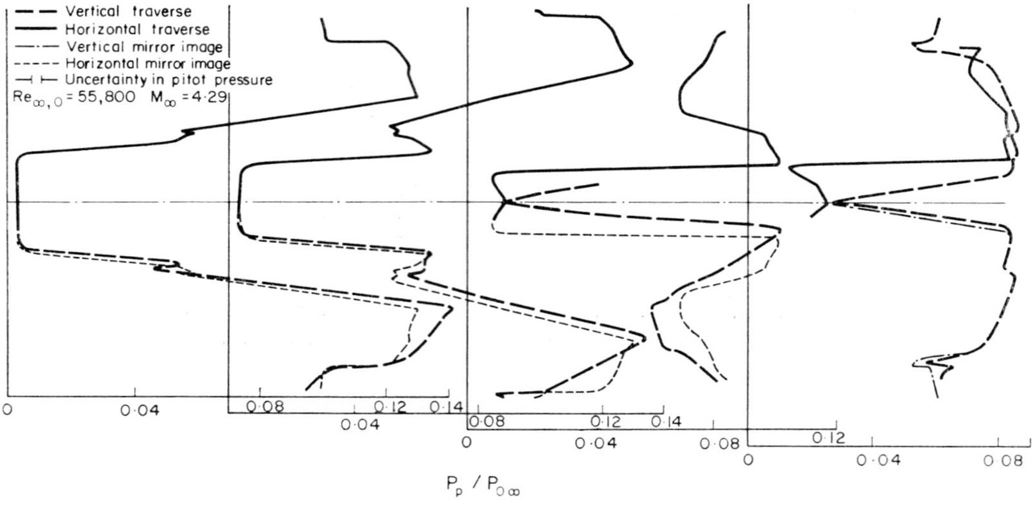

FIG. 47. Pitot pressure profiles showing wake bifurcation.

wake exhibits a notch (Fig. 47). In addition a technique for surface pressure telemetry from a magnetically suspended model has been reported in ref. 20. Finally note Crane (ref. 40) has measured air condensation effects on drag and wake properties. The correlation with static pressure measurements is excellent. The use of magnetic suspension for wake and flow field experiments represents a true free flight simulation and a practical alternate to ballistic range and free-flight aerodynamic testing techniques.

Conclusion

The data shown above indicates the magnetic balance system is of the same versatility of standard balances. The accuracy of recent data (ref. 26) suggests that repeatability of 0.5% or less can be expected from this instrument, with angular position being known to within ± 0.015 deg.

References to Section IV

1. M. TOURNIER and P. LAURENCEAU, Suspension magnétique d'une maquette en soufflerie, *La Recherche Aeronautique*, no. 59, pp. 21–27, July/Aug. 1957.
2. J. Mirande, Mesure de la résistance d'un corps de révolution à $M_0 = 2{,}4$ au moyen de la suspension magnétique ONERA, *La Recherche Aeronautique*, no. 70, p. 24, May/June 1959.
3. E. E. COVERT and E. L. TILTON, Further evaluation of a magnet suspension and balance system for application to wind tunnels, ARL 63-226, December 1963 (Dynamic stability, static data on a family of double enders, ogives, ellipses at $M = 4.8$).
4. J. E. CHRISINGER, E. L. TILTON, W. J. PARKIN, J. B. COFFIN and E. E. COVERT, Magnetic suspension and balance system for wind tunnel application, *Journal of the Royal Aeronautical Society*, vol. 67, November 1963, p. 635.
5. A. B. COPELAND, E. E. COVERT and T. STEPHENS, Recent advances in the development of a magnetic suspension and balance system for wind tunnels (Part III), ARL 65-114, June 1965. (Cone cylinder cone ($C_D C_L C_M$).)
6. A. B. COPELAND, E. E. COVERT and E. L. TILTON, Measured aerodynamic characteristics of the cone cylinder cone model with base separation at *M*-4.8, *Journal of Rockets and Spacecraft*, vol. 2, no. 6, November–December 1965, pp. 998–1000.

7. M. J. GOODYER, Some force and moment measurements using magnetically suspended models in a low speed wind tunnel, University of Southampton, Paper at ARL Symposium, ARL 66-0135, July 1966.
8. R. MOREAU, Use of magnetic suspension system in ONERA wind tunnel, ARL Symposium Paper, July 1966.
9. W. M. PHILLIPS, The measurement of low density sphere drag with a University of Virginia magnetic balance, ARL Symposium presentation, ARL 66-0135, July 1966.
10. A. B. COPELAND, E. E. COVERT, and R. A. PETERSON, Wind tunnel measurement at $M = 4.28$ of some static and dynamic aerodynamic characteristics of finned missiles suspended magnetically, *Journal of Spacecraft and Rockets*, vol. 5, no. 7, pp. 838–842, July 1968.
11. M. VLAJINAC, Wind tunnel measurements of the aerodynamic characteristics of the 2.75 wrap around fin rocket using a magnetic suspension system, MIT TR 150, December 1968. (C_{l_p}, $C_{m_q} + C_{m_{\dot\alpha}}$, C_D, C_L, C_M, $M = 4.25$).
12. K. R. SIVIER, A one-component, magnetic support-and-balance system, *Journal of Aircraft*, vol. 6, no. 5, p. 398, 1969.
13. M. JUDD, E. E. COVERT and M. VLAJINAC, Sting free drag measurements on ellipsoidal cylinders at transition Reynolds Numbers, *Journal of Fluid Mechanics*, vol. 47, part 2, pp. 353–364, 1971.
14. J. F. W. CRANE, J. G. WOODLEY and J. P. THOMPSON, Use of the RAE magnetic suspension system as a force balance, RAE TR 71140, July 1971 (20° Cone Mach 8.5 C_D, C_L).
15. M. VLAJINAC *et al.*, Subsonic static characteristics of slender wind configurations using a magnetic suspension and balance system, NASA CR-1796, July 1971.
16. W. M. PHILLIPS and A. R. KUHLTHAU, Transition regime sphere drag near the free molecule limit, *AIAA Journal*, vol. 9, no. 7, July 1971, pp. 1434–1435 ($M_\infty \sim 8$–10, *Re* 1–20. $M_\infty > 15$, $K_{n\infty} \sim 0.7$).
17. M. VLAJINAC *et al.*, Subsonic and supersonic static aerodynamic characteristics of a family of bulbous base cones measured with a magnetic suspension and balance system, NASA CR-1932, January 1972.
18. T. STEPHENS *et al.*, Recent developments in a wind tunnel magnetic balance, AIAA 10th Aerospace Sciences Meeting, January 1972, Paper No. 72–164.
19. A. B. COPELAND, Wind tunnel measurements of the roll aerodynamics of the Iroquois sounding rocket and the basic finner using a magnetic model suspension system, MIT TR-137, February 1967 (C_{l_p}).
20. G. DUBOIS and C. ROGUÉ, Sur une méthode de mesure de la pression de culot. Mesure et visualization sur un maquette cylindro-conique suspendue magnétiquement à $M_0 \simeq 7{,}6$, *La Recherche Aeronautique*, no. 79, Nov./Dec. 1960, p. 35.
21. O. E. VAS, E. M. MURMAN and S. M. BOGDONOFF, Studies of wakes of support-free spheres at $M = 16$ in helium, *AIAA Journal*, vol. 3, no. 7, p. 1237, July 1965.
22. E. M. MURMAN, Experimental studies of a laminar hypersonic cone wake, *AIAA Journal*, vol. 7, no. 9, September 1969. Sphere wakes at *M*-16.
23. D. K. MCLAUGHLIN, J. E. CARTER and M. FINSTON, Experimental investigation of the near wake of a magnetically suspended cone at $M = 4.3$, AIAA Paper 69-186.
24. D. K. MCLAUGHLIN, Experimental investigation of the mean flow and stability of the laminar supersonic cone wake, *AIAA Journal*, vol. 9, no. 3, March 1971, pp. 479–484.
25. F. K. BROWAND, M. FINSTON and D. K. MCLAUGHLIN, Some preliminary measurements behind cones magnetically suspended in a Mach number 4.3 stream, Presented at AGARD, Paris, France, May 1967. Vol. 1: *Fluid Physics of Hypersonic Wakes*, Preprint no. 19.
26. M. VLAJINAC and E. E. COVERT, Sting free measurements of sphere drag in laminar flow, *Journal of Fluid Mechanics*, vol. 54, pt. 33, pp. 385–392 (1972).
27. I. CHANTZ and R. T. GROVES, Dynamic and static stability measurements of the basic finner at supersonic speeds, NAVORD Report 4516, January 1960, U.S. Naval Ordnance Lab.
28. B. DAYMAN, Hypersonic viscous effects on free flight slender cones, *AIAA Journal*, vol. 8, 1965, p. 1391.
29. J. B. ADCOCK, Some experimental relations between the static and dynamic stability characteristics of sting mounted cones with bulbous bases, Paper presented at the 3rd Technical Workshop on Dynamic-stability Problems, Moffett Field, California, November 1968.
30. L. E. ERICSSON and J. P. REDING, Aerodynamic effects of bulbous bases, Lockheed Missiles and Space Co., Technical Report LMSC-4-17-68-4, November 1968.
31. E. C. POLHAMUS, A concept of the vortex lift of sharp-edge delta wings based on a leading-edge-suction analogy, NASA TN D-3767, 1966.
32. H. W. WENTZ and D. L. KOHLMAN, Wind tunnel investigations of vortex breakdown on slender sharp-edge wings, NASA CR-98737, November 1968.
33. D. H. PECKHAM, Low-speed-wind-tunnel tests on a series of uncambered slender pointed wings with sharp edges, R&M No. 3186, Brit., ARC, 1961.
34. F. W. ROOS and W. W. WILLMARTH, Some experimental results on sphere and disk drag, *AIAA Journal*, vol. 9, no. 2, Feb. 1971.
35. A. B. BAILEY and J. HIATT, Free-flight measurement of sphere drag at subsonic, transonic, supersonic, and hypersonic speeds for continuum, transition and near-free-molecular flow conditions, AEDC-TR-70-291, March 1971.

36. K. L. Goin and W. R. Lawrence, Subsonic drag spheres at Reynolds numbers from 200 to 10,000, *AIAA Journal* (Tech. Notes), vol. 6, no. 5, May 1968.
37. S. Hoerner, Tests of spheres with reference to Reynolds number, turbulence, and surface roughness, *NACA Technical Memorandum* No. 777, October 1935.
38. C. B. Millikan and A. L. Klein, The effect of turbulence, an investigation of maximum lift coefficient and turbulence in wind tunnels and in flight aircraft engineering, 1933.
39. F. J. Regan, Roll damping moment measurement for the basic finner at subsonic and supersonic speeds, NAVORD Report 6652, U.S. Naval Ordnance Lab., January 1964.
40. J. F. W. Crane, Air condensation effects at $M = 8.5$ measured on the drag and wake of a magnetically suspended 20° cone, RAE TR 70022, February 1970.

V. SPECULATION ON FUTURE APPLICATIONS

The results given in the previous sections demonstrate that magnetic suspension systems possess general utility as a data collecting instrument in a wind tunnel laboratory. This utility comes about because of the increased accuracy in measuring model orientation and aerodynamic forces and torques and flexibility in the nature of the tests that can be conducted.* Further, some individual components of a magnetic balance can be used separately. These applications will be considered in more detail below. This section will also contain a discussion of the application of magnetic balances to large tunnels and high q tunnels. The section will end with a short discussion of the applicability to short time, high Reynolds number facilities (such as shock tunnels, Ludwieg tubes) and of the limitations due to Lorentz forces.

Other Applications of Magnetic Balance System Components

Several of the ideas embodied in the use of magnetic fields for model suspension have other applications. The fact that the magnetically suspended body can be moved over a substantial distance in the wind tunnel test section can be used to advantage in measuring aerodynamic interference in a variety of situations. Further, the magnetic force has been shown to be a volume force. The weight of a mass in a gravity field is also a volume force. It follows that it may be possible to simulate gravity controlled trajectories on other than a one to one scale through application of this idea. Recently the techniques of ballistic range testing have been extended to wind tunnel testing by catapulting small models into the wind stream. Data reduction is hampered by the problems of defining the position of the free-flight model in the test section of the wind tunnel. An electromagnetic position sensing system appears to be well suited to this purpose.

Measurement of Aerodynamic Interference

Detailed knowledge is needed about the aerodynamic forces and torques resulting from the non-uniform flow field caused by adjacent bodies in a variety of applications ranging from positioning nacelles to separation of external stores, pilot escape capsules, and astronaut recovery towers. The magnetic balance is well suited to these experiments because, as noted above, the body in question can be positioned almost at will in the neighborhood of an axed, nonmagnetic body also located in the wind tunnel test section. Figure 48, which was taken by the authors, shows a droppable fuel tank near a model of the LTV A-7A. The aircraft model was obtained from a nearby hobby shop, was made of polyurethane plastic

* Magnetic balance systems may be useful in the future when applied to the problem of determining the aerodynamic characteristics of vehicles designed to "fly" in tubes.

FIG. 48. Wing body interference tests.

and was fixed to the tunnel side walls. The data taken from such a test could be used to compute trajectories of the fuel tank if it were to be released. Alternately it is possible to measure the forces and torques, feed them directly to a digital computer and determine the trajectory by using the wind tunnel as an aerodynamic function generator. In this connection note that this latter procedure has been carried out with standard balances* but never, as far as is known, with a magnetic balance.

Magnetic Simulation of Gravity

The usual laws of aerodynamic scaling, as applied to the problem of interpreting static wind tunnel data in the Mach number range of five or less, are satisfied if the fluids are identical, the model is geometrically similar to the full-scale aircraft, if the Mach number and Reynolds number are equal and if the ratios of temperature at corresponding points are equal. In other words, at the same angle of attack, C_L for the model is equal to C_L of the prototype, etc. Similarly, one can find appropriate dimensionless parameters for more general wind tunnel testing that include effects of power, aeroelasticity and dynamic parameters. In the case where the effect of gravity is important, the additional dimensionless parameter is called the Froude number (U^2/lg).† Here U is the relative speed between the wind and the model or aircraft, l the characteristic length and g is the local acceleration due to gravity. It is clear there is a problem here because g is g, independent of the model scale. This problem is overcome through use of non-uniform magnetic fields to generate "artificial" gravity. The scaling is based upon the same Mach number in the wind tunnel and flight. It is possible to determine coil configurations to simulate both level and inclined flight paths.[1,2,3] The geometry of the two cases is shown in Fig. 49a and 49b. Force field non-uniformities, whether due to imperfect coils‡ or due to other nearby jettisonable items on the model are shown to have a relatively small effect on the trajectories for short times. Naturally, the model of the carrier aircraft must be made of non-magnetic materials.

Figure 49b shows that climbing and diving flight requires a rotated acceleration vector. The process may be accomplished mechanically or electrically. Figure 50 shows a coil configuration that allows the acceleration field to be rotated electrically.[2] If the model core is a saturated sphere, then the required gradient along the rotated field line is

$$\frac{\partial B_n}{\partial n} = 0.019\left(\frac{g_{\text{model}}}{g_{\text{full scale}}} - 1\right)\frac{\text{weight of store model}}{\text{weight of magnetic material}} \text{ gauss/inch} \tag{90}$$

$$\frac{g_{\text{model}}}{g_{\text{full scale}}} = \frac{L_{\text{full scale}}}{L_{\text{model}}}\frac{T_{\text{model}}}{T_{\text{full scale}}} \text{ (Mach number scaling)} \tag{91}$$

where T is the static temperature of the fluid.

Figures 51 and 52 show distribution of typical acceleration and direction contours for a configuration similar to that in Fig. 50. These calculations are for a 48-in. transonic tunnel.

Clearly the fields are not uniform although the 10% error contour is always in a sphere of 12 in. radius. However, because the error in the acceleration only effects the trajectory at

* See ref. 1 for references related to these "captive flight" procedures.

† In the case for which the model density ρ_m is nearly the fluid medium density, ρ, the scaling law could be written in terms of the aerodynamic force per unit volume to the buoyancy force per unit volume, i.e. $\rho U^2/l \cdot (\rho_m - \rho)g$.

‡ It is not possible to design a "perfect coils system", because of limitations of Maxwell's equations, etc. $[\nabla \cdot B = 0,\ \nabla \times B = 0,\ F = (M \cdot \nabla)B]$. The goal of the design process is to minimize the non-uniformities.

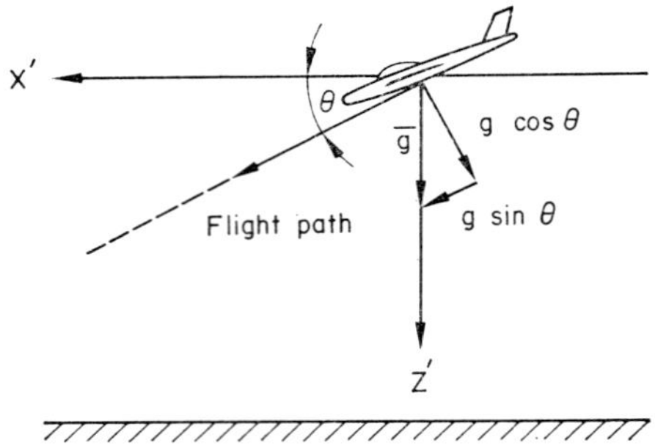

FIG. 49a. Prototype parent aircraft and store in earth-fixed reference frame.

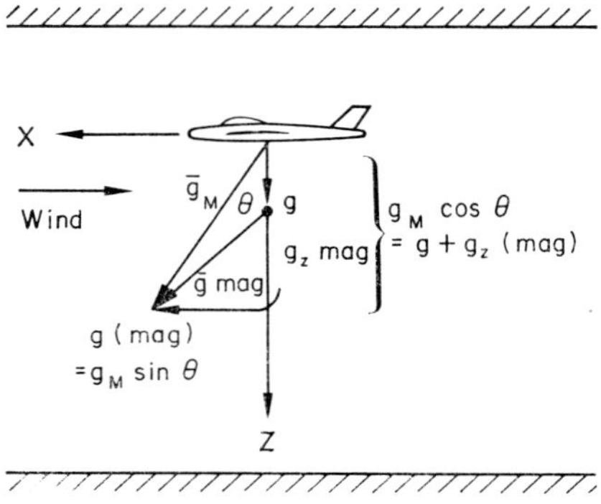

FIG. 49b. Model parent aircraft and store in wind-tunnel reference frame.

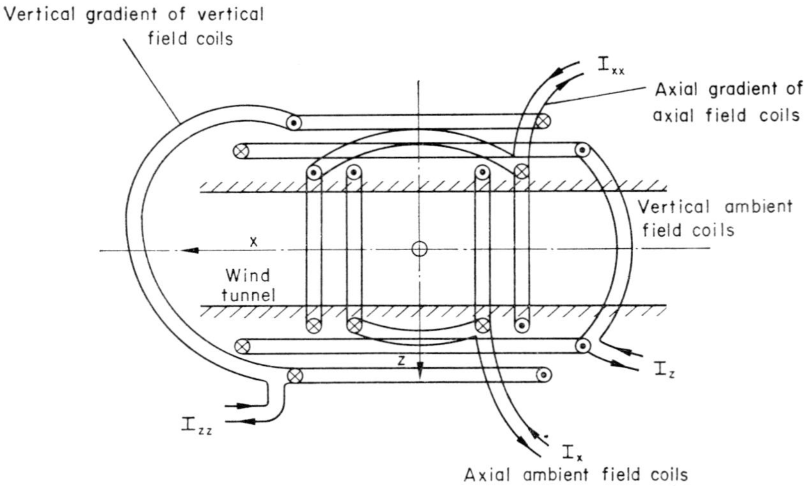

FIG. 50. Arrangement of coils to provide combined axial and vertical forces.

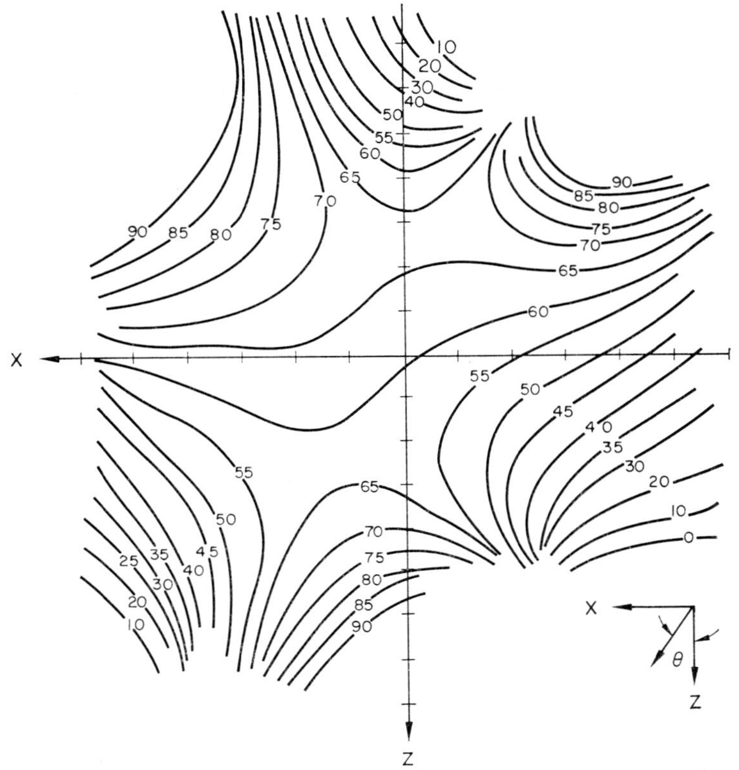

FIG. 51. Force direction result after Adams.[2]

a later time, the non-uniformities in acceleration are only important at larger distances. The integral for the fractional error in the trajectory suggests that the greater errors are incurred by these non-uniformities that occur for longer times. Hence, the 10% errors in force do not imply a 10% error in trajectory. The error is introduced by three factors:[3]

(a) The differences between actual and ideal gravity as it influences the trajectory directly.

(b) The change in angle of attack because the improper gravity induced improper velocities.

(c) The error in forces due to aerodynamic interferences, like buoyancy, flow inclination and flow curvature, that arises because the trajectory is in the incorrect plane.

If an iron sphere is released at the origin of the fields shown in Fig. 51 and Fig. 52 the trajectory computed for the non-uniform field may be compared with the ideal field trajectory (Fig. 53). The fact that the sphere moves into a stronger field is indicated by downward curved trajectory. The maximum error is 12% in z on a 24-in. fall. Note the force is nearly 30% too high over the latter phase of the trajectory.

Two additional trajectories were computed using an aerodynamic drag of $-\frac{1}{2}g$ in the x-direction. The errors here are comparable as shown in the table on page 98.

The $\frac{1}{2}g$ case follows the 60-deg contour (Fig. 51) quite closely and the errors are relatively small in z, but are larger in x because of increasing field strength.

This error is proportional to the magnitude of the non-uniformity and lags the variation of the field. Finally one can show that if multiple drop testing is desired, the spherical elements must be separated from each other three diameters to reduce the acceleration error to

	$-\frac{1}{2}g$		$-\frac{7}{8}g$	
z_1	$\frac{\delta x}{x}$	$\frac{\delta z}{z}$	x	$\frac{\delta z}{z}$
12.8 in.	0.08	0.0036	0.9 in.	+ 0.02

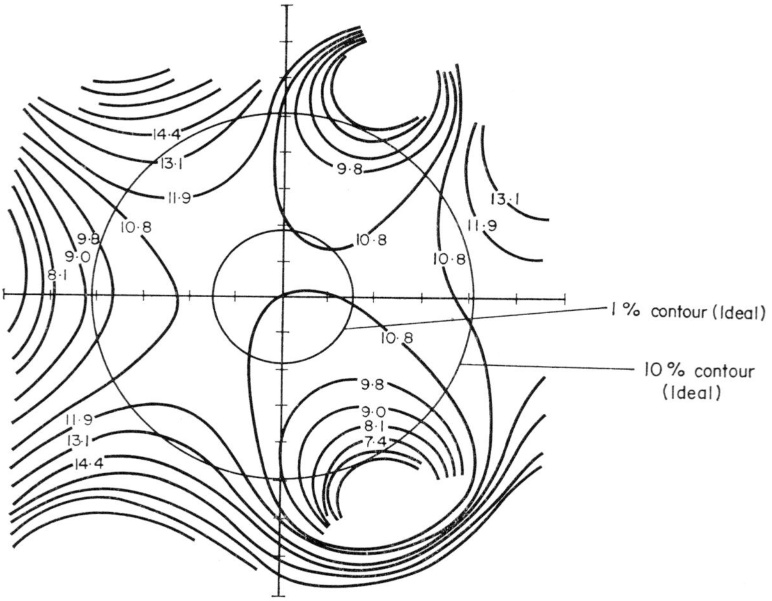

FIG. 52. Resultant force magnitude (typical result after Adams). [10.8 = 35 g's for an iron sphere; ortho 6.2-1. (*Note*: 10.8 = design-force.) Contour lines ± 10%, i.e. 11.9 = + 10%, 13.1 = + 20%, 9.8 = −10%, 9.0 = − 20%.]

10%. At four diameters the error is of the order of 3%. The direction of this erroneous acceleration depends upon position as shown in Fig. 54.

We conclude that such a method is a practical, feasible way to simulate separation of stores or other jettisonable items from aircraft.

Generalized Position Sensing

The data reduction problem for free-flight testing in wind tunnels is usually accomplished by measuring the position of the model in two photographs, taken simultaneously. One of these photographs depicts position in an arbitrary plane, and the other depicts the

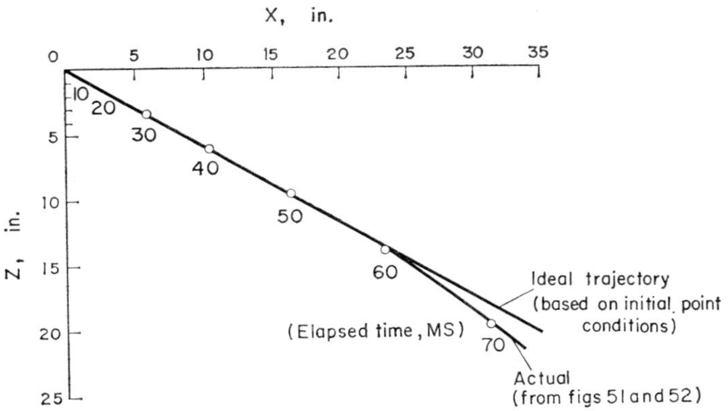

FIG. 53. Comparison of actual and ideal trajectories (fields of Figs. 51 and 52).

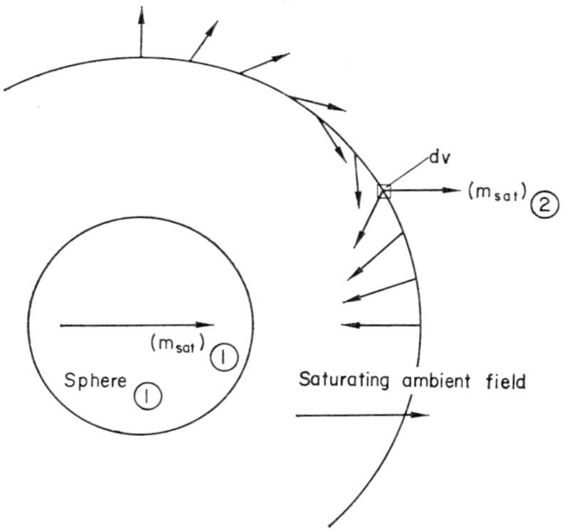

FIG. 54. Variation of strength and direction of magnetic force on a volume element *dv* due to a saturated sphere and a saturating ambient field, at a given radius.

position in a plane perpendicular to the first. The line of intersection of the two planes is usually along the wind axis. After the positions are read they are stored on punched cards and used for data analysis. It has been proposed that the electromagnetic position sensing system (EPS) be adapted to this same task. There is the further advantage that its data is in the form of an electric signal, and as such can be processed in a more straightforward way. It seems best to record the data on magnetic tape, and then use this as the primary source of data. The calibration can also be stored on tape for use in the data reduction. In as much as the model comes to rest in the critical part of its trajectory, the 20 kHz excitation discussed in section II seems to be suitable for this application. This application has yet to be tried, and must be treated as a speculation.

Limitations of Usage of Magnetic Balance and Suspension Systems

Any discussion of the limitations of magnetic balance and suspension systems seem to fall naturally into three distinct classes. First, there is the limitation of size. Second is the limitation of tunnel dynamic pressure, and third is the limitation of response time. Had this survey been prepared several years earlier there would have been a detailed discussion of power requirements.* However the successes achieved at the University of Virginia by Parker, Zapata and Moss[4] strongly suggest that such a limitation no longer exists.

Effect of Size

In discussing the consequences of size on magnetic balance or magnetic suspension systems for wind tunnel use, it is well to remember that the magnetic force is a volume force and thus changes with the cube of a change in the length scale, if everything else is fixed. The weight to be suspended is also a volume force, so that aspect presents no problems. Further the magnetic torque and the aerodynamic torque both are volume dependent. The aerodynamic forces are surface effect and so they change only as the second power of a change in dimension. Consequently the ratio of aerodynamic to magnetic forces increases as the reciprocal of the change in length. It is therefore more difficult to work with small objects than it is to work with larger objects, for a fixed size balance or suspension system. On the other hand if the magnetic field is held fixed the ampere turns required to develop this field increases linearly with the length scale. Similarly, for a fixed gradient, the ampere turn required to develop the gradient is proportional to the square of the length scale. Since the ampere turns for gradient increases faster than the ampere turns for magnetization the gradient coils require an increasing fraction of the total coil volume. If this law for the two values of ampere turns is followed the geometric relationship between the model and the field line in the test section remain exactly similar. If however the relative model size is reduced then the magnetizing current or the gradient current, or both, must be further increased. In pursuing these changes to the limit, three boundaries appear. The stresses in the coils become so large that they fail structurally. Bitter[5] has shown that a 250 kG solenoid $1\frac{1}{2}$ in. in diameter (3.8 cm) operates near the yield point of Cr–Cu coils. Figure 55 (adapted from Bitter) shows that stress is similar to size. Note that this figure also indicates the power required if the coils were copper and emphasizes the importance of the Parker and Zapata results.[4] Note too that cooling would not be a problem for these high-power coils. Hence one feels confident that large sized, high NI superconducting coils can be constructed readily though their mechanical details may evolve toward the high-power density Litz wire coils described by Haldeman *et al.*[6]

Second, the gradient field can become so large that the model is appreciably demagnetized on one-half its length. Under these circumstances the calibration of the model becomes increasingly difficult to handle, although the problem of non-linear calibration was faced successfully in the earlier systems.[7] Again a structural limit will be reached, this time in the

* All other things being about equal, the propulsion power in a wind tunnel increases as the square of a size change. That is, doubling the size causes four times the power to be needed. The power required to magnetize and to provide torques increases linearly with size, while the power required to balance forces increases as the cube of the size change. Hence the ratio of power for tunnel propulsion to the power for the balance tends to decrease with increasing tunnel size. This ratio is strongly Mach number dependent. At MIT the ratio is about 15:1 at a Mach number of 6. That is, propulsion takes fifteen times more power than the magnetic balance. At a Mach number of 0.5 these two powers stand in the ratio of 1:1.

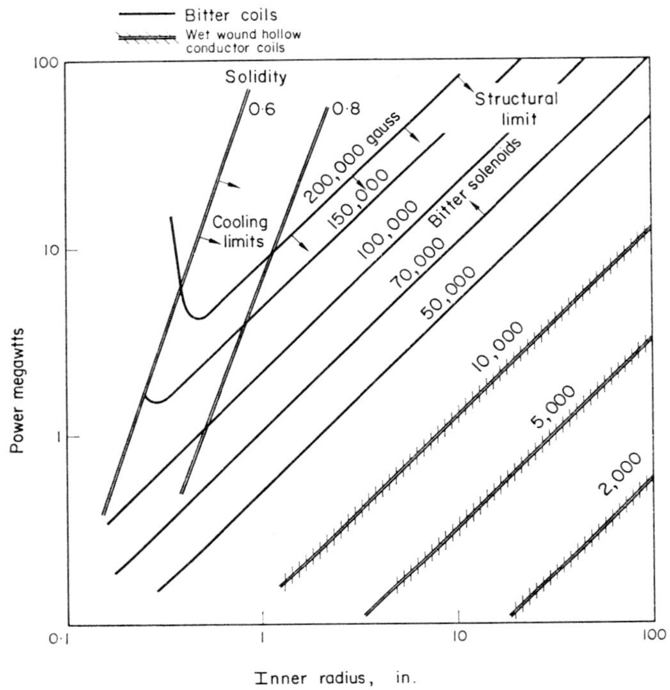

FIG. 55. Arrangement after Bitter.

model, particularly at the end of the model that is nearest the high intensity $\bar{B}$ field. If the model were fabricated from a permanently magnetized material it would be partially demagnetized under these high local field strength conditions. The third limit, the effect of local saturation in EPS, is less complicated and can be simply overcome by use of an optical position sensing system, for the EPS is less and less sensitive as the magnetization is increased well above saturation.

Effect of Dynamic Pressure

If the model size is fixed, then any increase in dynamic pressure must be accompanied by an increase in magnetization, or gradient, or both. This increase is easy to demonstrate by considering the ratio of the drag force to the model weight. This ratio can be expressed as g's of acceleration, if the model were free to move. The results are tabulated below for a sphere, together with the assumption of constant drag coefficient, here taken to be one-half. The assumption of constant C_d for the sphere is suitable for providing an estimate of the size of the terms involved. If the Reynolds number is so large that the boundary layer is turbulent the calculation can be in error by nearly a factor of 2. For a transonic Mach number a similar error in size but in the opposite direction is to be expected. The gradient (kG/in.) can be computed[2] for a saturated iron sphere magnetized to 21 kG as

$$B_{xx} = 0.019 \left(\frac{a}{g} - 1\right) \frac{w_{\text{mod}}}{w_{\text{mag}}} \tag{92}$$

TABLE 5. VARIATION OF SPHERE DRAG IN G'S WITH DYNAMIC PRESSURE

Radius (in.)	q (atm)				
	0.1	0.2	0.5	1.0	2.0
1.0	7.50	15.0	37.5	75.0	150.0
2.0	3.75	7.50	18.75	37.50	75.0
5.0	1.50	3.0	7.5	15.0	30.0
10.0	0.75	1.50	3.75	7.50	15.0

where w_{mod} is the overall model weight,
w_{mag} is the weight of magnetic material.
In this calculation two weights are equal. At 150 g's the gradient is 2.85 kG/in. This gradient is sufficiently large to unsaturate a sphere 8 in. in diameter. However, to require 150 g's on an 8-in. sphere implies a q of 15 atm, a condition not likely to be encountered. As indicated above the most severe condition is that of the highest q and the smallest sphere. Nevertheless a $\frac{1}{10}$-in. sphere will not demagnetize appreciably until 1500 g's are required. Thus this limit is not likely to be encountered. The dynamic pressure limit is usually determined by the power available. The use of superconducting coils would seem to move the q and size limit beyond practical needs for the immediate future.

Transient Wind Tunnel Limit

High Reynolds number testing techniques seem to be evolving toward the use of transient techniques to keep the wind tunnel power consumption to a minimum. This in turn raises the question of the usefulness of a magnetic balance in a short time facility. The time constant is largely governed by the ability of the power supply to add current to the coil. As this current is added an increased number of field lines must be created. This requires added electromotive force. The inductance is a geometric factor that, when multiplied by the rate change of electric current, defines the needed electromotive force. Experimental evidence[4] indicates that superconducting coils behave no differently from regular coils in this regard. The inductance is proportional to the square of the length scale, so if the coil size is doubled, the inductance is increased by a factor of four. Hence larger systems are slower. This difficulty can be overcome by use of a small capacitance in the circuit, as discussed in Section II. The problem here is the practical detail of constructing a capacitor to handle thousands of amperes. Further a high rate buildup of current of the magnitude requires high voltages. This technology was developed in connection with electro-magnetically driven shock tubes.[8] As with the problems of size and dynamic pressure discussed in this section, there does not seem to be a theoretical limit to the applications to short time facilities although one should do a pilot experiment to check for electric breakdown of the working fluid.

Effect of Lorentz Forces[9]

Lorentz forces come about because of the electromotive force induced by relative motion between a fluid and a magnetic field. This is the well-known $v \times B$ in magnetohydrodynamics. For high-temperature, low-density wind tunnel conditions, the magnetic fields introduce unwanted magnetohydrodynamic effects in the flow. An attempt is made here to obtain an estimate of the density and temperature limits to which the MHD effects limit the usefulness of the magnetic balance system.

In this calculation the magnetic force is held fixed and the lower limit of velocity is found for which a constant multiple of the magnetic force is equal to or less than the dynamic pressure effect. This constant is arbitrary, and has been chosen to be 0.001.

In aerodynamic analysis, the complexities of the basic processes frequently are correlated in terms of dimensionless parameters. One such parameter is the Reynolds number, *Re*, which can be derived in a qualitative sense by considering the ratio of inertial forces to the frictional forces, i.e.,

$$R_e = \frac{\rho v l}{\eta} \tag{93}$$

where ρ mass density of fluid,

v velocity of fluid,

l characteristic length of solid object,

η absolute viscosity of the fluid.

Similarly, in magnetohydrodynamic flows the square of the Hartmann number is the ratio of the magnetohydrodynamic forces to the viscous forces, i.e.

$$M^2 = \frac{B^2 l^2 \sigma}{\eta} \tag{94}$$

where B magnetic field intensity,

σ electrical conductivity.

Hence, the ratio of magnetohydrodynamic forces to the inertial forces is described by the ratio of the Hartmann number to the Reynolds number. Calling this N,

$$N = \frac{\sigma B^2 l}{\rho v}. \tag{95}$$

This ratio is also a measure of the ratio of induced volume force to dynamic pressure; thus, whenever N is small, it can be anticipated that there will be no appreciable interaction between the supporting field and the fluid. (The magnetic field induced by the magnetohydrodynamic effects is characterized by the magnetic Reynolds number, $R_m = \sigma \mu v l$; $\mu \cong$ permeability of free space. In the usual application of the suspension system, σ is small in the cases of interest and $\mu \cong 4\pi \times 10^{-7}$ in MKS units. R_m is very small for this problem from which it is inferred that the induced magnetic fields will be small compared to the fields of the suspension system.)

For the purposes of this discussion N will be assumed to be bounded by some value, say 0.1% or 1.0%, and the ratio v/l will be computed in terms of the constants (N) and (σ/ρ) to define the boundary of operation. In the Aerophysics Laboratory magnetic balance, B, at the pole face, rarely exceeds 5000 G. Thus as an upper bound let $B = 10{,}000$ G or 1.0 Wb/m^2. Hence, in the MKS system, eq. (95) gives

$$\frac{v}{l} \cong \frac{\sigma}{\rho} \frac{1}{N_0} \tag{96}$$

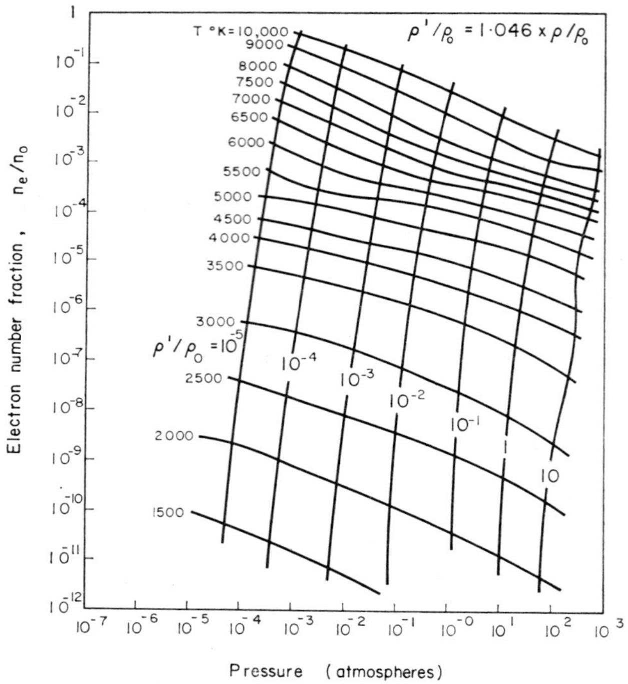

FIG. 56. Electron number fraction vs. pressure.

if N_0 is the reference value of N.

At low electron densities

$$\frac{\sigma}{\rho} \cong 0.566 \times 10^{18} \frac{n_e}{n_0} \cdot \frac{1}{\nu_c} \tag{97}$$

where n_0 number density of particles before ionization,

n_e number density of ionized particles,

n_g number density of unionized particles,

e charge of electron,

m_e mass of electron,

ν_c collision frequency,

m_g mass of "air" atom.

The value of n_e/n_0 may be found for the equilibrium conditions from charts; see Fig. 56 for example. It may be more reasonable, on a pragmatic basis, to use stagnation rather than the static temperature in this particular calculation. ν_c can be related to the density, temperature and a collision probability, P_c. Thus

$$\frac{\sigma}{\rho} \cong 1.08 \times 10^8 \frac{T^{3/4}}{P^{3/2}} \frac{e^{-(eV/2kT)}}{P_c}. \tag{98}$$

Substituting this value into eq. (96) gives

$$\frac{v}{l} \geqq 1.08 \times 10^8 \frac{T^{3/4}}{T^{3/2}} e^{-(eV/2kT)} \cdot \frac{1}{N_0} \frac{1}{P_c}. \tag{99}$$

Thus as $P \to 0$, v/l must become very large. Considering v/l as a function of pressure with temperature a parameter, the parameter a_0 ($a_0 = 1.08 \times 10^8 T^{3/4} e^{-eV/2kT}$) assumes the values tabulated in the following.

T	100	300	700	1000	3000	7000
$\frac{eV}{2kT}$	8.70×10^2	2.90×10^2	1.24×10^2	87	29	12.4
a_0	—	6.78×10^{-121}	1.84×10^{-44}	2.92×10^{-28}	1.09×10^{-2}	3.24×10^5

where $v/l \geq a_0/(P^{3/2} P_c N_0)$.

If l is taken as 1 m, then Table 6 gives the limiting value of velocity at which MHD effects are 0.1% of the dynamic pressure. According to this calculation at a temperature of 300°K, 700°K and 1000°K all reasonable velocities are acceptable. At 3000°K one must have velocities in excess of 10.9 m/sec if the ambient pressure is 1 atm.

In air at low velocities P_c is somewhat in excess of unity. However, this merely shifts the boundary towards operation at lower pressures.

It should be noted that when the pressure is below 10^{-8} atm the interaction is large at the high temperatures. However, these conditions correspond to circumstances where the dynamic pressure effects are only important for long-term operation, and can be estimated by the use of free molecule flow.

The results in Table 6 can be put in a different aspect by letting T represent stagnation temperature, and expanding the flow to 50°K static circumstances, the allowable wind tunnel operating regions are shown in Figs. 57 and 58. In Fig. 57 the maximum allowable velocity before condensation is plotted as a function of stagnation temperature. The values from Table 6 are shown plotted with constant pressure as a parameter. These lines represent the minimum velocity necessary for having $N_0 \leqq 10^{-3}$ at a given temperature and pressure. It should be noted that these lines are not vertical, but appear to be so due to the high sensitivity of the minimum velocity to temperature. The allowable operating region would always to be the left of the appropriate constant pressure line.

TABLE 6.* LIMITING VALUES OF (v/l) FOR SEVERAL P, T AT $N_0 = 10^{-3}$, $P_c = 1$

$P^{T.K}$	300°	700°	1000°	3000°	7000°
10^2	6.78×10^{-121}	1.84×10^{-44}	2.92×10^{-28}	1.09×10^{-2}	3.24×10^5
1	6.78×10^{-118}	1.84×10^{-41}	2.92×10^{-25}	1.09×10^1	3.24×10^8
10^{-2}	6.78×10^{-115}	1.84×10^{-38}	2.92×10^{-22}	1.09×10^4	3.24×10^{11}
10^{-4}	6.78×10^{-112}	1.84×10^{-34}	2.92×10^{-19}	1.09×10^7	3.24×10^{14}
10^{-6}	6.78×10^{-109}	1.84×10^{-32}	2.92×10^{-16}	1.09×10^{10}	3.24×10^{17}
10^{-8}	6.78×10^{-106}	1.84×10^{-19}	2.92×10^{-13}	1.09×10^{13}	3.24×10^{20}

N.B. If the fields were 10 times weaker, each entry would be 100 times smaller. Thus in this case at 3000°K, $P \sim 10^{-2}$, V must be greater than 109 m/sec.

* Due to uncertainties in ν_c, etc., the numerical factors should be regarded as qualitative. They may be in error by a factor of two or so. The entries on the left-hand side of the table have been included to show the strong influence of temperature on the velocity.

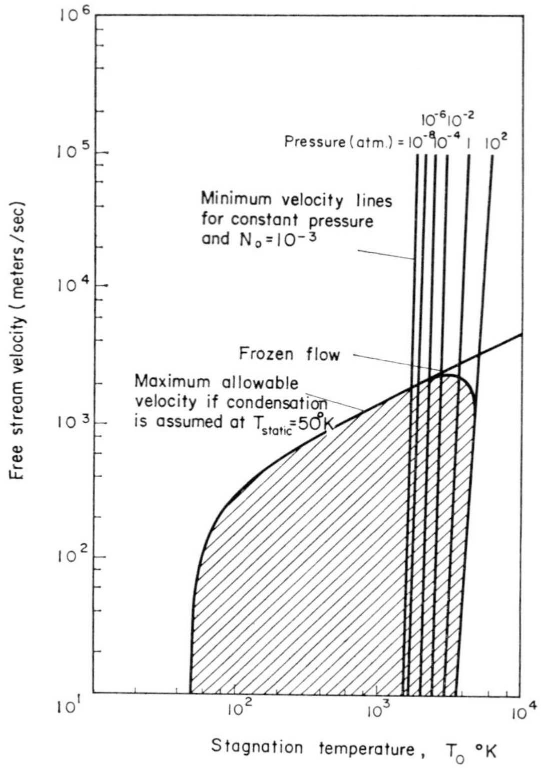

FIG. 57. Allowable wind tunnel operating region for $N_0 \leq 10^{-3}$, $B = 10^4$ G and $P_0 = 100$ atm.

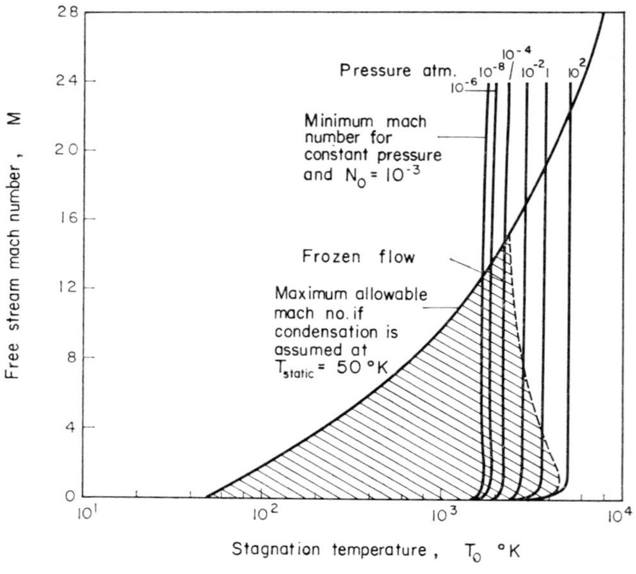

FIG. 58. Allowable wind tunnel operating region for $N_0 \leq 10^{-3}$, $B = 10^4$ G and $P_0 = 100$ atm.

For a frozen flow wind tunnel condition it is reasonable to assume that the stagnation temperature and the free stream static pressure are the limiting quantities. Then in Fig. 57 the constant pressure lines would represent the free stream static pressure. As an example a value of 100 atm is chosen as the wind tunnel stagnation pressure. The region of allowable wind tunnel operation is shaded in Fig. 57. The region is bounded on the left by the condensation line. On the right the region is bounded by a dotted line whose trajectory is determined by the static pressure at the given velocity and temperature; i.e., for a given velocity and stagnation temperature the corresponding static pressure must be equal to or greater than the pressure indicated by the constant pressure line running through that point. It should be kept in mind that the stagnation pressure is being held constant at 100 atm. If the stagnation pressure were lowered the dotted boundary would shift to the right. The dotted boundary asymptotically approaches the constant pressure line corresponding to the stagnation pressure but can never cross it since the static pressure can never exceed the stagnation pressure.

The lower right-hand region in Fig. 57 can be made more meaningful in terms of wind tunnel operation if the Mach number rather than the free stream velocity is the dependent variable. Figure 58 illustrates the allowable operating region in terms of Mach number and stagnation temperature for $N_0 \leqq 10^{-3}$ and a stagnation pressure of 100 atm.

When the flow in the wind tunnel is in equilibrium the free stream MHD effects are negligible. However, areas of MHD interaction could be present in the stagnation regions around the model in the case of either frozen or equilibrium flow since the velocity in these regions is low. The extent of the region around the model in which the MHD effects exceed a certain level are dependent on the shape of the model and the flow conditions. At this time a detailed solution to the problem of determining the nature of this region is not available.

References to Section V

1. E. E. Covert, Wind tunnel simulation of jettison of external store, *Journal of Aircraft*, vol, 6, no. 1, 1967, pp. 48–51.
2. Timothy Stephens and Ronald Adams, Wind tunnel simulation of store jettison with the aid of magnetic artificial gravity, MIT Aerophysics Laboratory Report 174, 1971.
3. E. E. Covert and Timothy Stephens, Magnetic simulation of gravity, *Proceedings of the 2nd International Symposium on Electro-magnetic Suspension*, Southampton, England, July 1971.
4. R. N. Zapata, The University of Virginia superconducting magnetic balance; H. M. Parker and J. R. Jancaitis, The use of iron and extended applications of the University of Virginia cold balance wind tunnel system. Both papers in *Proceedings of the 2nd International Symposium on Electro-magnetic Suspension*, Southampton, England, July 1971.
5. F. Bitter, Water cooled magnets, *Review of Scientific Instruments*, vol. 33, 1962, pp. 342–348.
6. C. W. Haldeman, J. Sullivan and E. E. Covert, A variable phase velocity traveling wave pump, *AIAA Journal*, vol. 9, 1971, pp. 1389–1395.
7. A. B. Copeland, E. E. Covert and E. L. Tilton III, Measured aerodynamic characteristics of a cone-cylinder-cone model, *Journal of Rockets and Spacecraft*, vol. 2, 1965, pp. 998–1000.
8. J. N. Bradley, *Shock Waves in Chemistry and Physics*, John Wiley & Sons, New York, 1962, pp. 127–128.
9. E. E. Covert and E. L. Tilton III, Recent advances in the development of a magnetic suspension and balance system for wind tunnel use, U.S. Air Force Aerospace Research Laboratories Report ARL 63-225, December 1963.

3

THEORY OF HYPERSONIC FLOW ABOUT A WING

A. L. Gonor
Institute of Mechanics, Moscow State University

NOTATION

r, θ, φ	conical coordinates	L_i	functionals
u, v, w	components of velocity	γ	dihedral angle
$\psi = \text{const}$	stream surface	v	volume of wing
ρ	density	s	wing planform area
p	pressure	l	length of body
ϵ	small parameter	c_p	pressure coefficient
$\theta^*(\varphi)$	angle of shock wave	C_x	drag coefficient
α	angle of attack	C_y	lift coefficient
2β	angle of apex wing	K	lift-to-drag ratio

INTRODUCTION

The possibility of using aerodynamic lift during re-entry of a vehicle into an atmosphere and for sustained hypersonic flight is attracting more and more attention from investigators. Studies are under way both in the direction of the investigation of hypersonic flow about a wing with a geometry close to the traditional configuration of supersonic wings and in the direction of searching for new shapes of vehicles in which the body simultaneously plays the role of wing producing lift in a nonsymmetrical flow. Besides, the configuration of a vehicle is often chosen so that the aerodynamic and thermal characteristics should assume the optimum value.

At present many results of great importance are obtained in both directions. Let us analyse those which are closely connected with the type of flow about a wing corresponding to the high values of lift-to-drag ratio.

The first solutions of the problem of hypersonic flow on the windward side of a triangular plate, by means of expanding the unknown functions in series of the order of a small parameter, have been suggested in the literature.[1–3] However, such an approach did not lead to a closed form solution for the problem of flow about a wing with supersonic leading edges; there were separate examples of numerical calculations made using the method of finite differences or the time-dependent method.[5–7] The generalization of the theory[2] concerning a wing of finite thickness has been considered in the literature.[8,9]

In their engineering applications the calculations of aerodynamic characteristics have followed a certain development on the basis of half-empirical approaches, assumed in refs. 10 and 11.

The qualitative investigation of the flow field is a special problem.[12] In this case it is assumed that there are complex singularities in the flow field near a delta wing. But the results are obtained making strong assumptions which are untrue.

The method of integral relations[13–16] has been widely used in the theory of flow about a wing at a high angle of attack. In this case the leading edge shock wave becomes detached and the boundary problem simplifies. On the basis of the first approximation of the integral relations method, G. G. Chernyi[16] carried out an analysis of all possible regimes of hypersonic flow about a wing.

Following the approach improved by Kennet,[13] a solution was obtained for the flow about a wing in the case of an attached shock wave.[17] However, it cannot be considered to be correct, as the analysis of the paper[17] has shown a break up of one of the boundary conditions at the junction with uniform streamlines.

The difficulty of the solution of the wing flow problem with an attached shock wave is explained by the necessity of a smooth junction of the disturbed flow with the uniform flow, along the Mach cone. To satisfy this condition a solution with a sufficient number of arbitrary functions is required. In this connection the use of the perturbation method was suggested.[18] This allowed the necessary matching of the uniform, potential and rotational flows in the flow field about a triangular plate.

The above-mentioned method of "two approximations" will be used in the first part of this paper for the research of the flow about a few classes of conical wings. Special attention is drawn to the investigation of hypersonic flow about conic wings having inverted V-shaped cross-sections which are often called caret wings.

The following parts of the paper are concerned with the numerical method based on the investigations referred to in the literature.[19–20] In the final part the solution of optimization problems is given. We are looking for the configuration of a conic body having a maximum lift-to-drag ratio, assuming that the distribution of pressure is determined by Newton's pressure law and that the coefficient of local friction is constant. Such a statement of the problem simplifies the investigation but it does not give effective results, as the optimization problem leads to a complex nonlinear equation with partial derivatives. Further simplification of the analysis can be made at the expense of additional restrictions on the class of considered surfaces. At present Strand's approach[21] has greatly progressed.

Strand was the first to solve the problem of the wing configuration with minimum drag provided the wing is thin in the longitudinal and lateral directions. Later the optimization of this class of surfaces was considered[22–24] under various additional conditions.

There was a detailed consideration of the thin wing problem with maximum lift-to-drag ratio.[23–24]

The shapes of the wing cross-section are known to influence markedly the values of the lift-to-drag ratio.[25–27]

Hence a more precise statement of the variational problem can give new results for optimum configurations.

The solution of the variational problem connected with the conic wing shape having maximum lift-to-drag ratio has been obtained.[28–31] Ferrari[32] considered a similar problem with two izoperimetric constraints (the given volume of the body and the lift). According to the author's statement the cross-section of the wing has a star-like configuration; a calculation of the shapes of the rays is in progress.

A similar problem has been considered[33] in which the local variation method is first used for the solution of optimum problems about a wing. Here the solution of the variational problem is stated by means of Newton's approach.[28–29]

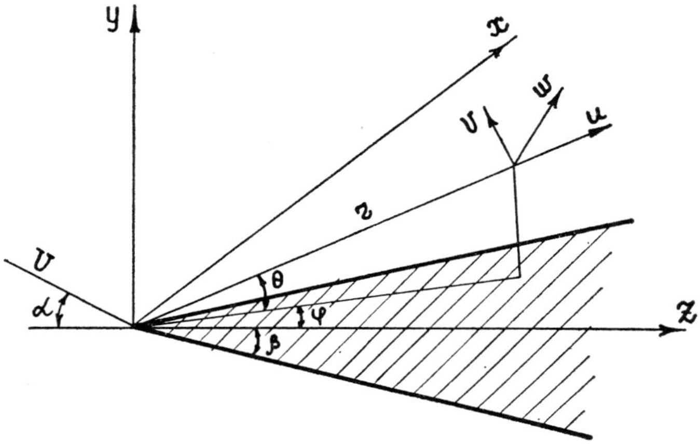

FIG. 1. Scheme of the flow about conical wing.

1. HYPERSONIC FLOW PAST A THIN WING WITH AN ATTACHED SHOCK WAVE

1. Let us consider a thin conical wing in hypersonic flow; in the coordinate system r, θ, φ shown in Fig. 1, the equations of conical flow for the ψ, φ variables have the following form:

$$\frac{w}{\cos\theta}\frac{\partial u}{\partial\varphi} - v^2 - w^2 = 0; \quad \frac{w}{\cos\theta}\frac{\partial v}{\partial\varphi} + uv + w^2\tan\theta = -\frac{1}{\rho\theta_\psi}\frac{\partial p}{\partial\psi}$$

$$\frac{x}{x-1}\frac{p}{\rho} + \frac{u^2+v^2+w^2}{2} = C; \quad \frac{\partial}{\partial\varphi}\frac{p}{\rho^x} = 0 \tag{1.1}$$

$$\frac{\partial}{\partial\varphi}\ln(\rho w\theta_\psi) + 2\frac{u}{w}\cos\theta = 0; \quad w\theta_\varphi = v\cos\theta.$$

In these equations all variables are dimensionless and refer to the velocity of undisturbed flow U, density ρ^0 and double dynamic pressure $\rho^0 U^2$. The variable ψ satisfies the relation $v\psi_\theta + w\psi_\varphi \sec\theta = 0$ and represents a stream surface in a conic flow.[1]

The solution will be sought assuming that the region of disturbed flow consists of a highly pressurized thin layer. Then according to traditional estimates of the shock layer theory (see ref. 1) it will be convenient to introduce the following transformation:

$$\theta = \epsilon\bar{\theta}, \quad v = \epsilon\bar{v}, \quad \rho = \epsilon^{-1}\bar{\rho}. \tag{1.2}$$

In new variables (the bar is dropped) one has the following system:

$$\frac{w}{\cos \epsilon\theta} \frac{\partial u}{\partial \varphi} - \epsilon^2 v^2 - w^2 = 0; \quad \frac{p^{1/x}}{\rho} = \delta(\psi)$$

$$\frac{xp}{(x+1)\rho} + \frac{u^2 + \epsilon^2 v^2 + w^2}{2} = C \tag{1.3}$$

$$\left(\frac{w}{\cos \epsilon\theta} \frac{\partial u}{\partial \varphi} - uv + w^2 \epsilon^{-1} \tan \epsilon\theta\right) \epsilon = -\frac{1}{\rho\theta_\psi} \frac{\partial p}{\partial \psi}$$

$$x = \frac{1+\epsilon}{1-\epsilon}; \quad \frac{\partial}{\partial \varphi} \ln (\rho w \theta_\psi) + 2\frac{u}{w} \cos \epsilon\theta = 0; \quad w\theta_\varphi = v \cos \epsilon\theta.$$

The fourth equation (1.3) allows us to represent pressure as a sum of two parts:

$$p = p_1(\varphi, \epsilon) + \epsilon p_2(\psi, \varphi, \epsilon). \tag{1.4}$$

Now let us simplify expressions (1.3) and omit terms of the order of ϵ^2 and higher in (1.3) and in the boundary conditions. Then the functions we are seeking will satisfy the following equations:

$$\frac{\partial u}{\partial \varphi} - w = 0; \quad w\frac{\partial v}{\partial \varphi} + uv + w^2\theta = -\frac{1}{\rho\theta_\psi} \frac{\partial p_2}{\partial \psi}$$

$$u^2 + w^2 = \Delta^2 (\psi); \quad \frac{\partial}{\partial \varphi} \ln (\rho w \theta_\psi) + 2\frac{u}{w} = 0; \quad w\theta_\varphi = v. \tag{1.5}$$

In this approximation the boundary conditions at the shock wave $\theta = \theta^*(\varphi)$ will have the following form:

$$\begin{aligned}
u^* &= \cos \alpha\,(\cos \varphi - \epsilon\theta^* \tan \alpha); \quad m_0 = 2/(x-1)\, M_\infty^2 \sin^2 \alpha \\
v^* &= -\,[\theta^*_\varphi \cos \alpha \sin \varphi + (1 + m_0) \sin \alpha + \epsilon \cos \alpha\, (\theta^* \cos \varphi \\
&\quad - \theta^*_\varphi \sin \varphi) + \epsilon \sin \alpha\, \theta^{*2}_\varphi] \\
w^* &= -\cos \alpha\,(\sin \varphi + \epsilon\theta^*_\varphi \tan \alpha) \\
p^* &= \sin^2 \alpha + \epsilon p_1^*; \quad \rho^* = (1 + m_0)^{-1} \\
p_1^* &= \sin 2\alpha\,(\theta^* \cos \varphi - \theta^*_\varphi \sin \varphi) - \sin^2 \alpha - M_\infty^{-2}.
\end{aligned} \tag{1.6}$$

Upon the surface of the wing defined by $\theta = \epsilon\theta^0(\varphi)$ one gets from the solid boundary condition:

$$v - w\, \theta^0_\varphi(\varphi) = 0. \tag{1.7}$$

While deriving the third of eqs. (1.5) we used the expression for pressure (1.6) and the second of eqs. (1.3) which can be written as follows:

$$\frac{1}{\rho} = \frac{1}{\rho'}\left[1 + \frac{\epsilon}{\sin^2 \alpha}\,(p_1^{*\prime} - p_1^* - p_2)\right]. \tag{1.8}$$

The primed variables correspond to the point of intersection of the shock wave with a streamline ($\varphi = \varphi'$ in Fig. 2). Pressure p_1 (φ) can be calculated directly from its value behind the shock wave (1.6) by putting $p_1 = p^*$; then p_2 is zero at the shock wave.

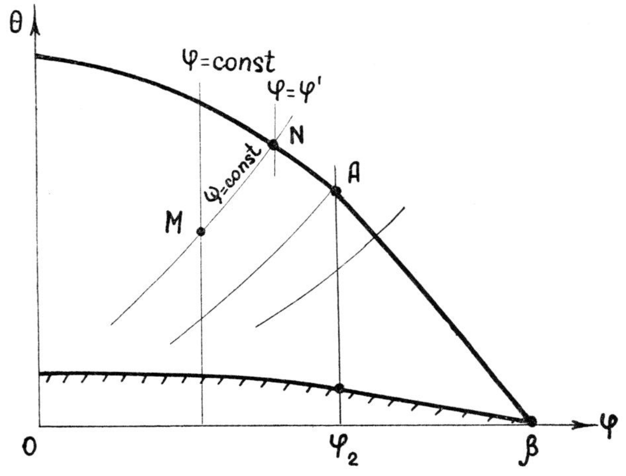

FIG. 2. Conical coordinate system.

The velocity components u and w follow from (1.5):

$$u = \Delta(\varphi') \cos[\varphi + \alpha(\varphi')]; \quad w = -\Delta(\varphi') \sin[\varphi + \alpha(\varphi')]$$

$$\Delta(\varphi') = \cos^2 \alpha - \epsilon \sin 2\alpha \,(\theta^{*\prime} \cos \varphi' - \theta^{*\prime}_{\varphi} \sin \varphi') \tag{1.9}$$

$$\alpha(\varphi') = \epsilon \tan \alpha \,(\theta^{*\prime}_{\varphi} \cos \varphi' + \theta^{*\prime} \sin \varphi') \tag{1.10}$$

The relations (1.9), (1.10) lead to an integral of the fifth equation of the system (1.5):

$$\frac{\rho\theta_\psi}{w} = \frac{\rho'\theta'_\psi}{w'}. \tag{1.11}$$

Integrating this equation with respect to ψ (or φ') from the wing surface along the line $\varphi = \text{const}$ yields the following expression for streamlines:

$$\theta = \theta^{\circ}(\varphi) + \int\limits_{\varphi^0(\varphi)}^{\varphi'} \frac{\rho' w}{\rho w'} \theta'_\psi \psi_{\varphi'} \, d\varphi'. \tag{1.12}$$

Here $\varphi^0(\varphi)$ is an arbitrary function, corresponding to the streamline lying upon the wing surface. As the wing is thin the function $\theta^0 \sim \epsilon_1$. Taking into consideration the last equation of the system (1.5) and the second condition (1.6) we find that

$$\theta'_\psi \psi_{\varphi'} \equiv \frac{d\theta'}{d\varphi'} - \left(\frac{\partial\theta}{\partial\varphi}\right)' = \frac{1}{w'}\left\{(1+m_0)\sin\alpha\left[1 + \frac{\epsilon}{1+m_0}(\theta^{*\prime}\cos\varphi' - \theta^{*\prime}_{\varphi}\sin\varphi')\right]\right\}. \tag{1.13}$$

Substituting (1.8) and (1.13) into (1.12) we get

$$\theta = \theta^0(\varphi) + (1+m_0)\sin\alpha \int\limits_{\varphi^0}^{\varphi'} \frac{w}{w'^2} R \, d\varphi'$$

$$R = 1 + \epsilon\left[\frac{(\theta^{*\prime}\cos\varphi' - \theta^{*\prime}_{\varphi}\sin\varphi')\cot\alpha}{1+m_0} + \frac{1}{\sin^2\alpha}(p_1^{*\prime} - p_1^{*} - p_2)\right]. \tag{1.14}$$

This expression at $\varphi' = \varphi$ yields the equation for the shock wave,

$$\theta^* = \theta^0(\varphi) + (1 + m_0) \sin \alpha \int_{\varphi^0}^{\varphi} \frac{w}{w'^2} R \, d\varphi', \tag{1.15}$$

where φ^0 and p_2 are unknown functions. From the solid boundary condition (1.7), taking into consideration (1.5) and (1.14), we get

$$\frac{d\varphi^0}{d\varphi} (w)_{\varphi' = \varphi^0} = 0. \tag{1.16}$$

This equation has two solutions

$$\begin{gathered} \varphi^0 = \text{const} \\ z'(\varphi^0) \cos \varphi^0 + z(\varphi^0) \sin \varphi^0 = - \varphi \cot^2 \alpha \\ z(\varphi) = \epsilon \cot \alpha \, \theta^*(\varphi). \end{gathered} \tag{1.17}$$

In an exact solution the wing surface is a stream surface, therefore $\varphi^0 = \beta$, β is the semi-angle at the apex of the wing. But as has been pointed out,[1] hypersonic solutions for conical bodies may not satisfy this condition. Omitting the analysis of all possible cases it should be noticed that in the theory of the wing developed now it is necessary to assume that $\varphi^0 =$ const, as there is a region of the uniform flow. Some singularities which can arise in this case are considered below. Let us find the second unknown function. It should be pointed out first of all that the pressure p_2 contains the multiplier ϵ and therefore it is enough to determine the main term.

After having integrated the second equation (1.5) along the coordinate $\varphi =$ const from the shock wave to the arbitrary point we shall determine the main term of the pressure:

$$p_2 = \sin \alpha \int_{\varphi'}^{\varphi} \frac{w^3}{w'^2} (\theta_{\varphi\varphi} + \theta) \, d\varphi'. \tag{1.18}$$

Differentiating (1.14) twice and substituting the results into (1.18) we obtain:

$$\begin{aligned} p_2 = \sin \alpha \, (\theta^0 + \theta^0_{\varphi\varphi}) \int_{\varphi'}^{\varphi} \frac{w^3}{w'^2} \, d\varphi' + 2 \sin^2 \alpha \, (1 + m_0) \Big\{ [\cos \varphi \, (z + z'') \\ + \sin \varphi \, (z' + z''')] \int_{\varphi'}^{\varphi} \frac{w^3}{w'^2} \, d\varphi' \int_{\beta}^{\varphi'} \left(\frac{w}{w'^2} \right)_{\varphi' = \xi} d\xi \\ - 2 \sin \varphi \, (z + z'') \int_{\varphi'}^{\varphi} \frac{w^3}{w'^2} \, d\varphi' \int_{\beta}^{\varphi'} \left(\frac{u}{w'^2} \right)_{\varphi' = \xi} d\xi \Big\}. \end{aligned} \tag{1.19}$$

The first term (1.19) characterizes the influence of the lateral contour curvature, the second one the influence of the curvature of the shock wave. The terms in (1.19) containing the values $\epsilon p_{2\varphi}$ and $\epsilon p_{2\varphi\varphi}$ are omitted. The omitted terms have the following form:

$$\delta_1 = \epsilon \int_{\varphi'}^{\varphi} \frac{w^3}{w'^2}\, d\varphi' \int_{\beta}^{\varphi'} \left(\frac{u}{w'^2}\right)_{\varphi'=\xi} p_{2\varphi}\, d\xi; \quad \delta_2 = \epsilon \int_{\varphi'}^{\varphi} \frac{w^3}{w'^2} \int_{\beta}^{\varphi'} \left(\frac{w}{w'^2}\right)_{\varphi'=\xi} p_{2\varphi\varphi}\, d\xi.$$

Suppose that $p_2 \sim a(\epsilon)$, we have $p_{2\varphi} \sim \epsilon^{-1/2}\, a(\epsilon)$; $p_{2\varphi\varphi} \sim \epsilon^{-1}\, a(\epsilon)$. Taking into consideration $u \sim 1$; $w \sim \epsilon^{1/2}$ (at $\varphi \lesssim \varphi_2$) and that the order of w' changes from 1 to $\epsilon^{1/2}$, we see that $\delta_1 \sim \delta_2 \sim \epsilon a(\epsilon)$. Hence in the region of the nonuniform flow where $\varphi \sim \sqrt{\epsilon}$ the pointed-out pressure gradients are small of higher order compared with the remaining terms.

As a result the main term of pressure is determined by formula (1.19). In the region of uniform flow p_2 is equal to 0.

Substituting p_2 in eq. (1.15) we obtain the following integro-differential equation for the shock-wave shape:

$$z + \delta[2L_2\,(z \cos\varphi - z' \sin\varphi) - L_1] + 2\epsilon(z^0 + z^{0\prime\prime})\, L_3 - z^0$$
$$= 2\epsilon\delta\, \{(z + z'')\,[2 \sin\varphi\, L_4 - \cos\varphi\, L_5] - (z' + z''')\sin\varphi\, L_5\}$$

$$z = \epsilon \cot\alpha\, \theta^*(\varphi); \quad z^0 = \epsilon \cot\alpha\, \theta^0(\varphi); \quad \delta = \epsilon \cos\alpha\, (1 + m_0)$$

$$L_1 = \int_{\beta}^{\varphi} \frac{w}{w'^2} \left\{1 + \frac{3 + 2m^0}{1 + m_0}\, [z(\varphi') \cos\varphi' - z'(\varphi') \sin\varphi']\right\} d\varphi'$$

$$L_2 = \int_{\beta}^{\varphi} \frac{w}{w'^2} d\varphi'; \qquad L_3 = \int_{\beta}^{\varphi} \frac{w}{w'^2}\, d\varphi' \int_{\varphi'}^{\varphi} \left(\frac{w^3}{w'^2}\right)_{\varphi'=\xi} d\xi \qquad (1.20)$$

$$L_4 = \int_{\beta}^{\varphi} \frac{w}{w'^2}\, d\varphi' \int_{\varphi'}^{\varphi} \left(\frac{w^3}{w'^2}\right)_{\varphi'=\xi} d\xi \int_{\beta}^{\xi} \left(\frac{u}{w'^2}\right)_{\varphi'=\eta} d\eta$$

$$L_5 = \int_{\beta}^{\varphi} \frac{w}{w'^2}\, d\varphi' \int_{\varphi'}^{\varphi} \left(\frac{w^3}{w'^2}\right)_{\varphi'=\xi} d\xi \int_{\beta}^{\xi} \left(\frac{w}{w'^2}\right)_{\varphi'=\eta} d\eta$$

The coefficients L_i are complex functionals of the form of shock waves, but in a number of cases (for instance in the case of flat shock wave) the integrals are easily estimated and all L_i are represented by closed form expressions. This is used later while constructing the solution. The third-order derivative present in the equation allows a smooth matching of a plane wave with a curvilinear one. This is according to eqs. (1.20) accompanied by a break of the curvature smoothness at point A (Fig. 2).

It should be noticed that though the system (1.5) and boundary conditions (1.6), (1.7) admit some errors of higher order, an exact solution of the approximate equations is being looked for.

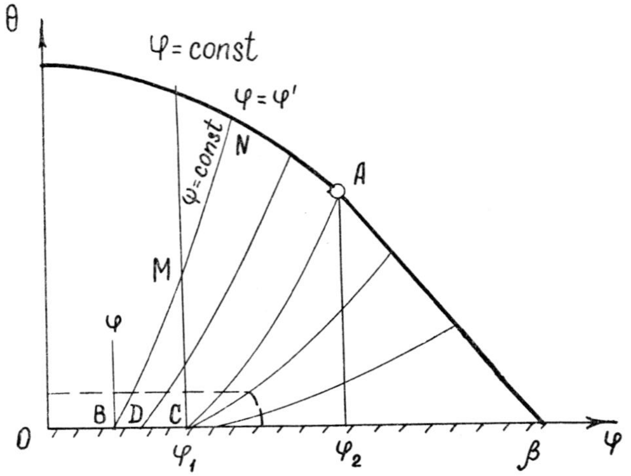

FIG. 3. Cross streamline pattern.

Let us now find the solution for eqs. (1.20). The equation of the shock wave surface in the region $0 \leq \varphi \leq \varphi_2$ will be sought for in form of the following series:

$$z(\varphi) = \sum_{n=0}^{\infty} \frac{z^{(n)}(\varphi_2)}{n!} (\varphi - \varphi_2)^n. \tag{1.21}$$

Values of functions $z(\varphi_2)$ and $z'(\varphi_2)$ are defined by the condition of smooth matching with the homogeneous flow. At the point φ_2 the coefficients L_i are continuous and known; therefore connection between $z'''(\varphi_2)$ and $z^{(4)}(\varphi_2)$ can be easily established from (1.20). By successive differentiation of eq. (1.20) one can find a similar connection between $z^{(4)}(\varphi_2)$, $z^{(5)}(\varphi_2)$, etc. As a result the series defined by eq. (1.21) will be dependent only upon an arbitrary constant which can be easily found from the condition in the plane of symmetry $z'(0) = 0$. The solutions for the different regions of homogeneous, potential and rotationary flows are matched continuously.

Another important result is the topology of stream surfaces.

A study of streamlines crossing a curvilinear shock shows that the former can cross the wing surface OC becoming singular in this region (Fig. 3). As in the case of the theory of hypersonic conical flows, there exists an entropy layer which, when taken into account, changes the velocity distribution in the region indicated in Fig. 3 by a dashed line (see section 2).

The solution of the equation (1.20) in the form of the series (1.21) is convenient for carrying out specific calculations, but it does not allow us to deduce the dependence of the solution from the parameters ϵ. This can be obtained from the perturbation method.

At $\epsilon = 0$ the solution $z \equiv 0$ satisfies the boundary condition and the equation (1.20). Taking the zero order approximation at $\epsilon \neq 0$ all the integrals L_i can be calculated. As a result the relation (1.20) becomes a differential equation and for small ϵ is reduced to Euler's equation of the third order. The solution of the latter one contains the terms $\exp(\epsilon^{-2/3} \ln \varphi)$, this shows the nonanalytical dependence from ϵ. The substitution of this solution into (1.19) allows us to find that the terms δ_1, δ_2 correspond to the above-mentioned evaluations.

2. The general relations obtained above can be applied to the flow past a triangular plate at an angle of attack. Let us denote the apex angle of the plate by 2β, and consider the Mach number to be equal to infinity ($m_0 = 0$). In this case the function $z^0 \equiv 0$ and coefficients L_i calculated along the plane wave are determined, $z = a \sin(\beta - \varphi)$, by relations

$$L_1 = \frac{1 + 3a \sin\beta}{\Delta_1} \frac{\sin(\beta - \varphi)}{\sin(\beta - \varphi_1)}; \qquad L_2 = \frac{\sin(\beta - \varphi)}{\Delta_1 \sin(\beta - \varphi_1)}$$

$$L_4 = \frac{\sin 2(\varphi - \varphi_1)}{2\Delta_1} d; \qquad L_5 = \frac{-\sin^2(\varphi - \varphi_1)}{\Delta_1} d$$

$$\Delta_1 = \cos\alpha\,(1 - 2a \tan^2\alpha \sin\beta)^{1/2}; \quad \varphi_1 = a \tan^2\alpha \cot\beta$$

$$d = \tfrac{1}{3}\cot^3(\beta - \varphi_1)\sin^3(\varphi - \varphi_1) + \tfrac{1}{2}\frac{\cot(\beta - \varphi_1)\sin(\beta - \varphi)\sin^2(\varphi - \varphi_1)}{\sin(\beta - \varphi_1)}$$

$$- \tfrac{1}{3}\cos^3(\varphi - \varphi_1). \tag{1.22}$$

In the given problem the coefficient L_3 drops out (the wing is plane); the coordinate φ_2 is determined by the intersection line of the Mach cone for the uniform flow behind the shock wave with the wing surface. Its value can be found from the expression:

$$\sin(\varphi_2 - \varphi_1) = \frac{1}{\Delta_1}\left(\epsilon \frac{1 + \epsilon}{1 - \epsilon} p^*\right)^{1/2}. \tag{1.23}$$

The formula (1.23) shows in particular that

$$\varphi_2 \sim \sqrt{\epsilon}.$$

Substituting the equation of the plane shock wave into (1.20) we have

$$a = \epsilon \frac{1 + \epsilon \sec^2\alpha}{\sin(\beta - \varphi_1)} \tag{1.24}$$

and the equation of the plane shock in physical coordinates can be written as

$$\theta^* = \epsilon \tan\alpha \frac{(1 + \epsilon \sec^2\alpha)\sin(\beta - \varphi)}{\sin(\beta - \varphi_1)}.$$

The pressure behind the shock wave in a homogeneous flow is determined by

$$p = p^* = \sin^2\alpha + \epsilon\left[\frac{\sin^2\alpha \sin\beta \tan\alpha}{\sin(\beta - \varphi_1)}(1 + \epsilon \sec^2\alpha) - \sin^2\alpha\right]. \tag{1.25}$$

The curvilinear section of the shock wave is found by the series (1.21). First of all we shall substitute the coefficients L_i calculated by means of the formula (1.22) at $\varphi = \varphi_2$ into the equation (1.20). Then at the point φ_1 we obtain the following relation:

$$z''' = -(z_2'' + z_2) K - z_2'; \quad z_2 = z(\varphi_2) \tag{1.26}$$

$$K = 2 \cot (\varphi_2 - \varphi_1) + \cot \varphi_2$$

Performing the differentiation of (1.20) analogous relations can be set between the derivatives of higher order $z^{(4)}$, $z^{(5)}$ and so on.

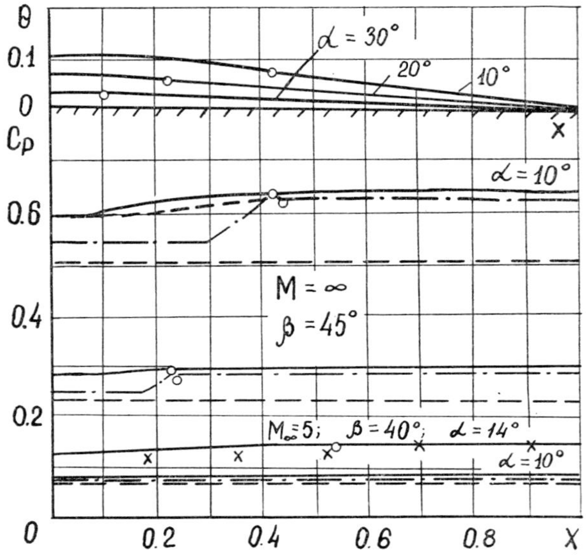

FIG. 4. Shock wave form and pressure distribution on delta wing.

Let us restrict the series (1.21) to four terms. Using the boundary condition $z'(0) = 0$ in this approximation we obtain for the second derivative the expression

$$z_2'' = \frac{z_2' (2 - \varphi_2^2) - K\varphi_2^2 z_2}{\varphi_2 (2 + K\varphi_2)}. \tag{1.27}$$

This formula allows us immediately to determine the order of z_2''. Indeed, $\varphi_2 \sim \sqrt{\epsilon}$, $z_2' \sim \epsilon$; consequently z_2'' will be $\sim \sqrt{\epsilon}$. From (1.26) we find that $z_2''' \sim 1$. It is easy to see that the shock wave is convex ($z_2'' < 0$). The relation (1.26) allows to put down the series (1.21) in the following way:

$$z(\varphi) = z_2 + z_2' (\varphi - \varphi_2) + \tfrac{1}{2} z_2'' (\varphi - \varphi_2)^2 - \tfrac{1}{6} [(z_2 + z_2'') K + z_2'] (\varphi - \varphi_2)^3 + \dots \tag{1.28}$$

Besides, the second derivative is excluded by using (1.27).

Having determined the shape of the shock wave we can easily find the remaining parameters of the flow field.

Let us consider some results of calculation of pressure distribution on the wing surface. Figure 4 shows the pressure coefficient on the wing surface calculated in the first approximation and accounting for p_2 (dashed line), and a comparison is made with corresponding values calculated[4] by the finite differences method and with Newton's theory (dashed solid line). On the same figure, a comparison is made with experimental[35] results at Mach number $M = 5$ and the shock wave configuration is shown (top). One can conclude from the shape of the curves that pressure changes slowly in the vicinity of the axis and that the pressure gradient normal to the plate is negative.

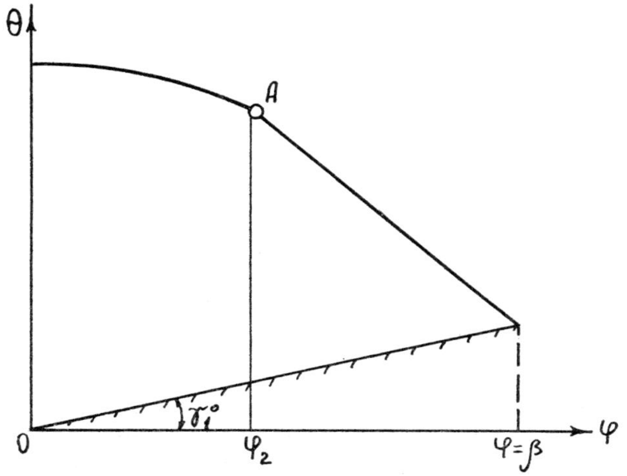

FIG. 5. Cross-section of V-shaped wing.

3. As a second example we shall investigate the flow over a triangular wing with the lateral cross section of V-shaped form. We shall denote the semiangle of the wing apex as β, and the dihedral angle in the lateral plane as γ_1 (Fig. 5).

Then the equation of the lateral contour is determined by $z^0 = \cot^2\alpha \sin\varphi$. Coefficients L_i calculated for the plane shock wave coincide with the expression (1.22). Coefficient L_3 can again be omitted and the formulae (1.23), (1.24), with the accepted accuracy, are not changed. As a result the equation of the plane shock in physical coordinates will be given

$$\theta^* = \gamma_1 \sin \varphi + \frac{\epsilon \tan \alpha \, (1 + \epsilon \sec^2 \alpha) \sin (\beta - \varphi)}{\sin (\beta - \varphi_1)}. \tag{1.29}$$

Let us determine the curvilinear part of the shock wave. At the point A equation (1.20) and the relevant coefficients coincide with the above-mentioned case of the plane delta wing. Therefore the relations (1.26), (1.27) can be also used for the V-shaped wing.

Consequently the series (1.28) will give the shock wave if the values z_2 and z_2' will be obtained from (1.29). The distribution of the pressure on the wing and the shape of the shock wave for some values of the dihedral angle are shown in Fig. 6. The diagrams obtained from the first approximation lead to the conclusion that the pressure along the wing span changes slightly. The main change is observed near the plane of symmetry where the growth of the pressure is taking place. Let us note that the theory is true for flows without inner shocks; therefore the dihedral angle can change within a narrow range.

2. CALCULATION OF THE ENTROPY LAYER ON THE WING SURFACE

1. From the proceeding analysis it follows that a singularity in the distribution of the circumferential velocity appears on the wing surface. Therefore the obtained solution is to be corrected in the region near the plane of symmetry of the wing.

This same case appeared first in the general theory of hypersonic conical flows when singularities exist on the body surface and a separate solution has to be built for the region called entropy layer.[36–42]

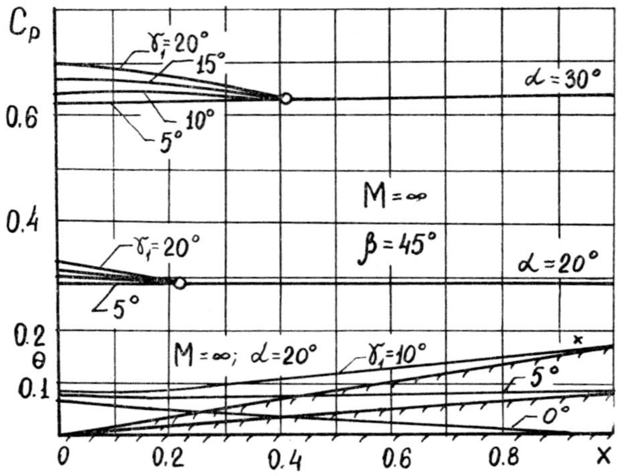

FIG. 6. Shock wave form and pressure distribution on V-shaped wing.

The calculation of the entropy layer on the wing has much in common with the above-mentioned investigations, but there is a significant difference connected with the type of outer solution. Below, the flow about a plane delta-wing is considered in detail, but this method can be extended to a general case.

Let us consider the projection of Euler's equation on the direction of the axis φ.

In variables φ, ψ we shall have

$$\frac{w}{\cos\theta}\frac{\partial w}{\partial\varphi}+uw-vw\tan\theta=-\frac{1}{\rho\cos\theta}\left[\frac{\partial p}{\partial\varphi}-\frac{\theta_\varphi}{\theta_\psi}\frac{\partial p}{\partial\psi}\right]. \tag{2.1}$$

Instead of u and w we shall introduce Δ and τ: $u=\Delta\cos\tau$; $w=\Delta\sin\tau$. Equation (2.1) using (1.1) in the new variables is written in the form:

$$\tan\tau\left(\frac{\partial\tau}{\cos\theta\,\partial\varphi}+1-\frac{v}{\Delta}\tan\theta\cos\tau+\frac{v^2}{\Delta^2}\right)=-B$$

$$B=\frac{1}{\rho\Delta^2\cos\theta}\left[\frac{\partial p}{\partial\varphi}-\frac{\theta_\varphi}{\theta_\psi}\frac{\partial p}{\partial\psi}\right]. \tag{2.2}$$

It is easy to determine that the gradient of pressure according to (1.3), (1.4) appears only with terms $\sim \epsilon^2$; therefore we shall have

$$\tan \tau \left(\frac{\partial \tau}{\partial \varphi} + 1\right) = 0 \tag{2.3}$$

where the terms $\sim \epsilon^2$ are omitted.

If using the shock wave conditions we take into account that the bracket (2.3) is zero, the solution given below corresponding to the relations (1.9) follows:

$$\tau = -\varphi - \alpha_1(\varphi'), \tag{2.4}$$

where α_1 (φ') is determined by the conditions on the shock wave.

The second possible solution contradicts the conditions on the shock wave, therefore the outer solution can be only given by (2.4), the transversal velocity having a singularity on the wing surface. In the flow field where $\tau = 0$ the small terms neglected deriving equation (2.3) are becoming comparable with the remaining terms.

The pressure gradient becomes essential along the wing span. This can be shown in the following way. Denoting

$$D = \left(\frac{v}{\Delta} \tan \theta \cos \tau - \frac{v^2}{\Delta^2}\right) \tan \tau \tag{2.5}$$

and supposing $\tau = \tau^0 + \epsilon^2 \tau^{(2)} + \ldots$ (further the top index k will denote the coefficient of ϵ^k), where τ^0 corresponds to (2.4), for $\tau^{(2)}$ we will receive the following relation:

$$\tan \tau^0 \left(\frac{\partial \tau^{(2)}}{\partial \varphi} - \frac{\theta^{(1)2}}{2}\right) = D^{(2)} - B^{(2)}$$

$$D^{(2)} = \left(\frac{v^{(1)}}{\Delta^{(0)}} \theta^{(1)} \cos \tau^0 - \frac{v^{(1)2}}{\Delta^{(0)2}}\right) \tan \tau^0 \tag{2.6}$$

$$B^{(2)} = \frac{1}{\rho^{(1)} \Delta^{(0)}} \left[\frac{\partial p^{(1)}}{\partial \varphi} - \frac{\theta_\varphi^{(1)}}{\theta_\psi^{(1)}} \frac{\partial p^{(1)}}{\partial \psi}\right].$$

It is seen from equation (2.6) that there exists a singularity at the points, where τ^0 is a zero. The main term is determined by the pressure gradient $B^{(2)}$.

For $B^{(2)} \neq 0$ this singularity surely exists; for $B^{(2)} = 0$ (the plane of symmetry) it cannot exist and a special investigation is required. Thus in the previously obtained solution (section 1), which will be called the outer solution, a term with a singularity appeared having the following form:

$$\tau^{(2)} \approx B^{(2)}_{\varphi = -\alpha_1(\varphi')} \ln \left[\alpha_1 (\varphi') + \varphi\right].$$

The non-uniformity of the solution indicates that there is a term, which is negligibly small in the outer flow and also small everywhere, is becoming the main term in the region of non-uniform convergence. It is easily seen that such term is a function of B.

2. Equation (2.3) has two solutions:

$$\frac{\partial \tau}{\partial \varphi} = -1 \text{ and } \tau = 0.$$

The first solution has the singularity near $\tau = 0$. Therefore the analysis of the second solution $\tau = 0$ becomes natural for the investigation of the non-uniform convergence.

We will transform the term τ_i (where i denotes an inner solution) into the form

$$\tau_i = \epsilon^2 \tau_i^{(2)} + \ldots \tag{2.7}$$

Having inserted the expression (2.7) into equation (2.2) we obtain

$$\tau_i^{(2)} = D_i^{(2)} - B_i^{(2)}. \tag{2.8}$$

By means of (2.6) we find that

$$D_i^{(2)} \sim 0\,(\epsilon^2)$$

$$B_i^{(2)} = \frac{1}{\rho_i^{(1)}\,\Delta_i^{(0)}} \left[\frac{\partial p_i^{(1)}}{\partial \varphi} - \frac{\theta_{i\varphi}^{(1)}}{\theta_{i\psi}^{(1)}}\,\frac{\partial p_i^{(1)}}{\partial \psi}\right]. \tag{2.9}$$

It is seen from here that for the determination $\tau_i^{(2)}$ it is necessary to determine $\rho_i^{(1)}$, $\Delta_i^{(0)}$; $p_i^{(1)}$. If we suppose that these values can be found, the problem of matching the inner and outer solutions arises. To prove that both solutions are compatible it is necessary to determine whether they have a common domain of existence where they are asymptotically equivalent.

On the basis of the second condition one may conclude that the inner and outer solutions do not match since they intersect.

Thus the first members of the expansion cannot be asymptotically equivalent at any common domain for two reasons.

Firstly, the scale of the independent variable is the same in every region; secondly, an arbitrary constant (or arbitrary function) by means of which the matching of the solutions could be accomplished is absent.

There are some methods by which the matching problem can be considered.[37–40]

The method used below is closely connected with the behaviour of the outer solution.

Analyzing the outer solution one could see that the term of the second order has a logarithmic singularity at $\varphi \to -\alpha_1(\varphi')$.

The analysis of the terms of higher order shows that they have even a more powerful singularity at a particular point. This divergence can be excluded by transforming the members of the series expansion indicated in the Poincare–Lighthill–Kuo method.[43]

In the PLK method both the dependent and independent variables are expressed by an auxiliary variable and they are expanded in series.

This gives an additional degree of freedom which to some extent can be used for controlling the behaviour of the expansion near the singularity. If we consider the equation (2.2) at small τ, the following approximation is introduced:

$$\tan \tau \sim \tau; \quad \tau = \varphi^0 + \epsilon^2 \tau^{(2)}$$

$$\cos \theta = 1 - \epsilon^2 \frac{\theta^{(1)2}}{2} + \ldots$$

where

$$\varphi^0 = -\varphi - \alpha_1(\varphi'),$$

and we obtain

$$(\varphi^0 + \epsilon^2 \tau^{(2)}) \frac{\partial \tau^{(2)}}{\partial \varphi^0} = B^{(2)} - D^{(2)} + \varphi^0 \frac{\theta^{(1)2}}{2}.$$

This kind of equation makes it possible to apply the PLK method.

First of all we shall rewrite the equations (2.2) in the form:

$$\frac{\partial \tau}{\partial \varphi} = -\frac{(\tan \tau + C) \cos \theta}{\tan \tau}, \quad C = B - D.$$

Let us introduce the auxiliary function $f(z)$ and the auxiliary variable z and substitute the preceding equation for the equivalent equation system

$$f(z)\frac{\partial \tau}{\partial z} = -(\tan \tau + C)$$

$$f(z)\frac{\partial \varphi}{\partial z} = \frac{\tan \tau}{\cos \theta}.$$

We choose the function $f(z)$ so that the expansions for τ and φ begin with terms $\tau^0 = -z - \alpha_1(\varphi')$ and z. Thus $f(z) = \tan \tau^0$. The main equation system becomes

$$\tan \tau^0 \frac{\partial \tau}{\partial z} = -(\tan \tau + C)$$

$$\tan \tau^0 \frac{\partial \varphi}{\partial z} = \frac{\tan \tau}{\cos \theta}. \tag{2.10}$$

All the variables may be presented as

$$C = \sum_{n=0}^{\infty} \epsilon^n C^{(n)}$$

$$\tau = \tau^0 + \epsilon^2 \tau^{(2)} + \ldots$$

$$\varphi = z + \sum_{n=1}^{\infty} \epsilon^n \varphi^{(n)}(z).$$

The functions $C^{(n)}$ and $\cos \theta$ may be expressed as z:

$$C^{(n)}(\varphi) = C^{(n)}(z) + \epsilon \frac{dC^{(n)}(z)}{dz} \varphi^{(1)} + \ldots$$

$$\cos \theta = 1 - \epsilon^2 \frac{\theta^{(1)2}}{2} + \ldots$$

$$\theta^{(1)}(\varphi) = \theta^{(1)}(z) + \epsilon \frac{d\theta^{(1)}(z)}{dz} \varphi^{(1)} + \ldots$$

and

$$C^{(0)}(\varphi) = 0; \quad C^{(1)}(\varphi) = 0,$$

as the main pressure term is constant.

Let us substitute these expansions into eqs. (2.10). After some transformations and integration we shall have:

$$\tau^0 = -z - \alpha_1(\varphi'); \quad \varphi^{(1)} = 0$$

$$\tau^{(2)} = \tan[z + \alpha_1(\varphi')] \int_{\varphi'}^{z} \frac{C^{(2)}(z)}{\tan^{(2)}[z + \alpha_1(\varphi')]} dz \tag{2.11}$$

$$\varphi^{(2)} = -\int_{\varphi'}^{z} \frac{\tau^{(2)} + \dfrac{\theta^{(1)2}}{2} \sin[z + \alpha_1(\varphi')] \cos[z + \alpha_1(\varphi')]}{\sin[z + \alpha_1(\varphi')] \cos(z + \alpha_1(\varphi')]} dz.$$

It is seen from the expressions (2.11) and (2.6) that for the determination of $\tau^{(2)}$ and $\varphi^{(2)}$, in which we are interested, it is quite enough to know the outer solution. Thus we have

$$\tau = -z - \alpha_1(\varphi') + \epsilon^2 \tan[z + \alpha_1(\varphi')] \int_{\varphi'}^{z} \frac{C^{(2)}(z)}{\tan^2[z + \alpha_1(\varphi')]} dz.$$

$$\varphi = z - \epsilon^2 \int_{\varphi'}^{z} \frac{\tau^{(2)} + \frac{1}{2}\theta^{(1)2}(z) \sin[z + \alpha_1(\varphi')] \cos[z + \alpha_1(\varphi')]}{\sin[z + \alpha_1(\varphi')] \cos[z + \alpha_1(\varphi')]} dz. \tag{2.12}$$

The obtained relations allow us to extend the uniform convergence of the first solution in a small region near the singularity of the outer solution.

If we introduce new variables z and $d = \rho w \theta_\psi$, then

$$\frac{\partial \varphi}{\partial z} = \frac{\tan \tau}{\cos \theta} \cot \tau^0$$

and the continuity equation (1.1–5) can be written down as:

$$\frac{\partial d}{\partial z} + 2d \cot \tau^0 = 0.$$

Therefore we obtain

$$d(\varphi', z) = d' \frac{\sin^2[\tau^0(z)]}{\sin^2 \tau^{0'}}$$

where (′) denotes that the value of the variable is taken from the shock wave or

$$\rho w \,\theta_\psi = \frac{\rho' w' \theta'_\psi \sin^2 \tau^0(z)}{\sin^2 \tau^0(\varphi')}$$

where for example, $w' = w|_{z=\varphi'}$.

After integration one can find the corrected expression for the streamline, as follows:

$$\theta = \epsilon \sin \alpha \int_{\beta}^{\varphi'} \frac{w^0}{w^{0,2}} R \frac{\sin [\tau^0(z)]}{\sin [\tau^0(z) + \epsilon^2 \tau^{(2)}]} d\varphi'$$

$$R = 1 + (\theta^{*\prime} \cos \varphi' - \theta_{\varphi}^{*\prime} \sin \varphi') \cot \alpha + \frac{\epsilon}{\sin \alpha} (p_1^{*\prime} - p_1^* - p_2); \quad w^{0\prime} = w(\tau^0) |_{z=\varphi'}.$$

The found solution is uniformly convergent over the outer region including the singularity surface and the small region near it.

To complete the solution for the whole region a total inner solution is to be found.

3. The main term for τ in the inner region has the following form:

$$\tau_i = \epsilon^2 \tau_i^{(2)} = - \epsilon^2 C_i^{(2)}. \tag{2.13}$$

Consequently $$w = 0 (\epsilon^2).$$

Applying these relations to the exact equation (1.1) for the variables of the inner region, we shall obtain

$$\frac{\partial d_i}{\partial \varphi} + \frac{2d_i}{\epsilon^2 \tau_i^{(2)}} = 0$$

$$\frac{\partial p_i}{\partial \psi} = 0(\epsilon^2) \tag{2.14}$$

$$\Delta_i = \Delta(\varphi') + 0(\epsilon^2).$$

It is seen from eqs. (2.14–2) that the pressure is constant across the entropy layer within a small order correction. Consequently, $C_i^{(2)}$ can be determined by the value of the pressure in the outer solution, since $D_i^{(2)}$ on it $\sim \epsilon^2$ and there remains only the term $B_i^{(2)}$ with the pressure gradient.

These considerations may be applied to $C^{(2)}$. If we consider the asymptotic value of $\tau^{(2)}$ (2.11) on the singularity surface,

$$\tau^{(2)} \approx - C^{(2)} [\alpha_1(\varphi')].$$

So the outer solution, formed by the PLK method, asymptotically matches the inner solution.

From the equation (2.14–1) we obtain:

$$d_i = d_r \exp \left[- \frac{2}{\epsilon^2} \int_{\varphi_r}^{\varphi} \frac{d\varphi}{\tau_i^{(2)}} \right].$$

Later on we find:

$$\theta_i = \frac{1}{\epsilon^2} \int_{\beta}^{\psi} \frac{d_r}{\rho \Delta \tau_i^{(2)}} \exp \left[- \frac{2}{\epsilon^2} \int_{\varphi_r}^{\varphi} \frac{d\varphi}{\tau_i^{(2)}} \right] d\psi \tag{2.15}$$

where d_r is determined from the matching condition with an outer solution:

$$d_r = d' \exp \left[- 2 \int_{\varphi'}^{z_r} \cot \tau^0 \, dz \right].$$

Substituting the expression for d_r in (2.15) we obtain

$$\theta_i = \frac{\sin\alpha}{\epsilon} \int\limits_{\beta}^{\varphi'} \frac{R}{\Delta\tau_i^{(2)}} \exp\left[-\frac{2}{\epsilon^2}\int\limits_{\varphi_r}^{\varphi} \frac{d\varphi}{\tau_i^{(2)}} - 2\int\limits_{\varphi'}^{z_r} \cot\tau^0\, dz\right] d\varphi' \tag{2.16}$$

and taking into account (2.13), (1.4) and (1.6) we have:

$$\tau_i^{(2)} = \frac{1}{\rho^{(1)}\,\Delta^{(0)}}\,[\sin 2\alpha \sin\varphi(\theta^* + \theta^*_{\varphi\varphi}) - p_{2\varphi}]. \tag{2.17}$$

An estimate of the expression (2.17) has shown that the value $\tau_2^{(2)}$ is negative and $p_{2\varphi}$ has a higher degree of smallness as compared with the first term in the square bracket.

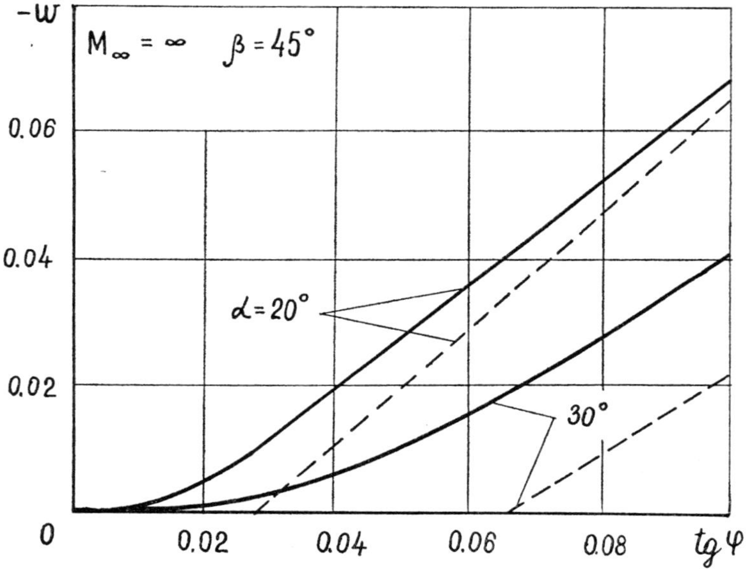

FIG. 7. Lateral component of velocity: ——— with entropy layer, – – – without entropy layer.

Now keeping in mind (2.16), (2.17), we can conclude that $\theta_i > 0$ and $\theta_i, w_i \to 0$ at $\varphi \to 0$. So, a solution consisting of two asymptotically matched solutions is formed, which applies to the whole region of the flow perturbation and fully satisfies Ferri's scheme. The diagrams of the lateral component of velocity w at $\beta = 45°$, $M_\infty = \infty$, $x = 1, 4$; $\alpha = 20°$ and $30°$, are given in Fig. 7. The values are calculated from expressions (2.12), (2.17), solid line, and those calculated using (1.9–2), dashed lines.

3. FINITE THICKNESS DELTA WING IN A HYPERSONIC FLOW

The method suggested in section 1 applies to the flow over a wing of finite thickness which has a different structure to a thin wing.

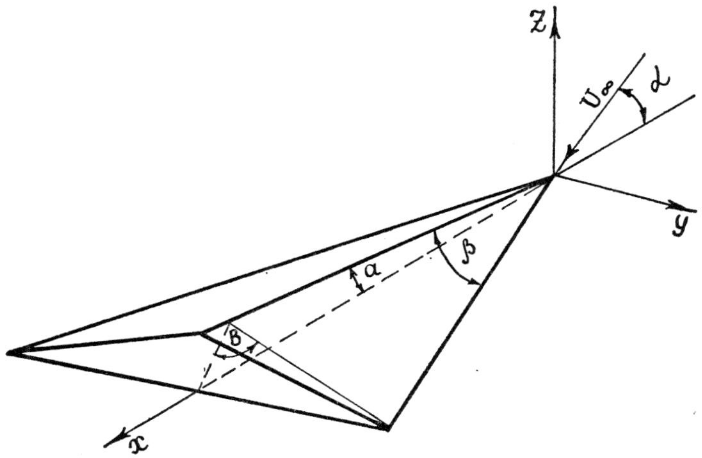

FIG. 8. Scheme of the flow about wing of finite thickness.

A. *Statement of the Problem and Basic Equations*

Let us consider the hypersonic flow past a delta wing (Fig. 8), assuming that the shape of the wing cross-section differs little from that of a diamond.

If we connect the system of coordinates (r, θ, φ) with one of the upper edges of the wing (Fig. 9), the conditions on the shock wave after the substitution (1.2) will read as follows (here as well as further on the bar is dropped):

$$\begin{aligned}
u^* &= \cos(a+\alpha)\,[\cos\varphi + \tan(a+\alpha)\cos B\sin\varphi - \epsilon\theta^*\tan(a+\alpha)\sin B] \\
v^* &= -\,\theta^*_\varphi\,[\cos(a+\alpha)\sin\varphi - \sin(a+\alpha)\cos B\cos\varphi] \\
&\quad -(1+m_0)\sin(a+\alpha)\sin B + \epsilon\cos(a+\alpha)(\theta^*_\varphi\sin\varphi - \theta^*\cos\varphi) \\
&\quad -\epsilon\sin(a+\alpha)\cos B\,[\theta^*_\varphi\cos\varphi + \theta^*\sin\varphi] - \epsilon\theta^{*2}_\varphi\sin(a+\alpha)\sin B
\end{aligned} \tag{3.1}$$

$$\begin{aligned}
w^* &= -\cos(a+\alpha)\,[\sin\varphi - \tan(a+\alpha)\cos B\cos\varphi + \epsilon\theta^*_\varphi\tan(a+\alpha)\sin B] \\
p^* &= \sin^2(a+\alpha)\sin^2 B + \epsilon p_1^* \\
p_1^* &= \sin 2(a+\alpha)\sin B\,(\theta^*\cos\varphi - \theta^*_\varphi\sin\varphi) \\
&\quad + \sin 2B\sin^2(a+\alpha)\,(\theta^*\sin\varphi + \theta^*_\varphi\cos\varphi) - \sin^2(a+\alpha)\sin^2 B - M_\infty^{-2} \\
\rho^* &= (1+m_0)^{-1}; \quad m_0 = \frac{1-\epsilon}{\epsilon\sin^2(a+\alpha)}\,M_\infty^{-2}
\end{aligned}$$

The meaning of coefficients a, b and α is indicated in Fig. 8. $\theta^*(\varphi)$ is the shock wave shape.

In the relations (3.1) members of the order ϵ^{1+m} have been dropped as compared with unity where $m > 0$; it is necessary for a to be of the order $\epsilon^{1/2+m}$.

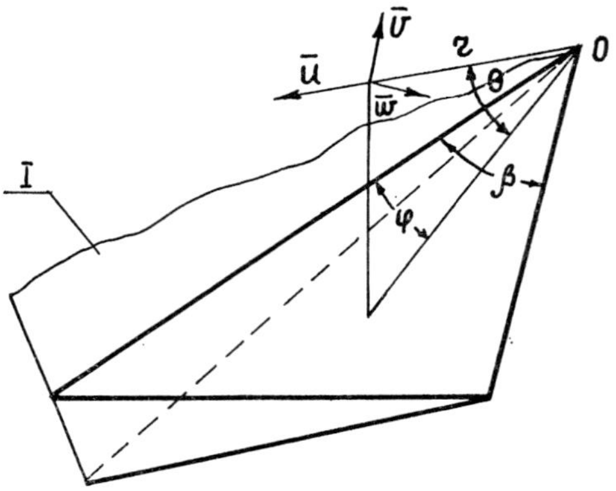

FIG. 9. Coordinate system.

Taking into account that the main member of the pressure (3.1) is constant along the wing, and dropping, in the system of equations (1.3), the members with an order higher than the first one, we shall again obtain the system of equations (1.5). If the shape of the wing cross-section differs from that of a diamond and is given by the equation $\theta = \epsilon\theta^0(\varphi)$, the boundary condition on the wing surface will be as in (1.7).

B. *The Solution of the Boundary Problem*

Let us introduce the new variables $\Delta(\psi)$ and $\tau(\varphi, \psi)$ and write them down as follows:

$$u = \Delta \cos \tau, \quad w = \Delta \sin \tau. \tag{3.2}$$

Then using (1.5–1), (1.5–3) and (3.1) we shall find that

$$\begin{aligned} \Delta^2(\varphi') &= \cos^2 (a + \alpha) [1 + \tan^2 (a + \alpha) \cos^2 B \\ &\quad + 2\epsilon \sin B \tan (a + \alpha) (\theta_\varphi^{*\prime} \sin \varphi' - \theta^{*\prime} \cos \varphi') \\ &\quad - \epsilon \sin 2B \tan^2 (a + \alpha) (\theta_\varphi^{*\prime} \cos \varphi' + \theta^{*\prime} \sin \varphi')] \\ \tau &= -[\varphi + f(\varphi')] \\ f(\varphi') &= a_1 + \epsilon B_1(\varphi') \\ a_1 &= -\arcsin \left[\frac{\tan (a + \alpha) \cos B}{\sqrt{1 + \tan^2 (a + \alpha) \cos^2 B}}\right] \\ B_1(\varphi') &= \frac{\tan (a + \alpha) \sin B}{1 + \tan^2 (a + \alpha) \cos^2 B} [(\theta_\varphi^{*\prime} \cos \varphi' + \theta^{*\prime} \sin \varphi') \\ &\quad + \cos B \tan (a + \alpha) (\theta_\varphi^{*\prime} \sin \varphi' - \theta^{*\prime} \cos \varphi')]. \end{aligned} \tag{3.3}$$

The primes indicate the corresponding magnitudes calculated for the point of the intersection of the shock wave and the streamline ($\varphi' = \varphi$) (Fig. 2).

The analysis of the expression for τ (3.3) makes it possible to conclude that w does not tend to zero at the point $\varphi = 0$ as is the case of a thin delta wing when the plane of symmetry is a separation line.

From (3.3) it follows that there exists a certain coordinate line $\varphi = \varphi_1$, different from $\varphi = 0$ on which the streamline with the zero cross-component velocity w is located. Thus the thickness of the wing a results in the formation of separation line not in the plane of symmetry (Fig. 10).

Instead of (1.8) we have an analogous relation

$$\frac{1}{\rho} = \frac{1}{\rho'}\left[1 + \frac{\epsilon}{\sin^2(a+\alpha)\sin^2 B}(p_1^{*\prime} - p_1^* - p_2)\right]. \tag{3.4}$$

The expressions (3.2) make it possible to find the integral of the fourth equation (1.5–4) in the form of (1.11).

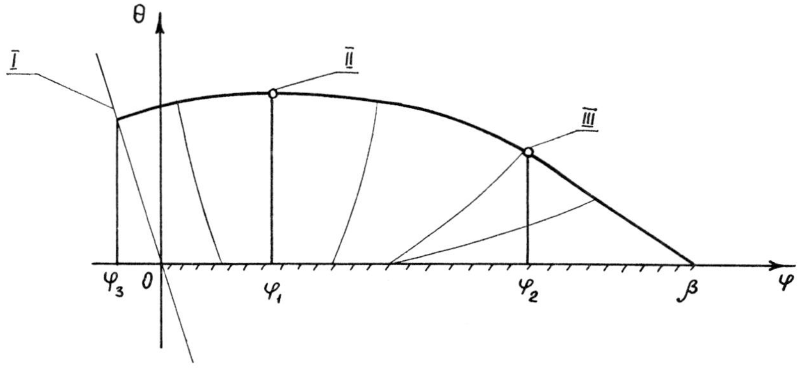

FIG. 10. Cross streamline pattern.

Integrating this equation with respect to ψ as it was done in section 1 from the wing surface along the line, $\varphi = \text{const}$ yields the following expression for the streamlines θ:

$$\theta = \theta^0(\varphi) + (1 + m_0)\sin(a+\alpha)\sin B \int_{\varphi^0(\varphi)}^{\varphi'} \frac{w}{w'^2} R\, d\varphi'$$

$$R = 1 + \epsilon\left[\frac{\cot(a+\alpha)}{(1+m_0)\sin B}(\theta^{*\prime}\cos\varphi' - \theta_\varphi^{*\prime}\sin\varphi')\right. \tag{3.5}$$

$$\left. + \frac{\cot B}{(1+m_0)}(\theta_\varphi^{*\prime}\cos\varphi' + \theta^{*\prime}\sin\varphi') + \frac{1}{\sin^2(a+\alpha)\sin^2 B}(p_1^{*\prime} - p_1^* - p_2)\right].$$

The boundary condition (1.7) presupposes one of the two conditions for the unknown function $\varphi^0(\varphi)$

$$\frac{d\varphi^0}{d\varphi} = 0 \tag{3.6}$$

$$w|_{\varphi' = \varphi^0(\varphi)} = 0,$$

of which we shall take the first $\varphi^0 = \text{const}$ which corresponds to Ferri's flow scheme (see Fig. 10).

If we choose one of the given conditions, there appear certain peculiarities connected with the entropy layer on the wing surface near the symmetry plane. The method of the correct solution of the problem in this region is given in section 2.

Thus, accepting the flow scheme given in Fig. 10, we must recognize that $\varphi^0 = \beta$ to the right of $\varphi = \varphi_1$ and $\varphi^0 = \varphi_3$ to the left of $\varphi = \varphi_1$ which corresponds to the idea of the physical nature of the flow in the shock layer and, as will be seen from what follows, it is necessary to give a closed solution of the problem (φ_3 is the streamline crossing the shock wave in the symmetry plane).

Integrating the second equation (1.5–2) from the shock wave, where $p_2 = 0$, along the line $\varphi = \text{const}$, and keeping the main members, we obtain:

$$p_2 = \sin(a + \alpha) \sin B \int_{\varphi'}^{\varphi} \frac{w^3}{w'^2} (\theta_{\varphi\varphi} + \theta) \, d\varphi'. \tag{3.7}$$

If we now differentiate eq. (3.5) and substitute in (3.7), p_2 will be:

$$\begin{aligned} p_2 = {} & \sin(a + \alpha) \sin B \, (\theta^0_{\varphi\varphi} + \theta^0) \int_{\varphi'}^{\varphi} \frac{w^3}{w'^2} \, d\varphi' \\ & + 2 \sin^2(a + \alpha) \sin^2 B \, (1 + m_0) \Big\{ [(\cos\varphi + \cos B \tan(a + \alpha) \sin\varphi)(z + z'') \\ & + (\sin\varphi - \cos B \tan(a + \alpha) \cos\varphi)(z' + z''')] \int_{\varphi'}^{\varphi} \left[\frac{w^3}{w'^2} \int_{\varphi^0}^{\varphi'} \frac{w}{w'^2} \, d\varphi' \right] d\varphi' \\ & - 2 (\sin\varphi - \cos B \tan(a + \alpha) \cos\varphi)(z + z'') \int_{\varphi'}^{\varphi} \left[\frac{w^3}{w'^2} \int_{\varphi^0}^{\varphi'} \frac{u}{w'^2} \, d\varphi' \right] d\varphi' \Big\}. \end{aligned} \tag{3.8}$$

In eq. (3.8), the terms having $\epsilon p_{2\varphi}$ and $\epsilon p_{2\varphi\varphi}$ as multipliers and which have a higher order of smallness as compared to the remaining members (see section 1) are omitted.

Substituting p_2 in eq. (3.5) and taking $\varphi' = \varphi$, we obtain the following integro-differential equation for the shock wave shape:

$$\begin{aligned} z + \delta \, \{ 2 \, [(z \cos\varphi - z' \sin\varphi) + \cos B \tan(a + \alpha)(z \sin\varphi + z' \cos\varphi)] \, L_2 - L_1 \} \\ + \epsilon (1 + m_0)(z^0 + z^{0\prime\prime}) \, L_3 - z^0 = 2 (1 + m_0) \, \epsilon\delta \, \{ (z + z'') \, [2 (\sin\varphi \\ - \cos B \tan(a + \alpha) \cos\varphi) \, L_4 - (\cos\varphi + \cos B \tan(a + \alpha) \sin\varphi) \, L_5] \\ - (z' + z''')(\sin\varphi - \cos B \tan(a + \alpha) \cos\varphi) \, L_5 \} \end{aligned} \tag{3.9}$$

where

$$z = \epsilon \frac{\cot(a + \alpha)}{\sin B} \theta^*(\varphi), \quad z^0 = \epsilon \frac{\cot(a + \alpha)}{\sin B} \theta^0(\varphi), \quad \delta = \epsilon (1 + m_0) \cos(a + \alpha)$$

$$L_1 = \int_{\varphi^0}^{\varphi} \frac{w}{w'^2} \left\{1 + \frac{3 + 2m_0}{1 + m_0} [z(\varphi') \cos \varphi' - z'(\varphi') \sin \varphi' \right.$$

$$\left. + \cos B \tan (a + \alpha) (z'(\varphi') \cos \varphi' + z(\varphi') \sin \varphi')] \right\} d\varphi'$$

$$L_2 = \int_{\varphi^0}^{\varphi} \frac{w}{w'^2} d\varphi'$$

$$L_3 = \int_{\varphi^0}^{\varphi} \left[\frac{w}{w'^2} \int_{\varphi'}^{\varphi} \frac{w^3}{w'^2} d\varphi'\right] d\varphi' \tag{3.10}$$

$$L_4 = \int_{\varphi^0}^{\varphi} \left[\frac{w}{w'^2} \int_{\varphi'}^{\varphi} \left[\frac{w^3}{w'^2} \int_{\varphi^0}^{\varphi'} \frac{u}{w'^2} d\varphi'\right] d\varphi'\right] d\varphi'$$

$$L_5 = \int_{\varphi^0}^{\varphi} \left[\frac{w}{w'^2} \int_{\varphi'}^{\varphi} \left[\frac{w^3}{w'^2} \int_{\varphi^0}^{\varphi'} \frac{w}{w'^2} d\varphi'\right] d\varphi'\right] d\varphi'.$$

Thus, we obtain the equation for the shock wave shape, where $\varphi^0 = \beta$ is to the right of $\varphi = \varphi_1$ and $\varphi^0 = \varphi_3$, to the left of $\varphi = \varphi_1$. Solving the equation for each region we are in a position to determine the parameters of the problem. The structure of the equation makes it possible to solve the Cauchy problem with three conditions. In other words, for the solution of the whole problem, six arbitrary conditions are available. Three conditions are known: two of them at the point $\varphi = \varphi_2$, a smooth gradual transition from the plane area of the shock wave to the curvilinear part, and one on the axis of symmetry (Fig. 8). There remain three arbitrary conditions and it might seem possible to link up $z^{(n)}(\varphi)$ for $n = 0, 1, 2$ at the point $\varphi = \varphi_1$. But it is known that the point where the separation line crosses the wing the pressure gradient $p_\varphi|_{\varphi=\varphi_1}$ is zero. The investigation of the conditions for reducing p_φ and w to zero makes it possible to find the exact value of $\varphi_1 = -a_1$, and ascertain that $z'(\varphi_1) = 0$.* Thus, the number of arbitrary conditions is reduced to two and $z^{(n)}(\varphi_1)$ can be linked up for $n = 0, 1$ at the point $\varphi = \varphi_1$.

Equation (3.9) can be solved to the right of the point $\varphi = \varphi_1$ as it was done in section 1: by expanding z in series at the point $\varphi = \varphi_2$,

$$z = \sum_{n=0}^{\infty} \frac{z^{(n)}(\varphi_2)}{n!} (\varphi - \varphi_2)^n \tag{3.11}$$

and determining the coefficients $z^{(n)}(\varphi_2)$. Now the functionals $L_i(\varphi_2)$ (3.10) can be easily calculated.

* Note that there is another possibility of satisfying these conditions if the shock wave curvature at the separation point is zero. This, however, will yield a trivial solution, corresponding to the uniform flow to the right of the separation line.

The solution of eq. (3.9) to the left of the point $\varphi = \varphi_1$ can be obtained by expanding z into a series at the point $\varphi = 0$:

$$z = \sum_{n=0}^{\infty} \frac{z^{(n)}(0)}{n!} \varphi^n. \tag{3.12}$$

Parameters $L_i(0)$ can be calculated with the necessary accuracy.

If, in particular, we confine ourselves to four terms in the expansions (3.11), (3.12), the problem will lead to the solution of the transcendental equation concerning $z''(0)$.

C. *The Calculation of the Diamond-shaped Wing Flow Line*

Let us examine the case $z^0(\varphi) = 0$ and $m_0 = 0$ which corresponds to $M_\infty = \infty$ and the cross-section of the diamond-shaped wing. Substituting the expression for the plane wave $z = a_2 \sin(\beta - \varphi)$ in eq. (3.9) we find:

$$a_2 = \delta \frac{1 + \delta \dfrac{N - L}{K \sin(\beta + a_1)}}{K \sin\left(\beta + a_1 - \dfrac{M\delta}{K \sin(\beta + a_1)}\right)}$$

$$N = \sin B - \cos B \tan(a + \alpha) \cos \beta$$

$$K = \cos(a + \alpha) \sqrt{1 + \tan^2(a + \alpha) \cos^2 B}$$

$$L = \frac{\sin^2 B \tan^2(a + \alpha) [\cos B \tan(a + \alpha) \cos \beta - \sin \beta]}{1 + \tan^2(a + \alpha) \cos^2 B}$$

$$M = \frac{\tan^2(a + \alpha) \sin^2 B}{1 + \tan^2(a + \alpha) \cos^2 B} [\cos \beta + \cos B \tan(a + \alpha) \sin \beta].$$

The functionals (3.10) will be:

$$L_1 = \frac{1 + 3a_2 N}{\Delta_1} \cdot \frac{\sin(\beta - \varphi)}{\sin(\beta + \sigma)}; \qquad L_2 = \frac{1}{\Delta_1} \cdot \frac{\sin(\beta - \varphi)}{\sin(\beta + \sigma)}$$

$$L_4 = \frac{\sin 2(\varphi + \sigma)}{2\Delta_1} d; \qquad L_5 = -\frac{\sin^2(\varphi + \sigma)}{\Delta_1} d$$

$$d = \tfrac{1}{3} \cot^3(\beta + \sigma) \sin^3(\varphi + \sigma) + \frac{\sin(\beta - \varphi) \sin 2(\varphi + \sigma) \cot(\beta + \sigma)}{2 \sin(\beta + \sigma)} - \tfrac{1}{3} \cos^3(\varphi + \sigma)$$

$$\sigma = a_1 - M a_2; \quad \Delta_1 = K[1 + a_2 L].$$

φ_2 will be determined from the equation

$$\sin(\varphi_2 + a_1 - M a_2) = \frac{1}{\Delta_1} \sqrt{\epsilon \frac{1 + \epsilon}{1 - \epsilon} p^*}$$

$$p^* = \sin^2(a + \alpha) \sin^2 B \{1 + 2 a_2 [\sin \beta - \tan(a + \alpha) \cos B \cos \beta] - \epsilon\}.$$

For the coefficients of the expansion (3.11) we obtain the following expressions:

$$z_2 = z(\varphi_2) = a_2 \sin(\beta - \varphi_2)$$

$$z_2' = -a_2 \cos(\beta - \varphi_2)$$

$$z_2''' = -(z_2 + z_2'')F - z_2'$$

$$z_2'' = \frac{z_2'[2 - (\varphi_2 + a_1)^2] - F(\varphi_2 + a_1)^2 z_2}{(\varphi_2 + a_1)[2 + F(\varphi_2 + a_1)]}$$

$$F = 2\cot(\varphi_2 + \sigma) + \frac{\cot\varphi_2 + \cos B \tan(a + \alpha)}{1 - \cos B \tan(a + \alpha)\cot\varphi_2}.$$

Before writing down the necessary relations on the left of the point $\varphi = \varphi_1$ we evaluate some magnitudes. For the shock wave slope at the point $\varphi_3 = -\tan(a + \alpha)\cos Bz(\varphi_3) + 0(\epsilon a^3)$ we obtain the relation

$$z'(\varphi_3) = \cot(a + \alpha)\cos B + 0(\epsilon a^3). \tag{3.13}$$

The order of the coefficients in the expansion (3.12) can be estimated as shown below, taking into account that

$$\varphi_3 \sim 0(\epsilon a); \quad \varphi_1 \sim 0(a)$$

$$z_0 \sim 0(\epsilon); \quad z_0' \sim 0(a); \quad z_0'' \sim 0(1); \quad z_0''' \sim 0(a^{-1}).$$

If we then leave out the members higher than ϵ^2 the functionals L_2 and L_1 will be:

$$L_2 = -\frac{1}{\cos(a + \alpha)(1 + Tz_0'')^2}\{Tz_0'' \ln[(1 - z_0'')\sin^2(a + \alpha)]$$
$$- 1 + (1 - z_0'')\sin^2(a + \alpha)\} \tag{3.14}$$

$$L_1 = (1 + 3z_0)L_2$$

$$T = \frac{\tan^2(a + \alpha)\sin^2 B}{1 + \tan^2(a + \alpha)\cos^2 B}.$$

The functionals L_4 and L_5, calculated at the point $\varphi = 0$ making analogous assumptions, are of the following order:

$$L_4 \sim 0(\epsilon a); \quad L_5 \sim 0(\epsilon^2 a^2). \tag{3.15}$$

Thus, taking into account (3.14)–(3.15), eq. (3.9) at the point of $\varphi = 0$ will read as follows:

$$z_0 + \frac{\epsilon(1 + z_0)}{(1 + Tz_0'')^2}\{Tz_0'' \ln[(1 - z_0'')\sin^2(a + \alpha)] - 1 + (1 - z_0'')\sin^2(a + \alpha)\} = 0.$$

By analogy, the condition (3.13) will appear in the form

$$z_0'' z_0 \tan(a + \alpha)\cos B - z_0' + \cot(a + \alpha)\cos B = 0.$$

At the point of $\varphi = \varphi_1$ we have $z'(\varphi_1) = 0$; $z(\varphi_1 - 0) = z(\varphi + 0)$, which, in the used notation, will take the form:

$$z_0' + z_0'' \varphi_1 + \tfrac{1}{2} z_0''' \varphi_1^2 = 0$$

$$z_0 + z_0' \varphi_1 + \tfrac{1}{2} z_0'' \varphi_1^2 + \tfrac{1}{6} z_0''' \varphi_1^3$$
$$= z_2 + z_2' (\varphi_1 - \varphi_2) + \tfrac{1}{2} z_2'' (\varphi_1 - \varphi_2)^2 + \tfrac{1}{6} z_2''' (\varphi_1 - \varphi_2)^3.$$

As can be easily checked, this system of relations is reduced to a transcendental equation for z''. Having found its solution the other coefficients of expansion (3.12) may be easily determined. The equation has three real roots, two of them being multiple.

The physical sense of the problem requires $z_0'' < 0$. If we choose the single root the analysis shows that with given a and α, $(z_0'')_1 > 0$ and therefore the problem is to find a multiple root.

The analysis of expression (3.8) shows that $p_2 < 0$. This proves the fact that the pressure on the axis of symmetry drops in the direction from a shock wave towards the body. Figure 11a, b shows the curves of the pressure distribution beyond the shock wave and the curves of pressure distribution on a plane delta wing (dashed lines), which makes it possible to estimate the influence of the wing thickness a upon the pressure distribution.

Figure 12 shows the curves of a shock distance in the plane of symmetry for various a ($20° \leq a \leq 27°$), $\beta = 45°$, according to α. The curves are limited on their left by $\alpha = 0$, and on their right by α corresponding to the detachment of the shock waves from the leading edges. The form of the curves shows that for a certain α the shock wave distance in the plane of symmetry begins to decrease. The shock layer over the wing edges is thicker (Fig. 13) ($M = \infty$; $a = 25°$).

Apparently, this behaviour of the shock wave can be explained by the simultaneous influence of two factors: the increase of the shock wave distance with the increase of the angle of attack, which we usually observe in the case of a thin delta wing on the other side, and the reduction of this distance due to the increase of the intensity of the gas flow down from the plane of symmetry, while the angle of attack increases.

Thus, for different a (the wing thickness) we have the corresponding angle of attack α, for which the second factor begins to dominate.

This solution makes it possible to examine the behaviour of the lift-to-drag ratio of the wing, made up of two wings of the type under discussion with $\alpha = 0$ (Fig. 14).

Our calculations were carried on for the different values of the area of the middle cross-section S. The friction factor C_τ was considered to be constant and equal to 2×10^{-3}. Note that the wing, whose upper thickness a_{00} is zero, has the maximum quality. As might be expected (Fig. 15), the curves of the maximum lift-to-drag ratio drawn according to the given solution lie higher than the curves of the maximum aerodynamic quality, according to Newton's formula, with the friction being taken into account.

4. HYPERSONIC FLOW ABOUT A CARET WING

The conical wing with an inverted V cross-section (caret wing) is considered a prospective configuration of the lifting body for sustained hypersonic flight. Such a configuration is of interest because theoretical estimates show that V-shaped schemes possess a higher lift-to-drag ratio in comparison with the conventional thin delta wings.

It is observed in experiments[25,34,44–47] that in the windward part of the flow about V-shaped wings a bifurcating system of shock waves exists. No methods for the calculation

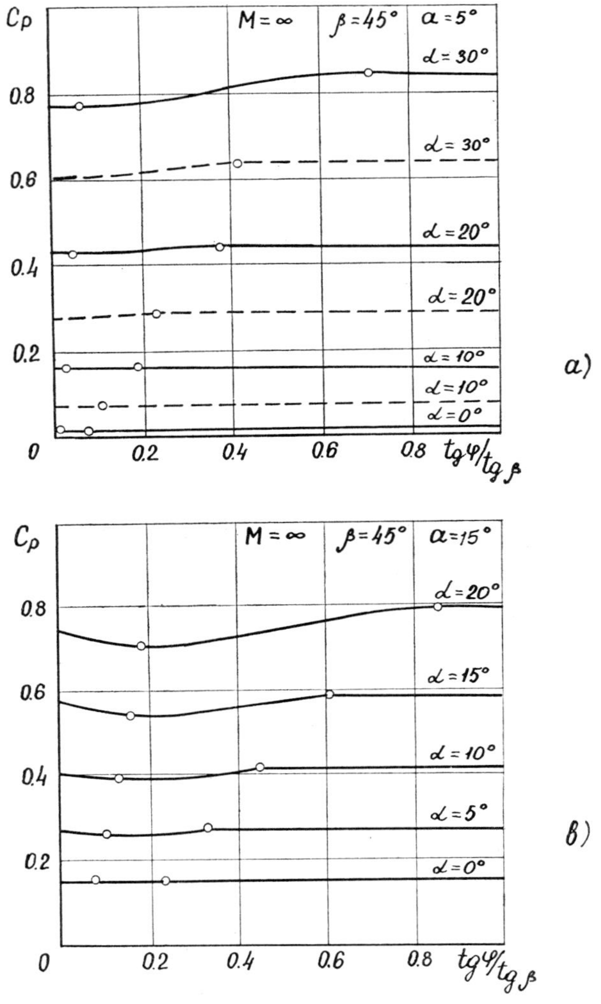

FIG. 11. Pressure distribution on a wing.

of such flows are available. Only a few exact solutions of the reverse problem are known for special geometric forms of the wing in the case of some simple flows.[48–51]

Below we describe the results obtained for hypersonic flow about V-shaped wings with a large dihedral when, according to experimental studies, a Mach configuration of curvilinear shocks is formed.

A. *Statement of the Problem and Basic Equations*

Let us consider a caret wing in a hypersonic flow (Fig. 16), with an angle of attack α. In the chosen system of coordinates (Fig. 17) the transformed equations of conical flow are those given in (1.1). After transformation into the new variables (1.2) and an appropriate evaluation of terms, neglecting the terms of order ϵ^{1+n} $(n > 0)$ in comparison with unity we obtain the following boundary conditions (4.1) at the attached shock wave I (Fig. 18).

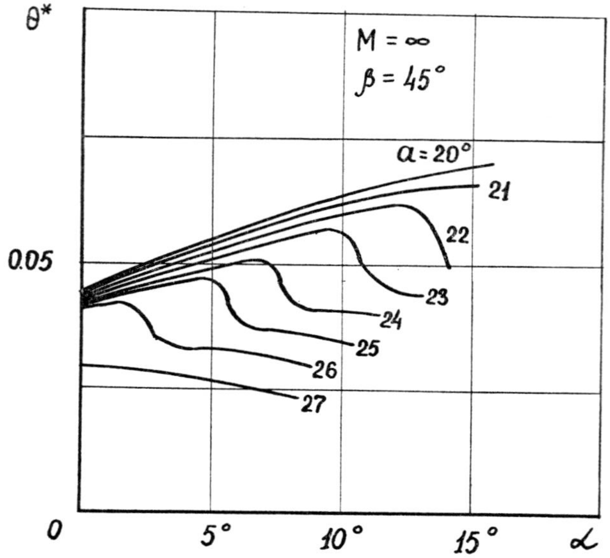

FIG. 12. Shock distance in the plane of symmetry.

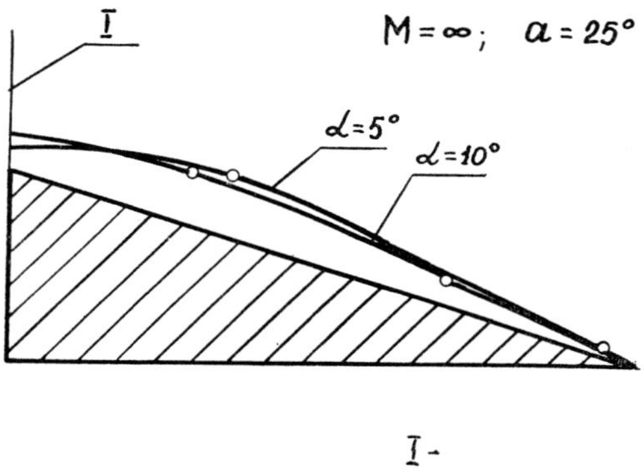

FIG. 13. Cross shock layer pattern.

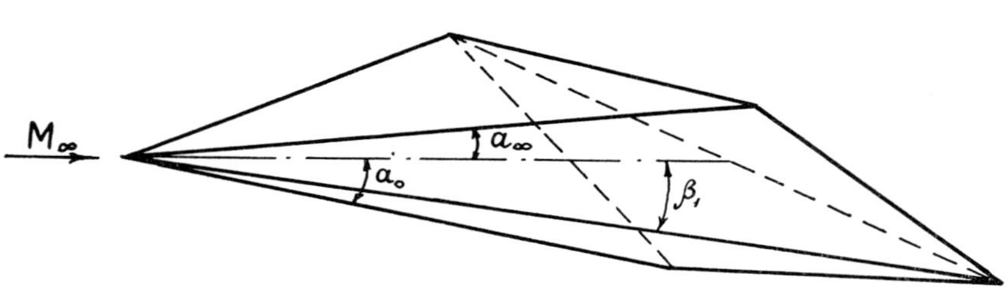

FIG. 14. Wing shape.

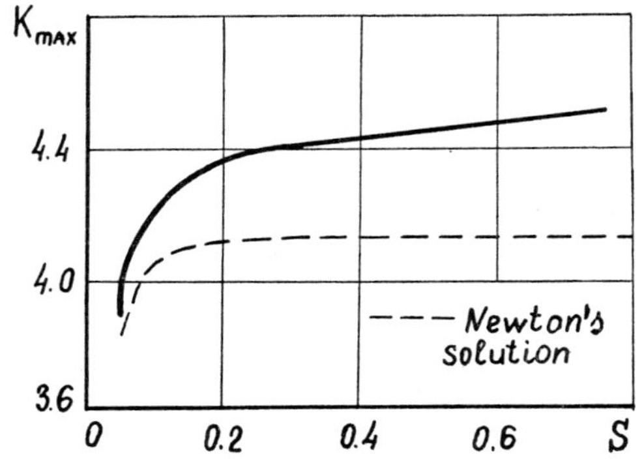

FIG. 15. Curves of the maximum lift-to-drag ratio.

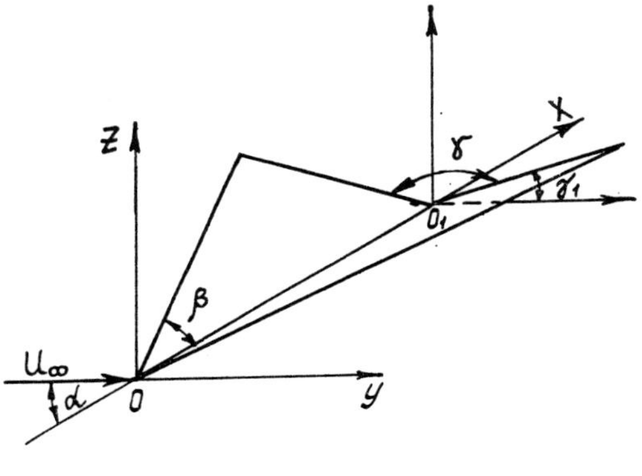

FIG. 16. Scheme of the flow about caret wing.

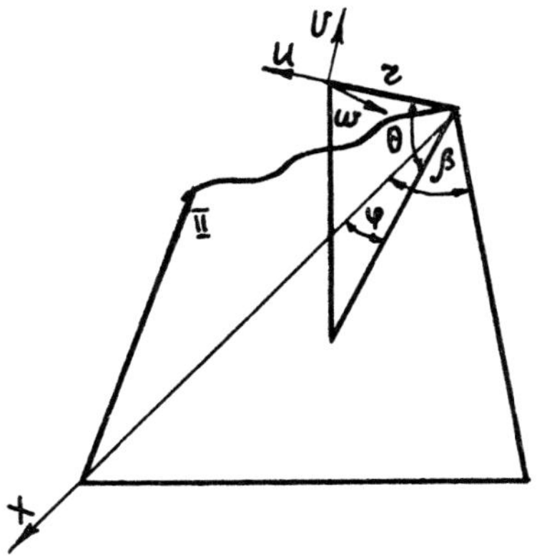

FIG. 17. Conical coordinate system.

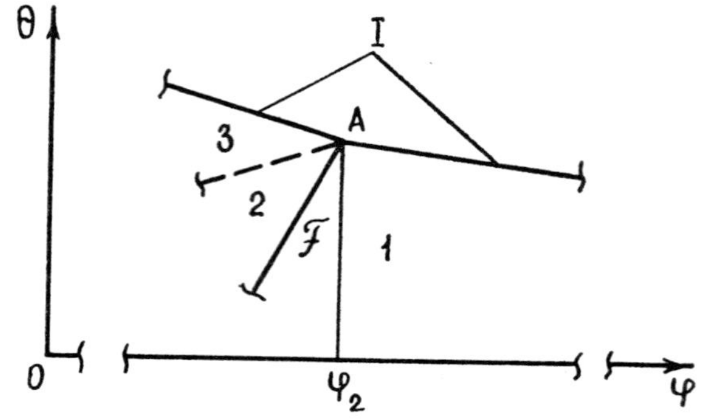

FIG. 18. Mach configuration of shocks.

$$
\begin{aligned}
u^* &= \cos\alpha\,[\cos\varphi - \tan\alpha\sin\gamma_1\sin\varphi - \epsilon\theta^*\tan\alpha\cos\gamma_1]\\
v^* &= -\,\theta^*_\varphi\,[\cos\alpha\sin\varphi + \sin\alpha\sin\gamma_1\cos\varphi] - (1+m_0)\sin\alpha\cos\gamma_1\\
&\quad + \epsilon\,(1+m_0)\cos\alpha\,[\theta^*_\varphi\sin\varphi - \theta^*\cos\varphi]\\
&\qquad + \epsilon\,(1+m_0)\sin\alpha\cos\gamma_1\,[\theta^*\sin\varphi + \theta^*_\varphi\cos\varphi] - \epsilon\theta^{*2}_\varphi\sin\alpha\cos\gamma_1\\
w^* &= -\cos\alpha\,[\sin\varphi + \tan\alpha\sin\gamma_1\cos\varphi + \epsilon\tan\alpha\cos\gamma_1\,\theta^*_\varphi]\\
p^* &= \sin^2\alpha\cos^2\gamma_1 + \epsilon p^*_1 \qquad (4.1)\\
p^*_1 &= \sin 2\alpha\cos\gamma_1\,[\theta^*\cos\varphi - \theta^*_\varphi\sin\varphi]\\
&\quad - \sin^2\alpha\sin 2\gamma_1\,[\theta^*\sin\varphi + \theta^*_\varphi\cos\varphi] - \sin^2\alpha\cos^2\gamma_1 - M_\infty^{-2}\\
\rho^* &= (1+m_0)^{-1}; \quad m_0 = \epsilon^{-1}(1-\epsilon)\,\mathrm{cosec}^2\alpha\sec^2\gamma_1\,M_\infty^{-2}.
\end{aligned}
$$

Here $\gamma_1 = 90° - \frac{1}{2}\gamma$, where $\frac{1}{2}\gamma$ is the angle between the plane of symmetry and the wing surface. $\theta^*(\varphi)$ describes the shock wave shape.

Bearing in mind that the main term of the pressure p^* (4.1) is constant along the wing surface and neglecting the terms of order higher than ϵ in the system of equations (1.3), we will obtain the system of equations (1.5). The boundary condition on the wing is

$$v = 0.$$

The system of equations (1.5) and the boundary conditions (4.1) differ from the exact ones only by terms of higher order, whereas the expansion in series of a small parameter would yield a system of equations which differ from the exact one by terms of the first order.

B. *The Solution of the Boundary Problem*

The first and third equations of the system (1.5) can be immediately integrated and using the boundary conditions (4.1) the components of the velocity u and w are written in the form:

$$
\begin{aligned}
u &= \Delta\cos\tau; \quad w = \Delta\sin\tau\\
\Delta^2(\varphi') &= \cos^2\alpha\,[1 + \tan^2\alpha\sin^2\gamma_1 + 2\epsilon\tan\alpha\cos\gamma_1\,(\theta^{*\prime}_\varphi\sin\varphi' - \theta^{*\prime}\cos\varphi')\\
&\quad + \epsilon\tan^2\alpha\sin 2\gamma_1\,(\theta^{*\prime}_\varphi\cos\varphi' + \theta^{*\prime}\sin\varphi')]
\end{aligned}
$$

$$\tau = -[\varphi + f(\varphi')]; \quad f(\varphi') = a_1 + \epsilon B_1(\varphi') \tag{4.2}$$

$$a_1 = \arcsin \frac{\tan \alpha \sin \gamma_1}{\sqrt{1 + \tan^2 \alpha \sin^2 \gamma_1}}$$

$$B_1(\varphi') = \frac{\tan \alpha \cos \gamma_1}{\sqrt{1 + \tan^2 \alpha \sin^2 \gamma_1}} [(\theta_\varphi^{*\prime} \cos \varphi' + \theta^{*\prime} \sin \varphi')$$

$$- \sin \gamma_1 \tan \alpha (\theta_\varphi^{*\prime} \sin \varphi' - \theta^{*\prime} \cos \varphi')].$$

The symbol (′) indicates that corresponding values are taken at the point of intersection of the shock wave with a streamline ($\varphi' = \varphi$). The formulae (4.2) are evidently true for every streamline passing through the leading shock wave *I*. The solution for the uniform flow behind the plane shock wave attached to the leading edge is given by

$$\Delta_1^2 = K^2(1 - 2a_2L); \quad f_1(\varphi') \equiv \sigma = a_1 + Ma_2$$

$$K^2 = \cos^2 \alpha (1 + \tan^2\alpha \sin^2 \gamma_1); \quad \rho^* = (1 + m_0)^{-1}$$

$$p_1^* = \sin^2 \alpha \cos^2 \gamma_1 (1 + 2a_2 N - \epsilon) - \epsilon M_\infty^{-2}$$

$$\theta_1 = (1 + m_0) \sin \alpha \cos \gamma_1 R_1 \sin (\varphi + \sigma) \sin (\beta - \varphi')/\Delta_1 \sin (\beta + \sigma) \sin(\varphi' + \sigma)$$

$$R_1 = 1 + a_2N$$

$$V_1 = -u_1 \theta_1$$

$$= -(1 + m_0) \sin \alpha \cos \gamma_1 R_1 \sin 2 (\varphi + \sigma) \sin (\beta - \varphi')/2 \sin (\beta + \sigma) \sin (\varphi' + \sigma)$$

$$a_2 = \delta[1 + \delta(N + L)/K \sin (\beta + a_1)]/K \sin [\beta + a_1 + M\delta/\sin (\beta + a_1)]$$

$$N = \sin \beta + \tan \alpha \sin \gamma_1 \cos \beta$$

$$L = \tan^2 \alpha \cos^2 \gamma_1/(1 + \tan^2\alpha \sin^2 \gamma_1)N$$

$$M = (\tan \alpha \sin \gamma_1 \sin \beta - \cos \beta)L/N$$

$$\delta = \epsilon \cos \alpha (1 + m_0).$$

We construct a closed form solution of the problem providing there is a Mach reflection of the plane shock wave from the plane of symmetry (Fig. 18). Inserting the internal shock $F(\varphi)$, the boundary conditions for region 2 (Fig. 18) on the discontinuity are written in the form:

$$u_2^* = u_1$$

$$v_2^* = -\epsilon^{-1} \Delta_1 \{\sin (\varphi + \sigma) F_\varphi (1 - \epsilon) (1 - M_{1n}^{-2})$$

$$+ \cos (\varphi + \sigma) F [1 - (1 - \epsilon) (1 - M_{1n}^{-2}) + F_\varphi^2)]\}/(1 + F^2 + F_\varphi^2)$$

$$w_2^* = -\Delta_1 \{\sin (\varphi + \sigma) [1 + F_\varphi^2 [1 - (1 - \epsilon) (1 - M_{1n}^{-2})]]$$

$$+ FF_\varphi \cos (\varphi + \sigma) (1 - \epsilon) (1 - M_{1n}^{-2})\}/(1 + F^2 + F_\varphi^2) \tag{4.3}$$

$$p_2^* = p_1^* [(1 + \epsilon) M_{1n}^2 - \epsilon]; \quad \rho_2^* = (1 + m_0)^{-1} [1 - (1 - \epsilon) (1 - M_{1n}^{-2})]^{-1}$$

$$M_{1n}^2 = \Delta_1^2 [F_\varphi \sin (\varphi + \sigma) - F \cos (\varphi + \sigma)]^2/a_1^{*2} (1 + F^2 + F_\varphi^2).$$

M_{1n} is the Mach number of the flow normal to the shock in the region 1, a_1^* the velocity of sound in the region 1.

The variables Δ and τ which determine u and w through formulae (4.2) in the region 2 are written as follows:

$$\Delta_2^2 = \Delta_1^2 \{1 + [\sin(\varphi_1+\sigma) F'_\varphi - \cos(\varphi_1+\sigma) F']^2 (1-\epsilon) \times$$
$$\times (1 - M_{1n}^{-2}(\varphi_1)) [(1-\epsilon)(1 - M_{1n}^{-2}(\varphi_1)) - 2]/(1 + F'^2 + F_\varphi'^2)$$
$$\tau_2 = -[\![\varphi + \sigma - (1-\epsilon)(1 - M_{1n}^{-2}(\varphi_1)) [\sin(\varphi_1+\sigma) F'_\varphi - \cos(\varphi_1+\sigma) F'] \times$$
$$\times \{F' \sin\varphi_1 + F'_\varphi \cos\varphi_1 + [\sin(\varphi_1+\sigma) F'_\varphi - \cos(\varphi_1+\sigma) F'] \times$$
$$\times [1 - \tfrac{1}{2}(1-\epsilon)(1 - M_{1n}^{-2}(\varphi_1))]\}/\cos\sigma\,(1 + F'^2 + F_\varphi'^2)]\!]. \quad (4.4)$$

Here $\varphi_1 = \varphi$ on the shock $F(\varphi)$. We consider such a compressed layer flow structure when $F_\varphi(\varphi_2) > 0$.* In this case the following interactions of the incident shock with the wall are possible: (1) Mach reflection (or in a particular case a regular one), (2) no reflection $F_\varphi(\varphi_4) = \infty$ (Fig. 19). Further we consider only the second case. There are known many cases[4,7] when reflected shocks do not appear.

Note further that we expect the parameter φ_1 which corresponds to the streamlines intersecting the plane shock wave and the shock F and going through to region 2 to be within the limits $\varphi_4 \le \varphi_1 \le \varphi_2$. Such a condition makes it easy to integrate the system (1.5) because now the implicit connection between φ_1 and φ' through the function describing the shape of the shock $F(\varphi_2 \le \varphi' \le \beta)$ is eliminated.

The relations (4.2) allow us to find an integral of the fourth equation of the system (1.5),

$$\frac{\rho\theta_\psi}{w} = \frac{\rho^*\,\theta_\psi^*}{w^*}.$$

The symbol (*) denotes that corresponding values are calculated on the leading shock wave in region 3 or on the shock F. Integrating this equation with respect to ψ in the interval $\varphi_4 \le \varphi_1 \le \varphi_2$ from the shock F along the line $\varphi =$ const yields the following expression for the streamlines in region 2 ($\theta_1 = \epsilon^{-1} F$):

$$\theta_2 = \theta_1 + \int_\varphi^{\varphi_1} \frac{\rho'_2\, w_2}{\rho_2\, w'_2}\, \theta'_{2\psi}\, \psi_{\varphi_1}\, d\varphi_1.$$

Taking into consideration

$$\theta'_{2\psi}\,\psi_{\varphi_1} = \frac{d\theta'_2}{d\varphi_1} - \left(\frac{\partial\theta_2}{\partial\varphi}\right)'; \quad \frac{d\theta'_2}{d\varphi_1} \equiv \theta'_{1\varphi}$$

we find

$$\theta'_{2\psi}\,\psi_{\varphi_1} = -\Delta_1 [1 - (1-\epsilon)(1 - M_{1n}^{-2}(\varphi_1))] [\sin(\varphi_1+\sigma)\,\theta'_{1\varphi} - \cos(\varphi_1+\sigma)\,\theta'_1]/w'_2. \quad (4.5)$$

* The proposed approach allows us to consider such a flow structure where $F_\varphi < 0$. But an investigation in this case yields an internal shock having a weak intensity which can be degenerated to a Mach wave.

From the system of equations (1.5) it follows that the expression for pressure in region 2 is

$$p_2 = p_3^*(\varphi) + \epsilon p_{32}(\varphi,\varphi_2) + \epsilon p_{22}(\varphi,\varphi_1). \tag{4.6}$$

Here the first index denotes the number of the region, the second one indicates an additional term for the pressure as given by the second equation of the system (1.5).

Using the fourth equation of the system (1.5) and relations (4.4) and (4.6) we find

$$\frac{\rho_2'}{\rho_2} = [(1+\epsilon)\, M_{1n}^2(\varphi_1) - \epsilon]\left[1 + \frac{\epsilon}{\sin^2\alpha\cos^2\gamma_1}(p_{11}^* - p_{31}^*(\varphi) - p_{32}(\varphi,\varphi_2) - p_{22}(\varphi,\varphi_1))\right]. \tag{4.7}$$

Substitution of (4.5) and (4.7) into the expression for θ_2 yields

$$\theta_2(\varphi, \varphi_1) = \theta_1(\varphi) - \int_{\varphi}^{\varphi_1} \frac{w_2}{w_2'^2} F_1 R_2\, d\varphi_1$$

$$F_1 = \Delta_1 \,[1 + \epsilon\,(M_{1n}^2(\varphi_1) - 1) + \epsilon\,(1 - M_{1n}^{-2}(\varphi_1))]\,[\sin(\varphi_1 + \sigma)\,\theta_{1\varphi}' - \cos(\varphi_1 + \sigma)\,\theta_1'] \tag{4.8}$$

$$R_2 = 1 + \frac{\epsilon}{\sin^2\alpha\cos^2\gamma_1}\,[p_{11}^* - p_{31}^*(\varphi) - p_{32}(\varphi,\varphi_2) - p_{22}(\varphi,\varphi_1)].$$

For streamlines in the region 3

$$\theta_3(\varphi, \varphi') = \theta_2(\varphi, \varphi_2) + \int_{\varphi_2}^{\varphi'} \frac{\rho_3' w_3}{\rho_3 w_3'}\,\theta_{34}'\,\psi_{\varphi'}\,d\varphi'.$$

Taking into consideration

$$\frac{\rho_3'}{\rho_3} = 1 + \frac{\epsilon}{\sin^2\alpha\cos^2\gamma_1}(p_{31}^{*\prime} - p_{31}^* - p_{32})$$

$$\theta_{3\psi}'\,\psi_{\varphi'} = (1 + m_0)\sin\alpha\cos\gamma_1\left\{1 - \epsilon\left[\frac{\cot\alpha}{\cos\gamma_1}(\theta_{\varphi}^{*\prime}\sin\varphi' - \theta^{*\prime}\cos\varphi')\right.\right.$$
$$\left.\left. + \tan\gamma_1\,(\theta^{*\prime}\sin\varphi' + \theta_{\varphi}^{*\prime}\cos\varphi')\right]\right\}\Big/ w_3'$$

we get

$$\theta_3(\varphi, \varphi') = \theta_2(\varphi, \varphi_2) + (1 + m_0)\sin\alpha\cos\gamma_1 \int_{\varphi_2}^{\varphi'} \frac{w_3}{w_3'^2} R_3\, d\varphi' \tag{4.9}$$

$$R_3 = 1 - \epsilon\left[\frac{\cot\alpha}{\cos\gamma_1}(\theta_{\varphi}^{*\prime}\sin\varphi' - \theta^{*\prime}\cos\varphi') + \tan\gamma_1\,(\theta^{*\prime}\sin\varphi' + \theta_{\varphi}^{*\prime}\cos\varphi')\right.$$
$$\left. - \frac{1}{\sin^2\alpha\cos^2\gamma_1}(p_{31}^{*\prime} - p_{31}^* - p_{32})\right].$$

Integrating the second equation of the system (1.5) from the leading shock wave in region 3 where $p_{32} = 0$ along the line $\varphi = \text{const}$ and retaining the principal terms, we get

$$p_{32} = \sin\alpha\cos\gamma_1 \int_{\varphi'}^{\varphi} \frac{w_3^3}{w_3'^2} (\theta_{3\varphi\varphi} + \theta_3)\, d\varphi'. \tag{4.10}$$

Integrating the same equation in the region 2 from the streamline $\varphi_1 = \varphi_2$ where $P_{22} = 0$, we get

$$p_{22} = -\int_{\varphi_1}^{\varphi_2} \frac{w_2^3}{w_2'^2} (\theta_{2\varphi\varphi} + \theta_2)\, F_2\, d\varphi_1$$

$$F_2 = \Delta_1 (1 + m_0)^{-1} [\sin(\varphi_1 + \sigma)\, \theta'_{1\varphi} - \cos(\varphi_1 + \sigma)\, \theta'_1]. \tag{4.11}$$

Differentiating (4.9) and (4.8) twice and substituting the results into (4.10) and (4.11) we get expressions for p_{32} and p_{22} to the right and to the left of the point φ_4. In these expressions the main terms (section 1) are retained. Substituting p_{32} and p_{22} into (4.8), (4.9) and taking $\varphi' = \varphi$ in (4.9) we get the following integro-differential equation for the shock wave shape:

$$\begin{aligned} z + \delta\, \{&(1 + m_0)^{-1} (1 + 2a_2 N)\, L_1 + 2\, [(z\cos\varphi - z'\sin\varphi) - \tan\alpha\sin\gamma_1 \times \\ &\times (z\sin\varphi + z'\cos\varphi)]\, L_2 - L_3\} - 2\epsilon\delta\, \{[(\sin\varphi + \tan\alpha\sin\gamma_1\cos\varphi)(z''' + z') \\ &+ (\cos\varphi - \tan\alpha\sin\gamma_1\sin\varphi)(z'' + z)]\, L_4 \\ &- 2\,(\sin\varphi + \tan\alpha\sin\gamma_1\cos\varphi)(z'' + z)\, L_5\} = \eta z_1 + \eta\epsilon\, (z_1'' + z_1)\, \Phi(\varphi)\, L_6 \end{aligned}$$

$$z = \epsilon\theta^* \frac{\cot\alpha}{\cos\gamma_1}; \quad z_1 = \epsilon\theta_1 \frac{\cot\alpha}{\cos\gamma_1}$$

$$L_1 = \int_{\zeta}^{\varphi_2} \frac{w_2}{w_2'^2} F_{11}\, d\varphi_1; \quad L_2 = \int_{\varphi_2}^{\varphi} \frac{w_3}{w_3'^2}\, d\varphi' - (1 + m_0)^{-1} \int_{\zeta}^{\varphi_2} \frac{w_2}{w_2'^2} F_{11}\, d\varphi_1 \tag{4.12}$$

$$\begin{aligned} L_3 = \int_{\varphi_2}^{\varphi} \frac{w_3}{w_3'^2} \{1 + 3\, [&(z(\varphi')\cos\varphi' - z'(\varphi')\sin\varphi') \\ &- \tan\alpha\sin\gamma_1\, (z(\varphi')\sin\varphi' + z'(\varphi')\cos\varphi')]\}\, d\varphi' \end{aligned}$$

$$\begin{aligned} L_4 = &\int_{\zeta}^{\varphi_2} \frac{w_2}{w_2'^2} F_{11}\, d\varphi_1 \int_{\varphi_2}^{\varphi} \left[\frac{w_3^3}{w_3'^2} \int_{\varphi_2}^{\varphi'} \frac{w_3}{w_3'^2}\, d\varphi'\right] d\varphi' - (1 + m_0)^{-1} \left(\int_{\zeta}^{\varphi_2} \frac{w_2}{w_2'^2} F_{11}\, d\varphi_1\right)^2 \times \\ &\times \int_{\varphi_2}^{\varphi} \frac{w_3^3}{w_3'^2}\, d\varphi' + (1 + m_0)^{-1} \int_{\zeta}^{\varphi_2} \left[\frac{w_2}{w_2'^2} F_{11} \int_{\varphi_1}^{\varphi_2} \left[\frac{w_2^3}{w_2'^2} F_{21} \int_{\zeta}^{\varphi_1} \frac{w_2}{w_2'^2} F_{11}\, d\varphi_1\right] d\varphi_1\right] d\varphi_1 \\ &- (1 + m_0) \int_{\varphi_2}^{\varphi} \left[\frac{w_3^3}{w_3'^2} \int_{\varphi'}^{\varphi} \left[\frac{w_3^3}{w_3'^2} \int_{\varphi_2}^{\varphi'} \frac{w_3}{w_3'^2}\, d\varphi'\right] d\varphi'\right] d\varphi' \\ &+ \int_{\zeta}^{\varphi_2} \frac{w_2}{w_2'^2} F_{11}\, d\varphi_1 \int_{\varphi_2}^{\varphi} \left[\frac{w_3}{w_3'^2} \int_{\varphi'}^{\varphi} \frac{w_3^3}{w_3'^2}\, d\varphi'\right] d\varphi' \end{aligned}$$

$$L_5 = \int_{\zeta}^{\varphi_2} \frac{w_2}{w_2'^2} F_{11}\, d\varphi_1 \int_{\varphi_2}^{\varphi} \left[\frac{w_3^3}{w_3'^2} \int_{\varphi_2}^{\varphi'} \frac{u_3}{w_3'^2}\, d\varphi' \right] d\varphi' - (1 + m_0)^{-1} \int_{\zeta}^{\varphi_2} \frac{w_2}{w_2'^2} F_{11}\, d\varphi_1 \times$$

$$\times \int_{\zeta}^{\varphi_2} \frac{u_2}{w_2'^2} F_{11}\, d\varphi_1 \int_{\varphi_2}^{\varphi} \frac{w_3}{w_3'^2}\, d\varphi' + (1 + m_0)^{-1} \int_{\zeta}^{\varphi_2} \left[\frac{w_2}{w_2'^2} F_{11} \int_{\varphi_1}^{\varphi_2} \left[\frac{w_3^3}{w_2'^2} F_{21} \times \right.\right.$$

$$\left.\left. \times \int_{\zeta}^{\varphi_1} \frac{u_2}{w_2'^2} F_{11}\, d\varphi_1 \right] d\varphi_1 \right] d\varphi_1 + \int_{\zeta}^{\varphi_2} \frac{u_2}{w_2'^2} F_{11}\, d\varphi_1 \int_{\varphi_2}^{\varphi} \left[\frac{w_3}{w_3'^2} \int_{\varphi'}^{\varphi} \frac{w_3^3}{w_3'^2}\, d\varphi' \right] d\varphi'$$

$$- (1 + m_0) \int_{\varphi_2}^{\varphi} \left[\frac{w_3}{w_3'^2} \int_{\varphi'}^{\varphi} \left[\frac{w_3^3}{w_3'^2} \int_{\varphi_2}^{\varphi'} \frac{u_3}{w_3'^2}\, d\varphi' \right] d\varphi' \right] d\varphi'$$

$$L_6 = \int_{\varphi}^{\varphi_2} \frac{w_2}{w_2'^2} F_{11}\, d\varphi_1 \int_{\varphi_2}^{\varphi} \frac{w_3^3}{w_3'^2}\, d\varphi' - \int_{\varphi}^{\varphi_2} \left[\frac{w_2}{w_2'^2} F_{11} \int_{\varphi_1}^{\varphi_2} \frac{w_2^3}{w_2'^2} F_{21}\, d\varphi_1 \right] d\varphi_1$$

$$- (1 + m_0) \int_{\varphi_2}^{\varphi} \left[\frac{w_3}{w_3'^2} \int_{\varphi'}^{\varphi} \frac{w_3^3}{w_3'^2}\, d\varphi' \right] d\varphi'$$

$$F_{11} = \Delta_1^{-1} \frac{\cot \alpha}{\cos \gamma_1} F_1; \quad F_{21} = \Delta_1^{-1} \frac{\cot \alpha}{\cos \gamma_1} F_2$$

$$\Phi(\varphi) = 1 + \frac{\Delta_1\, [1 - (1 - \epsilon)(1 - M_{1n}^{-2})]\, [\sin(\varphi + \sigma) F_{\varphi} - \cos(\varphi + \sigma) F]}{\Delta_2(\varphi) \sin[\varphi + \sigma + f_2(\varphi)]\, [F_{\varphi\varphi} + F]} \times$$

$$\times \{\cot[\varphi + \sigma + f_2(\varphi)] f_{2\varphi}(\varphi) + \Delta_2^{-1} \Delta_{2\varphi}(\varphi) + [1 - (1 - \epsilon)(1 - M_{1n}^{-2})]^{-1} \times$$

$$\times (1 - \epsilon)(1 - M_{1n}^{-2})_{\varphi} + [(1 + \epsilon) M_{1n}^2 - \epsilon]^{-1} [(1 + \epsilon) M_{1n}^2 - \epsilon]_{\varphi}$$

$$- [\sin(\varphi + \sigma) F_{\varphi} - \cos(\varphi + \sigma) F]^{-1} \sin(\varphi + \sigma)(F_{\varphi\varphi} + F)\}.$$

Here $\eta = 1$, $\zeta = \varphi$ in the interval $\varphi_4 \le \varphi \le \varphi_2$, $\eta = 0$, $\zeta = \varphi_4$ in the interval $0 \le \varphi \le \varphi_4$.

Because of the method used for integration of the system (1.5) the equation (4.10) contains a compatibility condition along the slip line φ_2 passing through the triple point A (Fig. 19). The second necessary condition in the intervals $[\varphi_4, \varphi_2]$ is that the pressure p_2^*, calculated at transition through the shock F from region 1 to region 2, should be equal to the pressure $p_2(\varphi, \varphi_1)|_{\varphi_1 = \varphi}$ from (4.6):

$$p_3^*(\varphi) + \epsilon p_{32}(\varphi, \varphi_2) + \epsilon p_{22}(\varphi, \varphi) = p_1^* [(1 + \epsilon) M_{1n}^2 - \epsilon]. \tag{4.13}$$

In this the integro-differential equation for the shock wave shape in the interval $[\varphi_3, \varphi_4]$ is obtained, namely eq. (4.12) at $\eta = 0$ and $\zeta = \varphi_4$. We find also a system of two equations for the leading shock wave and the internal shock F in the interval $[\varphi_4, \varphi_2]$. These equations are (4.12) at $\eta = 1$, $\zeta = \varphi$, and (4.13).

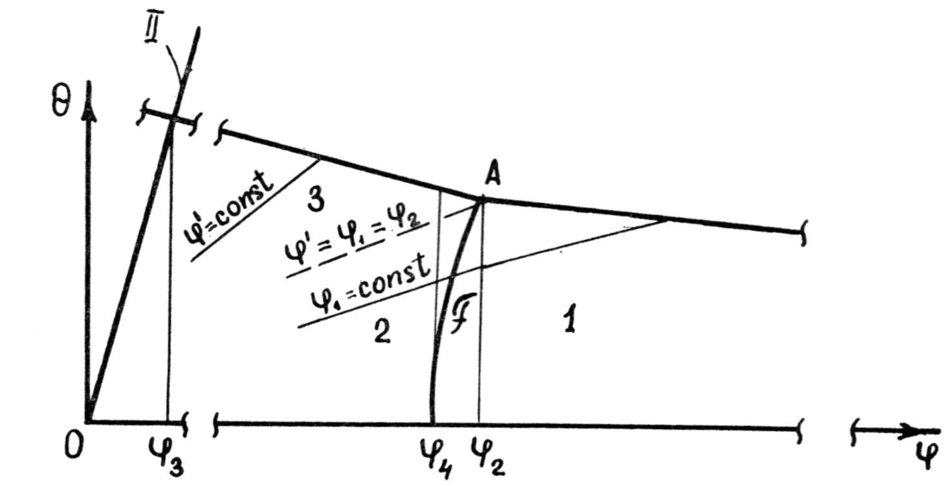

FIG. 19. Cross streamline pattern.

An investigation of p_{32} and p_{22} shows that for continuity of the pressure along the line $\varphi = \varphi_4$ we must have

$$(z_1'' + z_1)\, \Phi\,(\varphi)|_{\varphi=\varphi_4} = 0.$$

An analysis of the expression $\Phi(\varphi)$ (4.12) allows us to find the main term of the function $\Phi(\varphi)$:

$$\Phi(\varphi) \approx (1 - M_{1n}^{-2})$$

which is not zero. Thus it follows, that

$$(z_1'' + z_1)|_{\varphi=\varphi_4} = 0.$$

Having a differential operator of the third order for the function z in eq. (4.12) in the interval $[\varphi_3, \varphi_4]$ and of the third and the second order for functions z and z_1, respectively, in the system of equations (4.12) and (4.13) in the interval $[\varphi_4, \varphi_2]$, we have to apply eight restrictions to the unknown functions. Two unknown boundaries φ_3 and φ_2 may be found using the conditions of intersection of the shock waves with the plane of symmetry: $z\,(\varphi_3) = \cot \alpha \operatorname{cosec} \gamma_1 \sin \varphi_3$, and with the plane shock wave $z_2 = a_2 \sin(\beta - \varphi)$ attached to the leading edge:

$$z(\varphi_2) \equiv z_1(\varphi_2) = a_2 \sin(\beta - \varphi_2).$$

Two conditions are known for the function z_1 at the point φ_4 and the symmetry condition for the function z in the point φ_3

$$z'(\varphi_3) = -\tan \gamma_1 \cot \alpha \sec \gamma_1 + 0\,(\epsilon\gamma_1^2).$$

Thus there are five arbitrary conditions which may be used to connect $z^{(n)}(\varphi_4)$ for $n = 0, 1, \ldots, 4$. This means that on the line $\varphi = \varphi_4$ the functions p and θ as well as their derivatives are continuous.

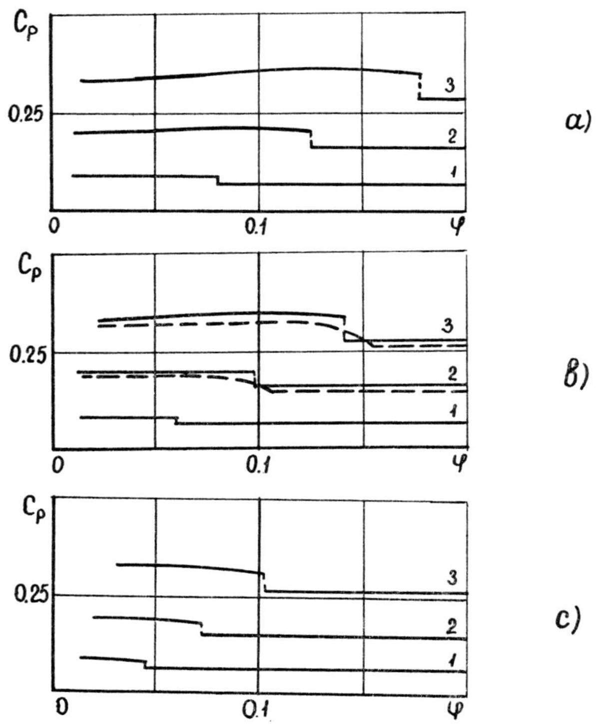

FIG. 20. Pressure distribution.

C. *Approximate Solution and Results of Calculations*

To find the approximate solution we shall use some simplifications. In the integrands the functions of φ_1 are considered to be constant; in the functions w_i only main terms are used assuming the shock F to be of small intensity. After integration of the functionals and some transformations the integro-differential equation for the function z in the interval $[\varphi_3, \varphi_4]$ is reduced to a Euler's equation, the general solution of which is known. The system of equations (4.12) and (4.13) for the functions z and z_1 in the interval $[\varphi_4, \varphi_2]$, after similar transformations, is reduced to an essentially nonlinear system of differential equations, no simple solution of which is known. So in the interval $[\varphi_4, \varphi_2]$ the functions z and z_1 were expanded in Taylor's series at the point φ_2. Four terms of the series were used. This allows us to link up, at the point φ_4, the functions z and their first and second derivatives only.

In Fig. 20a, b, c the curves of pressure distribution (solid lines) behind the shock wave are shown for $M_\infty = \infty$, $\beta = 45°$, $\gamma_1 = 10°$, $15°$ and $20°$. The curves 1, 2, 3 correspond to the angles of attack $\alpha = 10°$, $15°$ and $20°$.

In Fig. 20b is given a comparison of the pressure distribution calculated using the above theory with the pressure distribution (dashed line) resulting from the numerical computation described in section 5. There is good agreement between the shock positions. Some discrepancy in the pressure distributions may be explained by taking into consideration the somewhat smaller values of the pressure and density obtained by numerical computation behind the plane shock. Note that the velocity behind the shock F is subsonic. On the shock

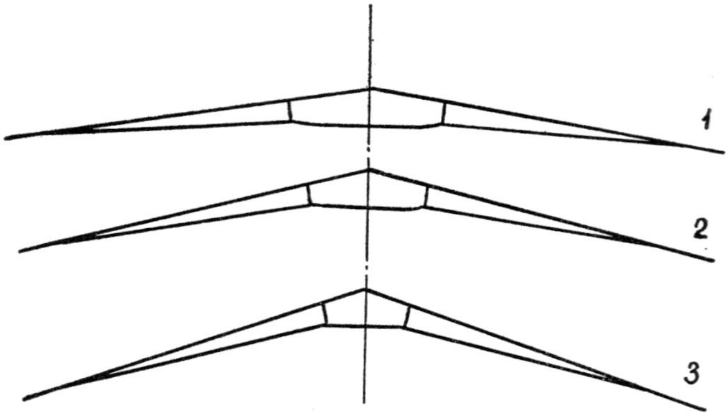

FIG. 21. Configuration of shock waves.

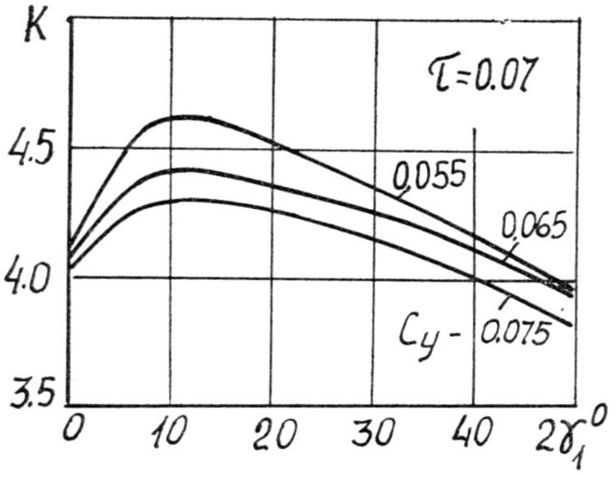

FIG. 22. Lift-to-drag ratio of caret wing.

wave in the interval $[\varphi_3, \varphi_2]$ there is a point of inflexion which moves to the left, as the angle γ_1 increases and then disappears. After this the curvature of the shock wave is positive. In Fig. 21 a general picture of the flow past a V-shaped wing is represented for $M_\infty = \infty$, $\beta = 45°$, $\alpha = 20°$; $\gamma_1 = 10°$, $15°$ and $20°$ in cases 1, 2 and 3 respectively.

The dependence of the lift-to-drag ratio K from the dihedral angle γ_1 at τ and C_y = const is given in Fig. 22. (Here $\tau = v/S^{3/2}$; v is the volume of the wing bounded on the top by two planes set along the flow, S the wing plan-form area.)

5. A NUMERICAL METHOD OF CALCULATION OF THE SUPERSONIC FLOW PAST A WING

In this section the time-dependent method is used where the difference scheme contains artificial viscosity but the role of time is replaced by the space of coordinates with respect to which the stationary system of the equations is hyperbolic.

A. *Basic Equations and Difference Scheme*

Let us consider the flow about a V-shaped wing with an angle of attack (Fig. 16). The coordinates used are

$$\xi = \ln x; \quad \eta = y/x; \quad \zeta = z/x.$$

The equations of motion in divergent form are

$$\frac{\partial f}{\partial \xi} + \frac{\partial}{\partial \eta}(F^y - \eta f) + \frac{\partial}{\partial \zeta}(F^z - \zeta f) + 2f = 0$$

$$f = \begin{Bmatrix} \Omega \\ R \\ S \\ T \\ E \end{Bmatrix}, \quad F^y = \begin{Bmatrix} S/u \\ S \\ R - \Omega u + S^2/\Omega u \\ TS/\Omega u \\ ES/\Omega u \end{Bmatrix}, \quad F^z = \begin{Bmatrix} T/u \\ T \\ TS/\Omega u \\ R - \Omega u + T^2/\Omega u \\ ET/\Omega u \end{Bmatrix} \tag{5.1}$$

$$\Omega = \rho u; \quad R = p + \rho u^2; \quad S = \rho uv; \quad T = \rho uw$$

$$E = \rho u\,(e + p/\rho + (u^2 + v^2 + w^2)/2).$$

u, v and w are components of velocity along the axes x, y, z; e the internal energy of gas. The system (5.1) is non-dimensional. The velocity components are referred to the velocity of the undisturbed flow, the pressure p to the dynamic pressure, the density ρ to the density of the undisturbed flow.

Consider a vector

$$\varphi = \begin{Bmatrix} u \\ v \\ w \\ p \\ \rho \end{Bmatrix}$$

which is connected with the components of the vector f by

$$u = x\,R/(x+1)\,\Omega + \{x^2\,p^2/(x+1)^2\,\Omega^2 - 2\,(x-1)\,[E/\Omega - (S^2 + T^2)/2\Omega]/(x+1)\}^{1/2} \tag{5.2}$$

$$v = S/\Omega; \quad w = T/\Omega; \quad p = R - \Omega u; \quad \rho = \Omega/u.$$

The axis x is chosen in such a way that $u > a$, where $a = \sqrt{xp/\rho}$ is the local sound velocity.

The boundary conditions: on the wing surface $\psi\,(\xi, \eta, \zeta) = \eta - g\,(\zeta) = 0$, the boundary condition is

$$(v - u\eta) - (w - \zeta u)\,g' = 0; \tag{5.3}$$

on the symmetry plane $\zeta = 0$,

$$w = 0; \tag{5.4}$$

in the undisturbed flow,

$$u = \cos\alpha; \quad v = -\sin\alpha; \quad w = 0 \tag{5.5}$$

$$p = \frac{1}{xM_\infty^2}; \quad \rho = 1.$$

The problem (5.1), (5.3)–(5.5) is solved by the time-dependent method where the space coordinate plays the role of the time. In the calculations Rusanov's difference scheme[52] was used.

Denoting

$$\Delta\xi = \tau; \qquad \Delta\zeta = h_1; \qquad \Delta\eta = h_2; \qquad h = \sqrt{h_1^2 + h_2^2}$$

$$h_1 = h \cos\chi; \qquad h_2 = h \sin\chi; \qquad \chi = \frac{\tau}{h_1}; \qquad \chi = \sqrt{\chi_1^2 + \chi_2^2}$$

The value of a quantity A at the mesh point with coordinates $(n\tau_1, lh_2, mh_1)$ is denoted by $A^n_{l,m}$.

We consider the case when the surfaces of the V-shaped wing are plane and the line intersection of the wing with the plane $x = \text{const}$ goes through the mesh points along the diagonals of the mesh cells (Fig. 23).

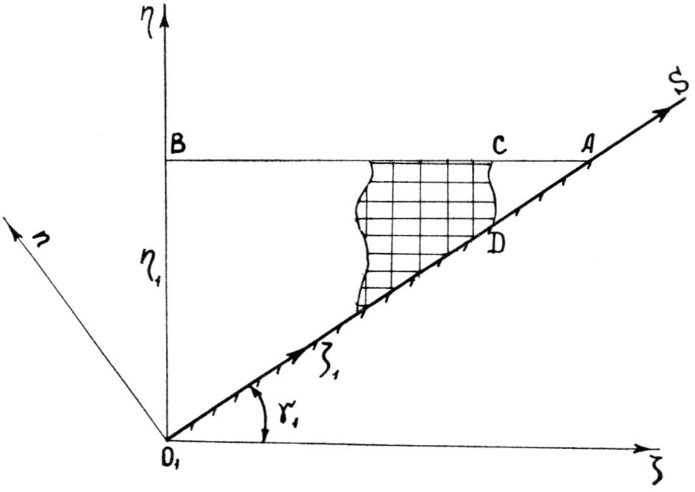

FIG. 23. Coordinate system.

The system of equations (5.1) is approximated in accordance with[52]

$$f^{n+1}_{l,m} = f^n_{l,m} - \frac{\chi_2}{2}\{(F^y - \eta f)_{l+1,m} - (F^y - \eta f)_{l-1,m}\}^n - \frac{\chi_1}{2}\{(F^z - \zeta f)_{l,m+1}$$
$$- (F^z - \zeta f)_{l,m-1}\}^n + \tfrac{1}{2}\{\varphi^y_{l+\frac{1}{2},m} - \varphi^y_{l-\frac{1}{2},m} + \varphi^z_{l,m+\frac{1}{2}} - \varphi^z_{l,m-\frac{1}{2}}\} - 2\tau^n f^n_{l,m}$$
$$\varphi^y_{l+\frac{1}{2},m} = \tfrac{1}{2}(\beta_{l+1,m} + \beta_{l,m})(f_{l+1,m} - f_{l,m}); \quad \beta_{l,m} = \chi\omega\sigma_{l,m}\cos^2\chi$$
$$\varphi^z_{l,m+\frac{1}{2}} = \tfrac{1}{2}(\alpha_{l,m+1} + \alpha_{l,m})(f_{l,m+1} - f_{l,m})$$
$$\alpha_{l,m} = \chi\omega\sigma_{l,m}\sin^2\chi; \quad \omega = \text{const}$$
$$\sigma_{l,m} = \{[u(v^2 + w^2)^{1/2} + a(u^2 + v^2 + w^2 - a)^{1/2}]/(u^2 - a^2)\}_{l,m}.$$

The stability analysis of the difference scheme was carried out by Fourier's method. The stability conditions are

$$\sigma_0 < \omega < (1 - \tau)/\sigma_0; \quad \sigma_0 = \max_{l,m} \sigma_{l,m}; \quad \sigma_0 < 1.$$

B. *The Calculation Formulae*

(1) The point O_1 (Fig. 23) of the plane of symmetry is on the wing surface. It follows from (5.3) and (5.4) that $T_{0,0} = S_{0,0} = 0$ and at this point it is necessary to find the components of the vector f, R and E only. To derive the calculation formulae the following new system of coordinates is used (Fig. 23):

$$\xi = \xi_1; \quad \eta = \eta_1 + \zeta \sin \chi; \quad \zeta = \zeta_1 \cos \chi.$$

The formula is

$$\frac{1}{\chi^n}\{f_{0,0}^{n+1} - (1 - 2\tau^n) f_{0,0}^n\} = -\tan \chi\, (F^z - \zeta f)_{1,1}^n - \cos \chi\, (F^y - \eta f)_{1,0}^n + \sin \chi F_{1,0}^{z,n}$$
$$+ \frac{\omega \cos^2 \chi \sin^2 \chi}{4} (\sigma_{1,1} - \sigma_{0,0})^n (f_{1,1} - f_{0,0})^n.$$

(2) A point of the plane of symmetry $\zeta = 0$. From the boundary condition (5.4) it follows that $T_{e,0} = 0$ and here all components of the vector f other than T should be found:

$$\frac{1}{\chi^n}[f_{l,0}^{n+1} - (1 - 2\tau^n) f_{l,0}^n]$$
$$= -\sin \chi\, (F^z - \zeta f)_{l,1}^n - \tfrac{1}{2} \cos \chi\, \{(F^y - \eta f)_{l+1,0}$$
$$- (F^y - \eta f)_{l-1,0}\}^n + \tfrac{1}{2}\omega \sin^2 \chi\, (\sigma_{l,0} + \sigma_{l,1})\, (f_{l,1} - f_{l,0})^n$$
$$+ \tfrac{1}{4}\omega \cos^2 \chi\, \{(\sigma_{l+1,0} + \sigma_{l,0})\, (f_{l+1,0} - f_{l,0}) - (\sigma_{l,0} + \sigma_{l-1,0})\, (f_{l,0} - f_{l-1,0})\}^n.$$

(3) A point on the wing surface. In the system of coordinates ξ_1, S, n (Fig. 23)

$$\xi_1 = \xi; \quad S = \zeta \cos \chi + \eta \sin \chi; \quad n = -\zeta \sin \chi + \eta \cos \chi.$$

The calculation formula is

$$\frac{1}{\chi^n}[\tilde f_{l,l}^{n+1} - \tilde f_{l,l}^n (1 - 2\tau^n)] = -\tfrac{1}{2} \sin \chi \cos \chi\, \{(\tilde F^z - \tilde\zeta \tilde f)_{l+1,l+1} - (\tilde F^z - \tilde\zeta \tilde f)_{l-1,l-1}\}^n$$
$$- \{(\tilde F^y - \tilde\eta \tilde f)_{l,l-1} \sin^2 \chi + (\tilde F^y - \tilde\eta \tilde f)_{l+1,l} \cos^2 \chi\}^n$$
$$+ \tfrac{1}{4}\omega \sin^2 \chi \cos^2 \chi\, \{(\sigma_{l+1,l+1} + \sigma_{l,l})\, (\tilde f_{l+1,l+1} - \tilde f_{l,l})$$
$$+ (\sigma_{l,l} + \sigma_{l-1,l-1})\, (\tilde f_{l,l} - \tilde f_{l-1,l-1})\}^n.$$

All quantities marked by the sign ($\sim$) are calculated using the above formulae for corresponding quantities without such a sign but with u, w, η, ζ replaced by:

$$\tilde v = -w \sin \chi + v \cos \chi; \quad \tilde w = w \cos \chi + v \sin \chi; \quad \tilde \eta = -\zeta \sin \chi + \eta \cos \chi;$$
$$\tilde \zeta = \zeta \cos \chi + \eta \sin \chi.$$

The third component of the vector $\tilde f_{ll}$ is not calculated as it follows from (5.3) that $\tilde S_{ll} = 0$.

When $f_{l,l}$ is known the vector $f_{l,l}$ is found by formulae

$$\Omega_{l,l} = \tilde\Omega_{l,l};\ R_{l,l} = \tilde R_{l,l};\ S_{l,l}, = \tilde T_{l,l} \sin \chi;\ T_{l,l} = \tilde T_{l,l} \cos \chi; E_{l,l} = \tilde E_{l,l}.$$

(4) An internal point of the flow region:

$$\frac{1}{\chi^n}[f_{l,m}^{n+1} - (1 - 2\tau^n) f_{l,m}^n] = -\tfrac{1}{2}\sin\chi\,\{(F^z - \zeta f)_{l,m+1} - (F^z - \zeta f)_{l,m-1}\}^n$$

$$-\tfrac{1}{2}\cos\chi\,\{(F^y - \eta f)_{l+1,m} - (F^y - \eta f)_{l-1,m}\}^n$$

$$+\tfrac{1}{4}\omega\sin^2\chi\,\{(\sigma_{l,m+1} + \sigma_{l,m})(f_{l,m+1} - f_{l,m}) - (\sigma_{l,m} + \sigma_{l,m-1})(f_{l,m} - f_{l,m-1})\}^n$$

$$+\tfrac{1}{4}\omega\cos^2\chi\,\{(\sigma_{l+1,m} + \sigma_{l,m})(f_{l+1,m} - f_{l,m})$$

$$-(\sigma_{l,m} + \sigma_{l-1,m})(f_{l,m} - f_{l-1,m})\}^n.$$

The calculated region O_1DCB is shown in Fig. 23. On the boundary BC, all quantities correspond to the undisturbed flow parameters; on the boundary DC, to the quantities obtained behind an oblique shock wave attached to the leading edge; on the wing surface O_1D, the non-permeable surface condition; on the plane of symmetry O_1B, $w = 0$. For the initial flow parameters the values behind an oblique shock wave attached to the leading edge of the V-shaped wing were taken. The calculation was terminated when

$$\max_{l,m}\{(f_{l,m}^{n+1} - f_{l,m}^n)/f_{l,m}^n\} \leqslant \epsilon$$

for every component of the vector f (in the calculations $\epsilon = 10^{-2}$ to 10^{-3}).

C. *The Results*

To check the efficiency of the method a comparison of the results of calculations with experimental data[47] was made. In Fig. 24 the calculated pressure distribution $C_p = (p - p_\infty)/\frac{1}{2}\rho_\infty V_\infty^2$ along the wing surface is shown (solid lines), together with experimental data for $\beta = 29°30'$, $M_\infty = 3.95$, $Re = 6.8 \times 10^6$. There is a good correlation between experimental and calculated data, excluding the region in front of the shock adjacent to the wall where the experimental pressure is greater than the calculated pressure. Such a pressure overshot was mentioned in ref. 53 and may be explained[54] by the interaction of the shock with the boundary layer. In Fig. 25a, b, the shapes of the isobars indicate flow regimes with Mach and regular reflection of the shock coming from the leading edge at the plane of symmetry. The isobars in Fig. 25b indicate quite a few reflections of the shock at the plane of symmetry and the wall. The dashed line shows the exact position of the plane shock wave attached to the leading edge.

Note that the pressure and the density behind the calculated plane shock differ from the exact values by 3%. The width of the calculated shock is not more than 5–6 mesh intervals.

In conclusion we shall consider some results of the flow past the leeward side of the V-shaped wing (the region of rarefied flow). The plane delta wing ($\gamma = 180°$) is known to be flown with inner shocks on the leeward side. These shocks can also occur in case of V-shaped wings ($\gamma < 180°$). The investigation of the flow field over the leeward side is carried out by means of the numerical method mentioned above. The results of the pressure calculation on the leeward side are shown in Fig. 26a for $\gamma = 116°$. As can be seen from the diagram, a suspension shock in the region of the flow may appear. Let us study the distribution of the streamlines. In Fig. 26b the streamlines in the plane $x = 1$ are given for the delta wing (Fig. 26b, $\gamma = 180°$) and for the V-shaped wing (Fig. 26c, $\gamma = 116°$). It is seen that all streamlines are convergent at the point O, the latter being the sink point.

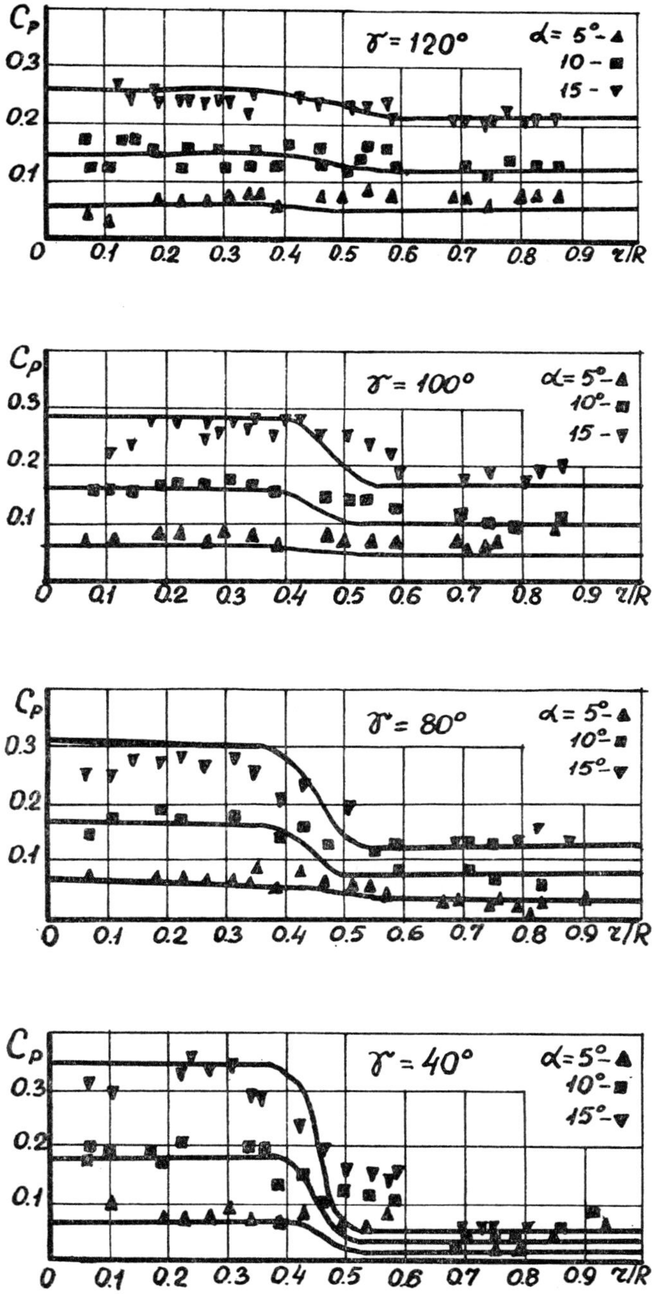

FIG. 24. Pressure distribution on caret wing.

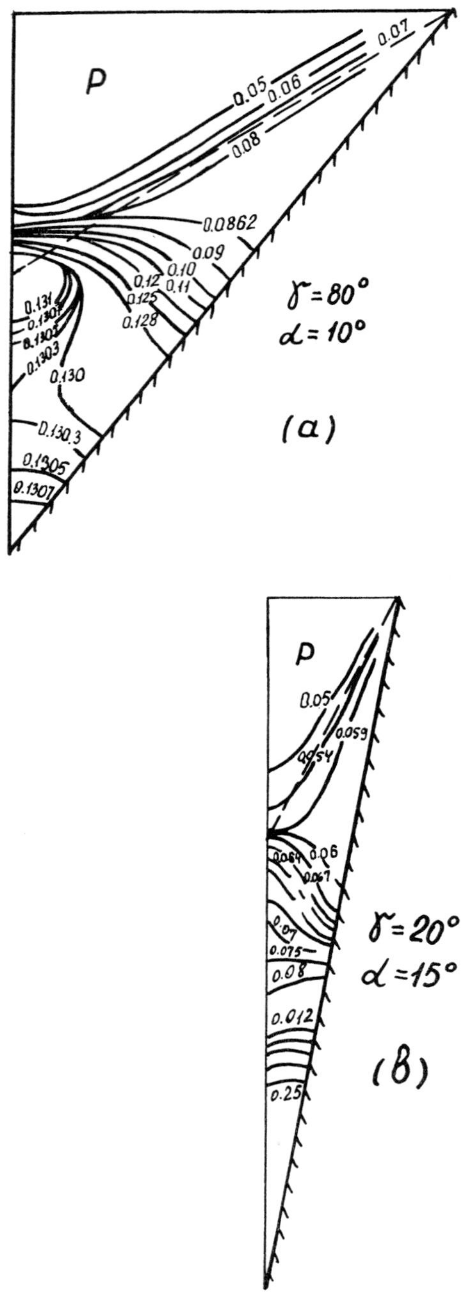

FIG. 25. Isobaric curves.

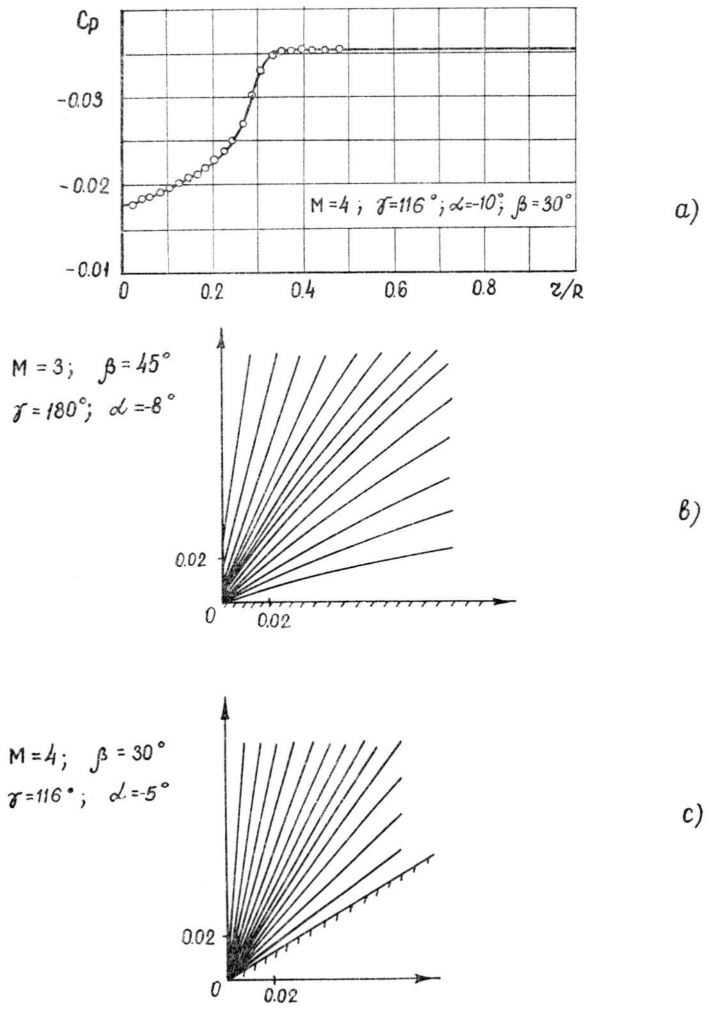

FIG. 26. (a) Pressure distribution. (b, c) Streamline pattern on leeward side wing.

The case $\gamma = 270°$, $\beta = 30°$, $M = 4$, for the angles of attack $\alpha = -5°, -10°$ was considered in detail. In Fig. 27a the streamline field at $\alpha = -5°$ (dashed lines) is given. Hence we can see that a part of the streamlines is almost parallel to the wing surface and their distortion begins at a certain distance from the wing.

It is difficult to conclude whether the sink point is on the wing surface or in the stream.

To define more accurately the position of the sink point the computation programme was changed in such a way as to isolate the region near the point of the plane of symmetry. New calculations have been made for this region.

This method allows us to decrease the step of the difference mesh and to raise the exactness of the results. In the case when the angle of attack $\alpha = -5°$, and the regions near the sink point are chosen, the step of the mesh is decreased by a factor of 5 (see Fig. 27a solid lines, and Fig. 27b dashed lines) and by a factor of 25 (Fig. 27b, solid lines).

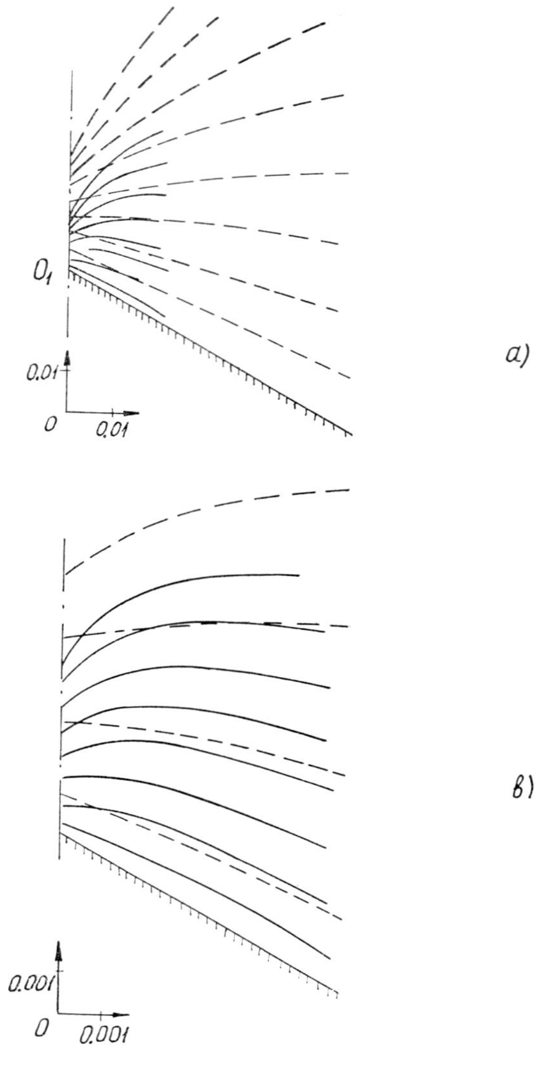

FIG. 27. (a, b, c) Streamline distribution near Ferri point.

It is seen that as the step of the mesh is decreased the size "of the spread" of the sink point becomes smaller and approaches the wing surface. For the angle of attack $\alpha = -10°$ the situation changes. The investigation has shown that a 5 times decrease of the mesh step reduces the size "of the spread" of the sink point, but its position does not change (Fig. 27c). On Fig. 27c are shown the streamlines at $\alpha = -10°$ and the mesh step h_0 (dashed lines) and $h_1 = 0.2\, h_0$ (solid lines). Thus as the angle of attack increases, the sink point leaves the wing surface and moves into the region of flow along the plane of symmetry.

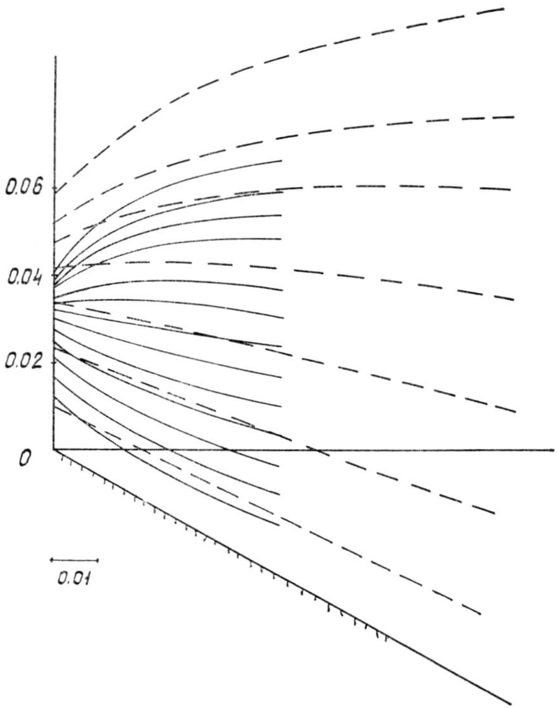

FIG. 27 (c).

6. THE FLOW REGIMES OF CARET WINGS WITH SUPERSONIC LEADING EDGES

In this section the shock wave system for the flow about inverted V-shaped wings (caret wing) was investigated by means of numerical calculations made using the method described in section 5.

The analysis has allowed us to discover some flow regimes about a caret wing depending on the dihedral angle, with qualitatively different systems of shock waves: from one shock wave to four shock waves and more. In particular the calculations carried out have shown that there are flow regimes with a strong shock wave and a strong reflected shock wave from the plane of symmetry starting at the leading edge.

As is known, a strong shock wave cannot be realized in the case of two-dimensional flow.

The notation used below is shown in Fig. 16. The results of calculations made for $\beta = 29°30'$; $M_\infty = 3.95$, angles γ from 100° to 150°, $\alpha = 15°$ and $\gamma = 160°$, $\alpha = 10°$ are shown in Fig. 28a, b.

The last regime is near to the flow regimes with the detached shock wave. In Fig. 28a are shown the configurations of the shock waves, and in Fig. 28b the distributions of the pressure coefficient $C_p = (p - p_\infty)/q_\infty$ on the wing surface, where p is the pressure, q_∞ the dynamic pressure head of the uniform flow. On the diagram, r is the distance from the plane of symmetry to the arbitrary point on the wing surface, R distance along the wing from the plane of symmetry to the leading edge.

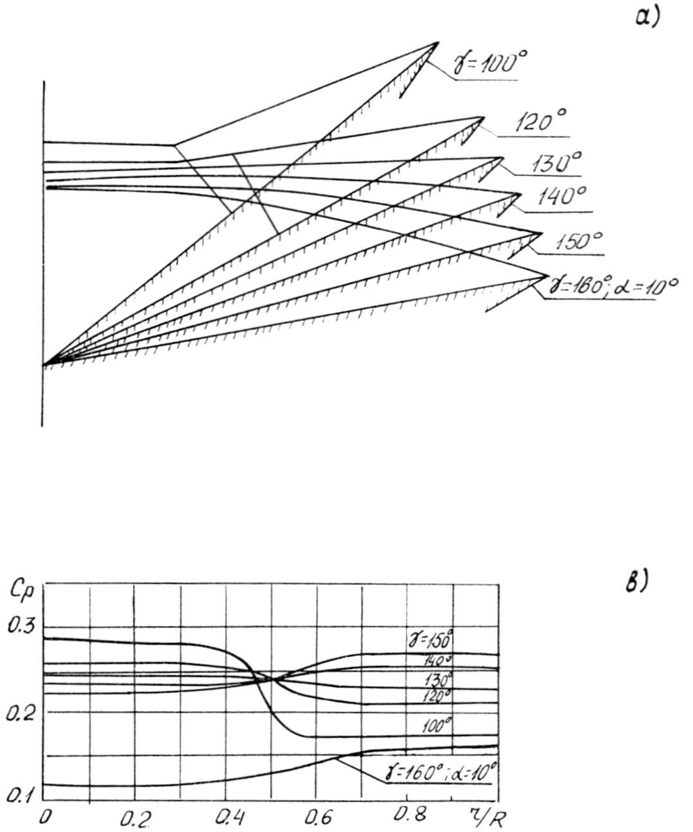

FIG. 28. (a) Flow regimes. (b) Pressure distribution on caret wing.

All the values shown in diagrams and tables are dimensionless, viz. the components of the velocity vectors are referred to the modulus of the velocity vector of the uniform flow, the pressure is referred to the double dynamic head of the uniform flow, and density is referred to the density of the uniform flow.

While adapting the results of numerical calculations, the position of the shock wave was determined as the region of high values of the gradients of the parameters.

The dependence of $C_p = f(r/R)$ is shown in Fig. 28b and the analysis of the flow field for various angles γ at $M_\infty = 3.95$ shows that with the reduction of the angles γ there is a transition from the flow with the convex shock wave through the plane shock wave to the flow with the concave shock wave.

In the case of the concave shock wave, the pressure near the wing symmetry plane increases when the angle $\gamma \lesssim 120°$. The increase of the pressure can be identified with the presence of the inner shock wave, i.e. the type of flow of the V-shaped wing with the Mach wave configuration. It is not clear whether the inner shock wave forms at once at the transition from the flow pattern with the plane shock wave to the flow with the concave shock wave. As an example, in Fig. 29a the family of isobars for $\gamma = 120°$, $\alpha = 15°$, $M_\infty = 3.95$ is shown. The position of the isobaric curves indicates the flow with the Mach wave configura-

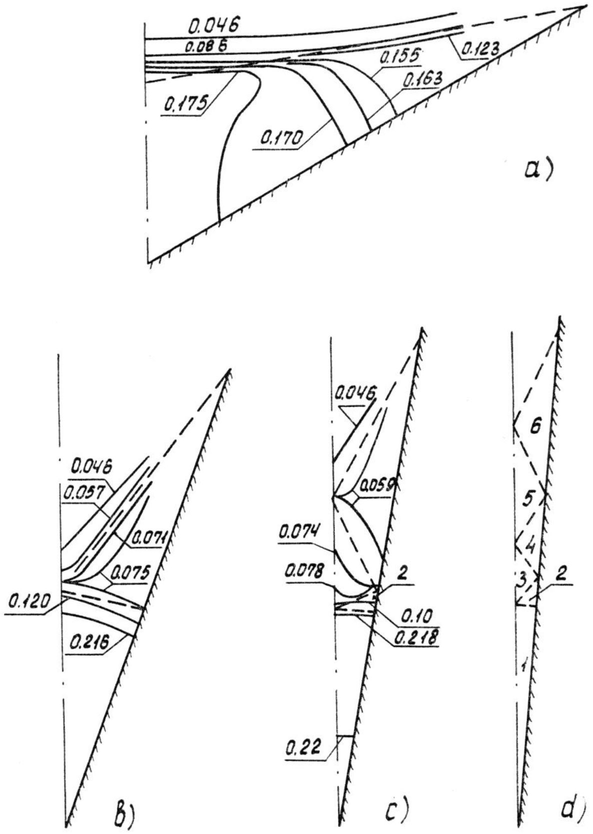

FIG. 29. (a, b, c, d) Isobaric curves.

tion. On the same figure the position of the plane shock wave attached to the leading edge is given as a dashed line.

When the angle γ is reduced to 40° (Fig. 29b) the shock wave being attached to the leading edge has a regular reflection from the wing plane of symmetry according to the position of the isobars.

The results obtained by numerical methods were compared with the calculations of the plane shock wave system for the determination of the reflected shock waves shapes (weak or strong shock wave) as well as for a more precise type of reflection (regular or Mach reflection). This analysis has shown that for $\gamma = 40°$, $\alpha = 15°$ the plane shock wave attached to the leading edge has been regularly reflected as a strong shock from the wing plane of symmetry. Besides, the values of density and pressure obtained from the numerical calculations and from the relations for the plane shock wave differed by less than 2%. For this case the system of the shock waves is shown in Fig. 29b (dashed lines).

With the reduction of the angle γ the system of shock waves changes and at the angle $\gamma = 20°$ the family of isobaric curves shows that the shock wave reflected from the wing surface appears (Fig. 29c).

The comparison of the results with calculations obtained from the relations for the plane shock has shown that the system of shocks takes an aspect shown in Fig. 29c (dashed lines), viz. the shock wave attached to the leading edge is reflected regularly as a weak shock wave from the wing plane symmetry and the latter is regularly reflected from the wing surface as a secondary weak shock wave, which in its turn is being reflected from the plane of symmetry as a Mach wave configuration.

The difference in the values of velocity, density and pressure obtained from numerical calculations and from the relations of the plane shock wave did not exceed 2%.

Reducing further the angle ($\gamma = 10°$) the number of the reflections from the wing surface and the plane of symmetry increases. Besides this, the last reflection from the plane of symmetry becomes a regular strong reflected shock. In this case all the remaining reflections are weak reflected shocks. The system of the shock waves corresponding to this case is shown in Fig. 29d. The difference in the values of velocity, density and pressure obtained from numerical calculations and from the relations of the plane shock wave did not exceed 2%. In the numerical methods the flow region (denoted in Fig. 29d) number 2 is badly determined since a small number of points of the calculation mesh falls in this region.

As an example the results of numerical calculations are given in Table 1, and the results obtained from the relations for the plane shock in Table 2.

TABLE 1

Number of the region	u	v	w	p	ρ
1	0.8946		0	0.223	3.004
2					
3	0.9500	−0.2143	0	0.0768	1.444
4	0.9538	−0.2278	−0.01994	0.0676	1.320
5	0.9574	−0.2394	0	0.0599	1.211
6	0.9617	−0.2494	−0.02183	0.0522	1.098

TABLE 2

Number of the region	u	v	w	p	ρ
1	0.8995	−0.00579	0	0.222	3.020
2	0.9472	−0.2017	−0.01759	0.0865	1.577
3	0.9507	−0.2169	0	0.0764	1.443
4	0.9544	−0.2283	−0.01998	0.0678	1.325
5	0.9576	−0.2403	0	0.0599	1.213
6	0.9617	−0.2494	−0.02184	0.0522	1.099

In these tables u, v, w are components of velocity along axes x, y, z; p, ρ are pressure and density. A comparison of the tables shows that the results of the calculations coincide quite well and allow the identification of the pattern of the shock waves. The numbers of regions given in the tables correspond to the numbers indicated in Fig. 29d.

The values of the component velocity v for the region 2 (Fig. 29d) are not given in Table 1 as v is monotonically decreasing to zero on the axis.

In Fig. 30 the diagrams $\bar{p} = p/q_\infty$ along the axis of the symmetry (1) and along the wing surface (2) obtained from the numerical calculations for $\gamma = 10°$, $\alpha = 15°$, $\beta = 29°30'$, $M_\infty = 3.95$ are given. y/H is the ratio, where the numerator is the distance from the intersection point of the wing walls to an arbitrary point lying on the axis of symmetry and the denominator is the length of the projection of the leading edge on the plane of symmetry. The system of shock waves shown in Fig. 30 (dashed lines) corresponds to the solutions obtained from the relations of the plane shock. In this figure the position of the curves confirms that the above system of shock waves takes place. The analysis of the diagrams

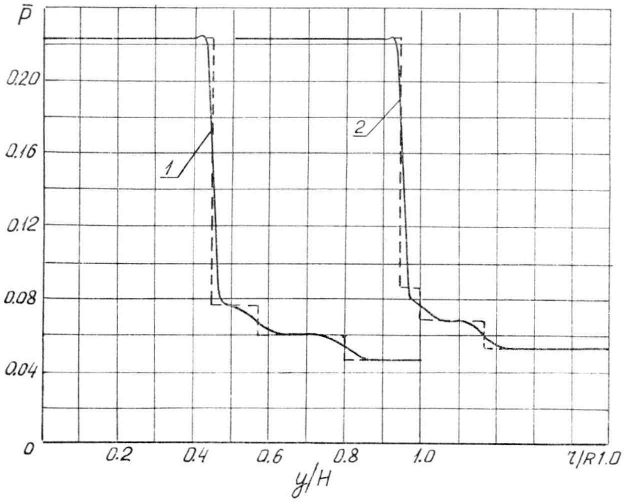

FIG. 30. Pressure distribution on the plane of symmetry (1) and along the wing surface (2).

(Fig. 30) shows a sufficient width of shock wave structure at the points of regular reflection that can be explained by the presence of two shock waves. The calculations of the flow over V-shaped wings at $\gamma = 140°$, $\alpha = 15°$ and different values of the number M_∞, have shown that with the increase of M_∞ there is a transition from the flow with the convex shock wave to the flow with the concave shock wave and Mach wave configuration, i.e. the flow is qualitatively the same as the flow when the angle γ is reduced at $M_\infty = \text{const}$ (if the angle γ is not small).

Thus the analysis shows that for large angles γ a Mach shock wave system takes place at an arbitrary Mach number (the calculations were carried out for $\beta = 29°30'$, $\gamma \geq 80° \div 100°$). The results of the numerical calculations compared with the experimental data[46] have shown good agreement. Let us consider the problem of uniqueness of the numerical solution. In the calculations as a zero approximation (initial field) we used the flow behind the plane shock wave attached to the leading edge of the wing. The shock is supposed to be weak in the sense that the projection of the velocity vector of the flow behind the shock on the normal plane to the leading edge is supersonic. There is another case: the flow velocity behind the shock wave is supersonic, but the projection of the vector velocity on the normal plane to the leading edge is subsonic. These two solutions will be called weak and strong shocks respectively. Thus, generally speaking, it is not clear which solution can be used as an initial field and how it depends on the choice of the initial approximation. To clear up this problem, flow calculations of V-shaped wing at $\gamma = 100°$, $\beta = 29°30'$, $\alpha = 15°$ and with $M_\infty = 3.95$, while using three different versions of initial data, were carried out.

In the first version the flow with a weak shock wave attached to the leading edge was used as an initial field. In the second one the flow behind the strong shock wave was used. The incoming flow was used as an initial field in the third version.

Let us consider briefly the schemes of calculation.

For the first version the calculated region $O_1A'B'C'$ (Fig. 31) was chosen so that the influence of the plane of symmetry of the wing was inside the region $O_1A'B'C'$.

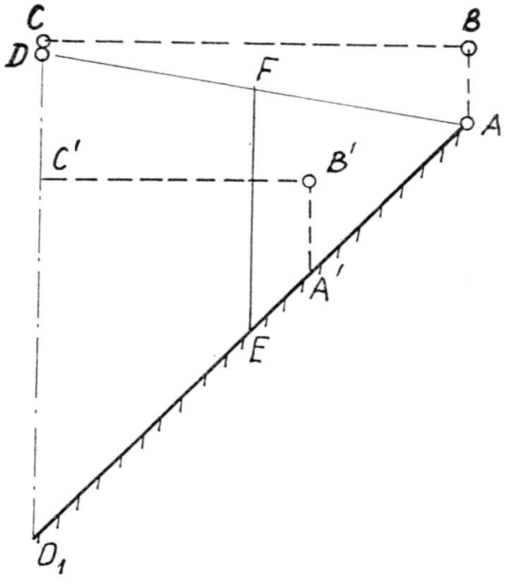

FIG. 31. Calculated region.

In the case of the strong shock attached to the leading edge the whole region of the disturbed flow is subject to the influence of the plane symmetry.

Therefore the calculated region O_1ABC was chosen as shown in Fig. 31, where at the point A the parameters of the flow are supposed to correspond to the strong shock wave. In all the remaining points AB the parameters correspond to the uniform flow. The segment BC is placed so that the line of the strong shock was below the segment BC (Fig. 31).

The values of the flow parameters on the segments AB and BC remained invariable.

If we assume (for the initial field) the values of the parameters of the uniform flow, the calculated region will be the same as in the version 2 (Fig. 31). Besides, on the segment AB (including the point A) the values of the parameters of the undisturbed flow were given. On the segments BC and AB the conditions during the calculations did not change. It is supposed that the point A has a weak influence upon the numerical solution, as the point A is in the undisturbed flow. The results of the calculations are given in Figs. 32–34. In the diagrams the number 1 corresponds to the first version, while the number 2 refers to the second one. The small circles in the figure correspond to the third version. u, v, w are the components of the vector velocity along the axis of the coordinates, p the dimensionless pressure related to the doubled dynamic pressure, ρ the dimensionless density related to the density of the undisturbed flow, and s the entropy function ($s = p/\rho^x$). In the calculations $x = 1, 4$ was

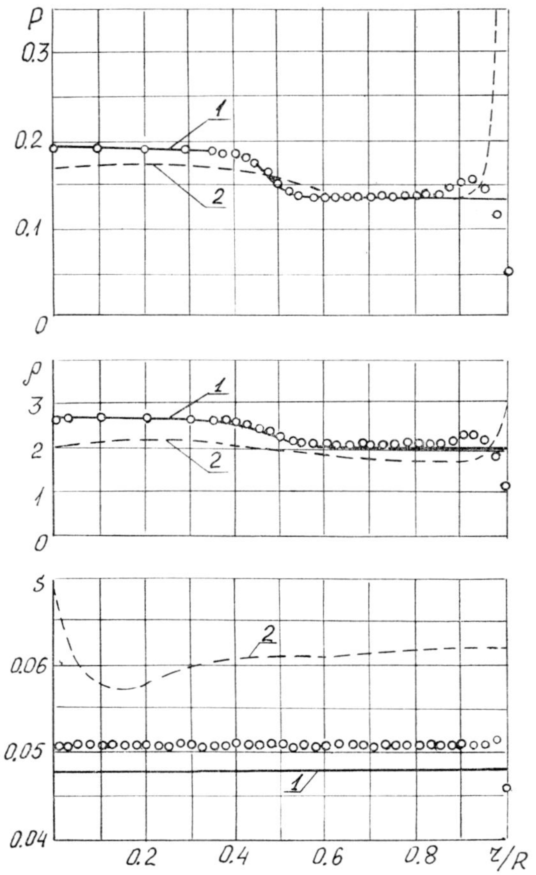

FIG. 32. p, ρ, s—distribution along the wing span.

assumed. In Fig. 32 the curves p, ρ, $s = f_i(r/R)$ along the wing span are given; r is the distance from the point on the wing surface to the plane of symmetry, R the distance from the leading edge to the plane of symmetry. In Fig. 33 the diagrams of dependence of p, ρ, $s = f_i(y/h)$ on the plane of symmetry are shown. y is the distance from the point O_1 to the arbitrary point, h the distance from the projection of the point of intersection of the leading edge with the plane ($x = 1$) on the plane of symmetry, to the point O_1. The diagrams of $f_i(y/h)$, along the straight line EF (Fig. 31), which lies in the region of undisturbed flow behind the weak plane shock wave, are shown in Fig. 34; h is the same as in Fig. 33, y the distance from the point E to the arbitrary point of the straight line EF.

Following from the above-mentioned diagrams, the calculations of the versions $N1$ and $N3$ coincide. There is a certain discrepancy in values of density and s. The diagrams in Fig. 34 show that the results of the calculations for the versions $N1$ and $N3$ in the region of the uniform flow coincide well. As seen from Fig. 33, the position of the shock waves for the versions $N1$ and $N3$ also coincide well, following the coincidence of the points on the diagrams in the region of the large gradients.

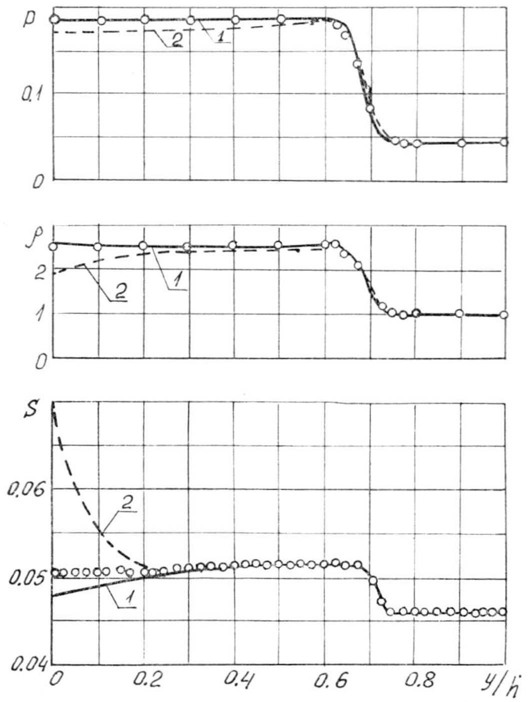

FIG. 33. p, ρ, s—distribution in the plane of symmetry.

The diagrams for the $N2$ version point out that there is a great difference, especially on the wing surface, between the $N2$ and the $N1$ and $N3$ cases. Let us note that although the values of p, ρ, s in the region of influence of the plane of symmetry on the wing surface differ considerably, as seen from the data of $N1$ and $N3$ the results of the calculation on the head wave coincide for all three versions (excluding the small region near the leading edge for the $N2$ version).

It is noticeable that the lateral velocity in the region of point A (excluding the point A itself, in which the flow parameters during the calculations did not change) was supersonic as well as in the $N1$ and $N3$ versions. The values p, ρ shown in Fig. 34 for the region of uniform flow behind the weak shock wave, attached to the leading edge in the $N2$ version, are close to the values, obtained in the $N1$ and $N3$ versions; the values of u, v, w for the $N2$ version differ greatly from the same values in $N1$ and $N3$; besides, there is a qualitative difference. The pointed out difference can be explained by the different values of density, since the pressure is the same for all three versions. In the diagrams of Figs. 33–34 it is seen that near the wing surface ($N2$ version) there is a narrow strip with a great entropy gradient. Thus the results of calculation in the $N2$ version lead to a new solution. But the latter is not the solution with a strong shock wave attached to the leading edge since the pressure and density on the wing surface in the vicinity of the point A falls off rapidly (in one mesh cell). In fact this solution corresponds to the conditions of the following problem.

Suppose that the flow of the leading edge with the detached shock wave takes place and the flow parameters behind it at point A are equal to the corresponding flow parameters

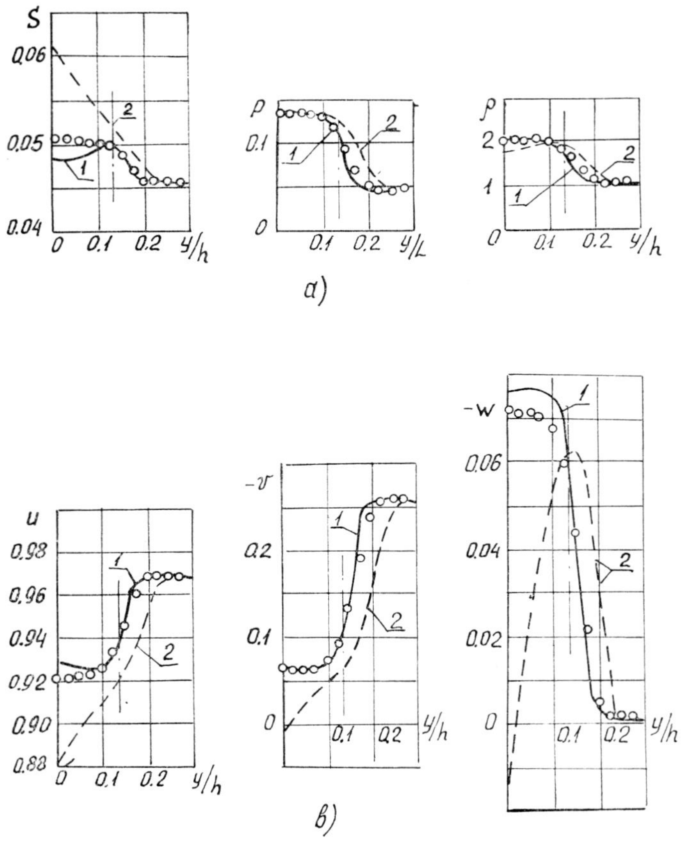

FIG. 34. (a, b) Flow parameters in the region of uniform stream.

behind the strong shock wave attached to the leading edge: then the solution obtained for the $N2$ version will be the solution of the problem of flow about a V-shaped wing with a slightly blunted leading edge.

This analogy helps to understand the peculiarities of the solution of the $N2$ version, viz. the existence of a narrow zone of large entropy gradients (entropy layer) and the coincidence of pressure on some part of the wing surface in all versions. All the above features are characteristic for the flow about bodies with a small bluntness. The experimental investigations of delta wings with pointed and blunted leading edges[59] have shown that when there is a small bluntness, the value of the pressure on the wing surface at rather small distances from the bluntness region becomes equal to the pressure on the wing with a pointed leading edge. The small bluntness is known not to have a marked influence upon the shape of the shock wave being formed near a conical body.

Both these facts are present in the calculations for the $N2$ version. Assuming this analogy with the flow about a V-shaped wing where the leading edge is slightly blunted, the calculations carried out yield some preliminary conclusions:

1. The shape of the bow shock wave does not depend on the blunted shape except for

the small region close to the leading edge and it coincides with the weak bow shock wave being formed for the flow about the same wing with the pointed leading edge.

2. The bluntness of the leading edge does not influence the value of the pressure behind the bow shock wave in the main flow.

3. The extension of the influence region of the plane of symmetry and also the value of the pressure in this region depend on the leading edge bluntness.

Let us note in conclusion, that for the $N1$ version the value s on the wing plane behind the weak plane shock wave attached to the leading edge did not differ practically from the exact solution—the difference did not exceed 0.25%. The values of the pressure and density differ from the exact ones by not more than 0.5%.

As is seen from the diagrams corresponding to the $N3$ version, the values of density on the wing surface are somewhat less (3.5%) than the exact values for the weak shock; the difference of the values of the pressure do not exceed 1%. In Fig. 32 for the $N3$ version the region of the high gradients of p is clearly observed in the vicinity of $(r/R) \simeq 1$ that corresponds to the weak shock wave attached to the leading edge.

The diagrams (Fig. 33) show a good coincidence of the results of calculations for the $N1$ and $N3$ versions at the plane of symmetry with the exception of the neighborhood of point O. The explanation for this is that point O is the Ferri point and the entropy value at this point is not unique. Numerical analysis used here does not allow us to recognize the non-uniqueness, since the singularity is "smeared out" in the calculations, but the reduction of the value s in the vicinity of the point O from the value s behind the shock wave at the plane of symmetry to the value s on the wing surface points to its presence.

The flow regime with a strong shock wave attached to the leading edge. As is known from the experiments the supersonic flow about a wedge, cone and delta wing has always been realized with weak shock waves. For a V-shaped wing a simple exact solution with plane shock wave can correspond to a strong shock wave in a plane orthogonal to the leading edge. In reference 53 such flow was studied by means of experiments. It was shown from experiments that the flow being close to the stream a strong wave takes place indeed. Below we shall consider the flow about a V-shaped wing at $\gamma = 100°$, $\beta = 29°30'$, $M_\infty = 3.95$ and different angles of attack up to detached shock wave ($\alpha = 41°$). Let us denote by index M the intermediate angle of attack corresponding to exact solution with a plane strong shock wave. In Fig. 35 the isobaric fields for $\alpha = 20°$ and $\alpha = 31°$ and the configurations of shock waves for various α are shown. In Fig. 35a the dashed line corresponds to weak shock waves attached to the leading edge. It follows from the above figures that the flow takes place with Mach shock wave configuration at $\alpha < \alpha_M$. With increasing angle of attack the region of flow restricted by central shock BC, plane of symmetry O_1C and inner shock BD is being increased (Fig. 35c). In addition the inner shock approaches the leading edge and the central shock transfers upwards. The calculations were carried out by using for the initial field the flow behind the strong shock as well as behind the weak shock. The results are given as the distribution of pressure coefficient C_p along the span of wing (Fig. 36a,b). As is seen from the diagrams, when the initial field is behind the strong shock a sharp decrease C_p at $\alpha < \alpha_M$ takes place (compare Fig. 32). If $\alpha > \alpha_M$, the value C_p approaches the corresponding value behind the strong shock; some oscillations of the pressure may be explained by errors arising from numerical method. It should be noted that when $\alpha = \alpha_M$ the numerical solution coincides with the exact solution corresponding to the strong shock with high accuracy. If we compare the diagrams $C_p\ (r/R)$ shown in Fig. 36a,b with the preceding results we can conclude that at $\alpha < \alpha_M$ a flow regime with a weak shock, as well as at $\alpha > \alpha_M$

a flow regime with strong shock, takes place. When $\alpha \to \alpha_M$ ($\alpha < \alpha_M$) the inner shock approaches the leading edge and the regime with the weak shock arises. Under further increase of α the inner shock forces out the weak shock attached to the leading edge and there arises a strong shock. The numerical calculation indicates that the shock wave is convex at $\alpha > \alpha_M$ (Fig. 35c). Consequently, the regime with the strong shock having an inflexion near the leading edge or the regime with detached shock takes place.

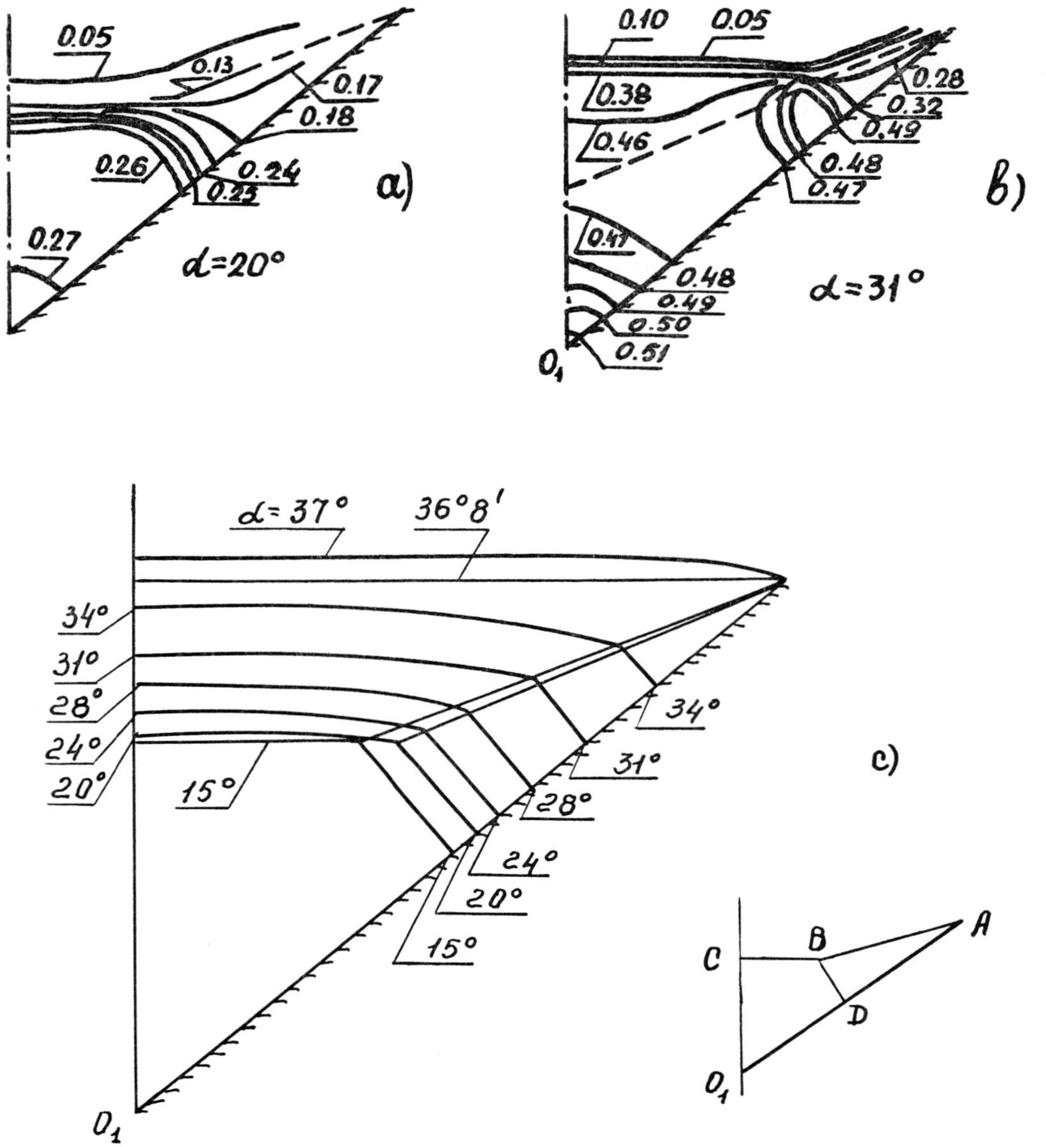

FIG. 35(a, b, c). Isobaric curves and configuration of shock waves.

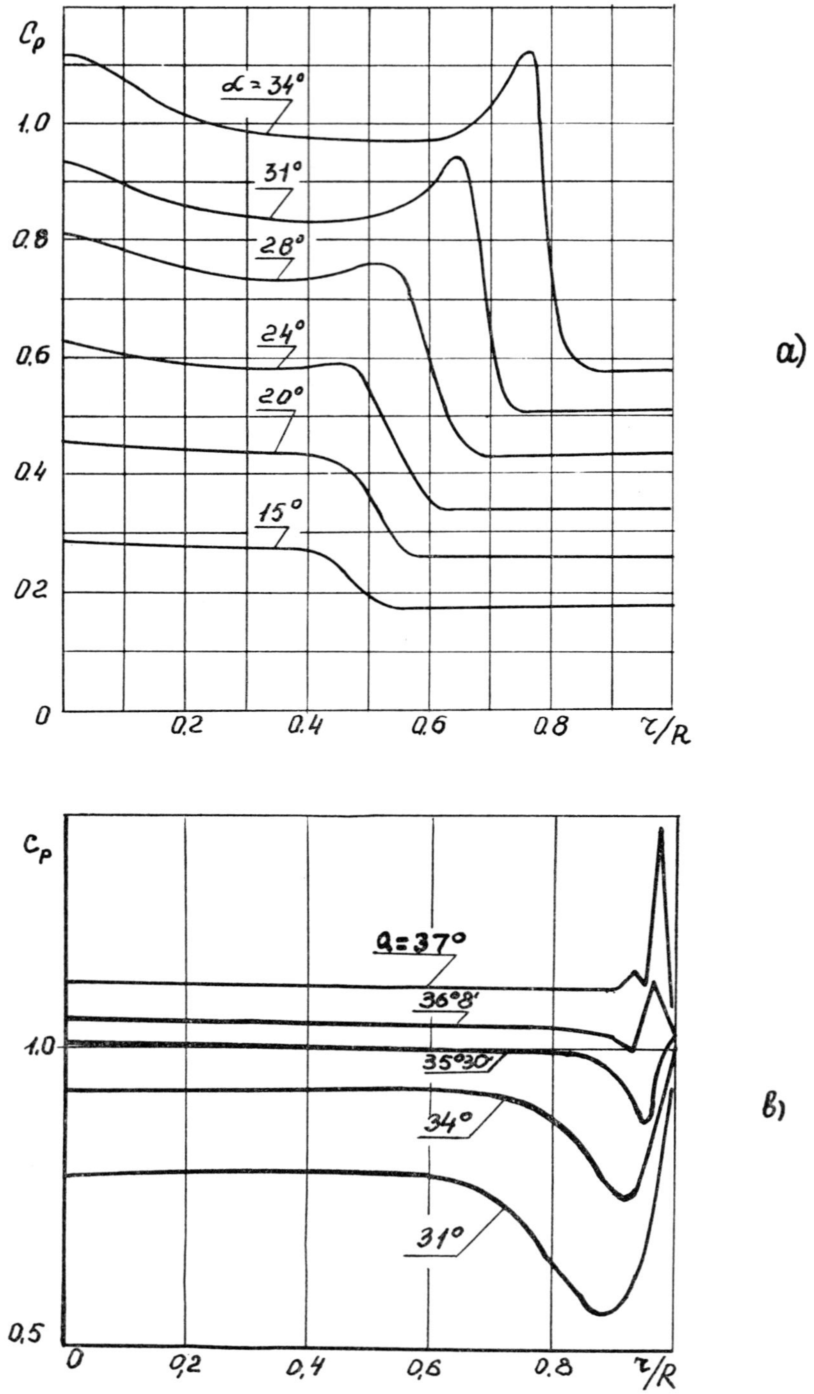

FIG. 36(a, b). Pressure distribution along the wing span.

7. THE CONICAL WING SHAPE WITH MAXIMUM LIFT-TO-DRAG RATIO

In this section the shape of the cross-section conical wing is determined by the condition under which the lift-to-drag ratio is a maximum.

The method is illustrated by the solution of optimum problems with one isoperimetric constraint (the wing volume is given).

The solution of the problem with two isoperimetric constraints (the volume and lift of the wing given) is in progress.[32]

A. *Statement of the Variational Problems*

Let us consider a conical body in hypersonic flow. In the system of coordinates chosen (Fig. 37) the surface of a conical body is written in the form

$$\rho = xr(\varphi)/l \tag{7.1}$$

where (ρ, φ, x) are cylindrical coordinates and V is the velocity of undisturbed flow. Below, the geometrical parameters are related to the length of the body, l being equal to 1. Let us assume that the total drag consists of wave drag and friction drag; for the calculations of the lift the friction is negligible.

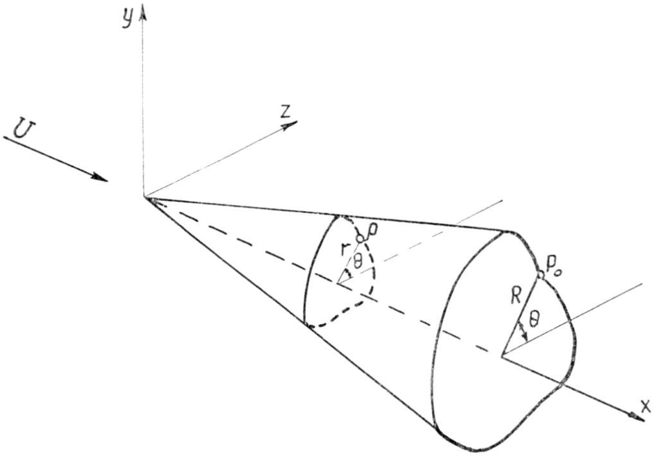

FIG. 37. Coordinate system.

The pressure distribution is Newtonian:

$$C_p = b \cos^2 (\bar{n}, \bar{v}) \tag{7.2}$$

where C_p is the pressure coefficient related to the free stream dynamic pressure, k is constant, $\bar{n}$ is normal to the body surface. The skin-friction coefficient is constant. Hence the coefficients of the aerodynamic forces can be written as

$$SC_x = \int_0^\pi \left\{ \frac{r^4}{1 + r^2 + r'^2/r^2} + ar\sqrt{1 + r'^2/r^2} \right\} d\varphi \tag{7.3}$$

$$SC_y = \int_0^\pi \frac{r^3 (\cos\varphi + r'/r \sin\varphi)}{1 + r^2 + r'^2/r^2} d\varphi \tag{7.4}$$

where

$$a = \frac{C_\tau}{b} \quad (b \cong 2),$$

$$S = \int_0^\pi r^2 \, d\varphi \text{ is the base area.}$$

The symmetry of the problem with respect to the plane xoy allows us to consider the "half-body". If the body is slender ($r^2 \ll 1$), the coefficient of aerodynamic forces simplifies to

$$I_1 = C_x a^{-2/3} \cdot S_1 = \int_0^\pi \left(\frac{y^6}{y'^2 + y^2} + \sqrt{y^2 + y'^2}\right) d\varphi \tag{7.5}$$

$$I_2 = C_y a^{-1/3} \cdot S_1 = \int_0^\pi y^4 \frac{y' \sin\varphi + y\cos\varphi}{y^2 + y'^2} d\varphi \tag{7.6}$$

$$S_1 = Sa^{-2/3} = \int_0^\pi y^2 \, d\varphi \tag{7.7}$$

where $r = a^{1/3} \cdot y$. The integrals (7.5), (7.6) do not contain the skin friction coefficient C_τ, which simplifies the investigation.

The statement of the problem. Let us find the function y for which the relation

$$\gamma = \frac{I_2}{I_1} \tag{7.8}$$

takes the maximum value if there is the constraint (7.7). The lift-to-drag ratio K is connected to the parameter λ by the relation

$$K = \lambda a^{-1/3}. \tag{7.9}$$

B. *The Algorithm of Numerical Solution*

The analytical investigation of this problem cannot be accomplished because Euler's equation is of second order and essentially non-linear. The direct method, called the "local variations method", is used for the solution of the problem.[56–58] Let us describe in brief its contents. All admitted variations must satisfy the isoperimetric constraint $S_1 = \text{const.}$ Beside this general variations are used in the calculations (Fig. 38a). The use of the particular variations (Fig. 38b) makes the convergence worse. The interval of integration is divided into equal cells τ:

$$\varphi_k = k\tau; \quad \tau = \frac{\pi}{N}; \quad k = 0, 1, \ldots, N$$

and the integrals (7.8) are approximated as

$$I_{i,k}(y_{k-1}, y_k) = \tau f_i \left(\frac{\varphi_k + \varphi_{k-1}}{2}, \frac{y_k + y_{k-1}}{2}, \frac{y_k - y_{k-1}}{\tau} \right) \tag{7.10}$$

$$I_i = \sum_{k=1}^{N} I_{i,k}, \quad i = 1, 2 \tag{7.11}$$

$$f_1 = \frac{y^6}{y'^2 + y^2} + \sqrt{y'^2 + y^2} \tag{7.12}$$

$$f_2 = y^4 \frac{y' \sin \varphi + y \cos \varphi}{y'^2 + y^2}. \tag{7.13}$$

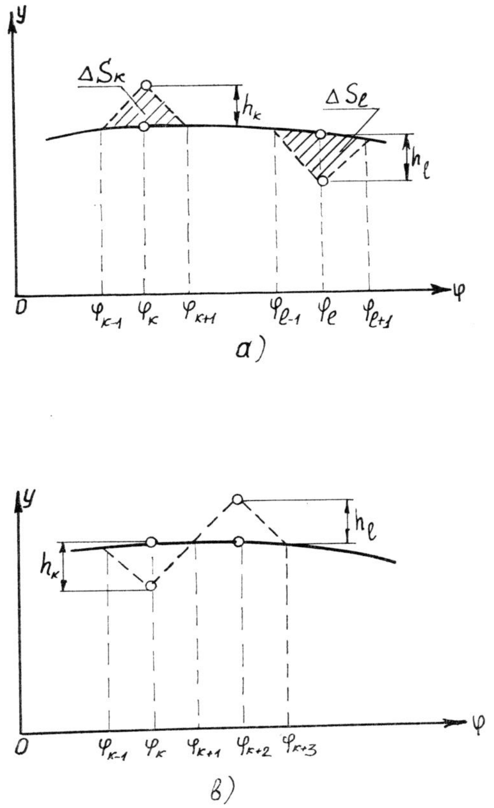

FIG. 38. (a, b) Calculated scheme.

The initial curve $y = y(\varphi)$ must satisfy the isoperimetric constraint. Suppose that the nth approximation is known. We shall show how to determine the $(n+1)$th approximation. Let us suppose that the first k $(0 \leq k \leq N-1)$ numbers $(n+1)$ approximations are found;

then we shall calculate the values

$$
\lambda^{\pm}_{k,l} = \frac{\displaystyle\sum_{n=1}^{k-1} I_{2,n} + I_2\,(y_{k-1},\, y_k \pm h_k) + I_2\,(y_k \pm h_k,\, y_{k+1}) + \sum_{n=k+2}^{l=1} I_{2,n}}{\displaystyle\sum_{n=1}^{k-1} I_{1,n} + I_1\,(y_{k-1},\, y_k \pm h_k) + I_1\,(y_k \pm h_k,\, y_{k+1}) + \sum_{n=k+2}^{l-1} I_{1,n}}
$$

$$
\frac{+\, I_2\,(y_{l-1},\, y_l + h_l) + I_2\,(y_l + h_l,\, y_{l+1}) + \displaystyle\sum_{n=l+2}^{N} I_{2,n}}{+\, I_1\,(y_{l-1},\, y_l + h_l) + I_1\,(y_l + h_l,\, y_{l+1}) + \displaystyle\sum_{n=l+2}^{N} I_{1,n}} \tag{7.14}
$$

$$
\lambda_{k,l} = \frac{\displaystyle\sum_{n=1}^{k-1} I_{2,n} + I_2\,(y_{k-1},\, y_k) + I_2\,(y_k,\, y_{k+1}) + \sum_{n=k+2}^{l-1} I_{2,n} + I_2\,(y_{l-1},\, y_l)}{\displaystyle\sum_{n=1}^{k-1} I_{1,n} + I_1\,(y_{k-1},\, y_k) + I_1\,(y_k,\, y_{k+1}) + \sum_{n=k+2}^{l-1} I_{1,n} + I_1\,(y_{l-1},\, y_l)}
$$

$$
\frac{+\, I_2\,(y_l,\, y_{l+1}) + \displaystyle\sum_{n=k+2}^{N} I_{2,n}}{+\, I_1\,(y_l,\, y_{l+1}) + \displaystyle\sum_{n=k+2}^{N} I_{1,n}}
$$

where the values $\lambda_{k,l}$ are known. Sign h_l is determined by sign h_k which is given before. The condition $\Delta S_k + \Delta S_l = 0$ determines the relation between h_k and h_l in the following way:

$$
h_l = -\tfrac{1}{2}\,(a - \sqrt{a^2 - 4b}) \quad \text{for inner points} \tag{7.15}
$$

$$
h_l = -\,(a - \sqrt{a^2 - 2b}) \quad \text{for boundary points} \tag{7.16}
$$

where $b = 2\Delta S_k$.

$$
a = y_{l-1} + 2y_l + y_{l+1} \quad \text{for inner points}
$$

$$
a = y_l + y_{l+1} \quad\quad\quad\quad \text{for boundary points.}
$$

The $(n+1)$ approximation is determined by the formulae:

$$
y = (y_k,\, y_l) =
\begin{cases}
(y_k + h_k,\, y_l + h_l) & \text{at } \lambda^{+}_{k,l} > \lambda_{k,l} \\[4pt]
(y_k - h_k,\, y_l + h_l) & \text{at } \lambda^{-}_{k,l} > \lambda_{k,l} \quad \lambda^{+}_{k,l} \leqslant \lambda_{k,l} \\[4pt]
(y_k,\, y_l) & \text{at } \lambda^{+}_{k,l} \leqslant \lambda_{k,l} \quad \lambda^{-}_{k,l} \leqslant \lambda_{k,l}
\end{cases} \tag{7.17}
$$

The iteration is considered to be complete when k is passing through O to N.

Having accomplished one iteration one can proceed to the following iteration and so on. The iterations at a given h_k value stop as soon as the difference is

$$
|\,\lambda^{(n+1)} - \lambda^{(n)}\,| < \delta. \tag{7.18}
$$

Using the expression for λ and supposing that the derivates $y(\varphi);\, f_y;\, f_{y',\,\varphi}$ are continuous, we can show[57] the equivalence of (7.18) to the following unequality:

$$
\frac{\partial \phi}{\partial y} - \frac{d}{d\varphi}\left(\frac{\partial \phi}{\partial y'}\right) < I_1\left\{\frac{\delta}{h\tau} + 0(\tau) + 0(h) + 0\!\left(\frac{h}{\tau^2}\right)\right\} \tag{7.19}
$$

where

$$\phi = I_2 - \lambda I_1. \tag{7.20}$$

The left part of (7.19) is the left part of Euler's equation at an arbitrary value of S_1. In the calculations S_1 is constant and one can expect that the method converges in this case too.

C. *The Results of the Calculations and Discussion*

The calculations were done on the computer BESM-4M. The step along φ was equal to $\tau = 4°$, the initial value of the variation radius $h_k = 0.01$, $0 \leq k \leq N$. For checking the convergence to the unique solution the initial contours with the same areas shown in Fig. 39 were chosen. We shall denote the angle corresponding to the maximum radius (y_{max}) φ^*. It was discovered that the angle remains as it was in the original approximation, though the number of iteration was large and h_k decreased by a factor of eight (except at the triangle contour, Fig. 39).

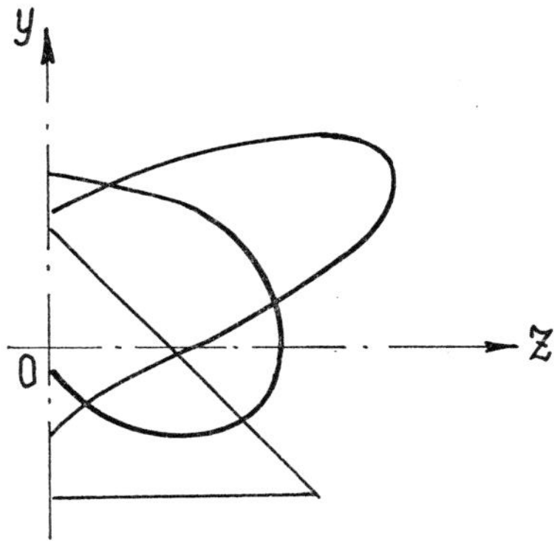

FIG. 39. Initial contours.

To determine the optimum angle φ^* there were several contours with equal area and similar in shapes to the optimum ones with different angles φ^* chosen. The calculations have shown that the lift-to-drag ratio has a weak dependence on φ^* in the vicinity of the optimum angle (see Table 3, $S_1 = \pi$).

TABLE 3

φ^*	72°	64°	60°	56°	48°	36°
λ	0.3904	0.3977	0.4018	0.4015	0.3963	0.3739

An analogous behaviour of the lift-to-drag ratio is observed for other values of S_1. It should be noted that with the increase of the iteration number the value K changes very little (of the order of four significant figures). Therefore the obtained dependence $\lambda(S_1)$ is of a high accuracy. The shape of the optimum contour is found with less accuracy. The dependence of the optimum lift-to-drag ratio K at $C_\tau = 10^{-3}$ from the volume parameter

$\tau = v/s^{3/2}$, where V is the volume body, s the plan form area, is given in Fig. 40. In the diagram the maximum values K for a V-shaped wing are marked with circles. They practically coincide with the values K for optimum bodies. Let us compare the above obtained data with the values of the lift-to-drag ratio of some investigated bodies.[30] The semicone with a base of semicircled shape at $l = \text{const}$, $C_\tau = 10^{-3}$, $v/l^3 = 0.00727$ has $K = 3.6$. For an elliptical base the maximum increment is 2%. For a base in the shape of sinusoidal line the increment is 12.5%. For the triangular form base, with the windward lateral sides, the lift-to-drag ratio approaches the value of optimum bodies.

If we consider the triangular windward base cross-section then its lift-to-drag ratio practically coincides with that of the optimum body (Fig. 41).

The cross-section optimum shapes are shown in Fig. 42 for various values of S_1. It should be noticed that the body is bounded on the top by surfaces representing two planes (with the accepted accuracy) set along the flow.

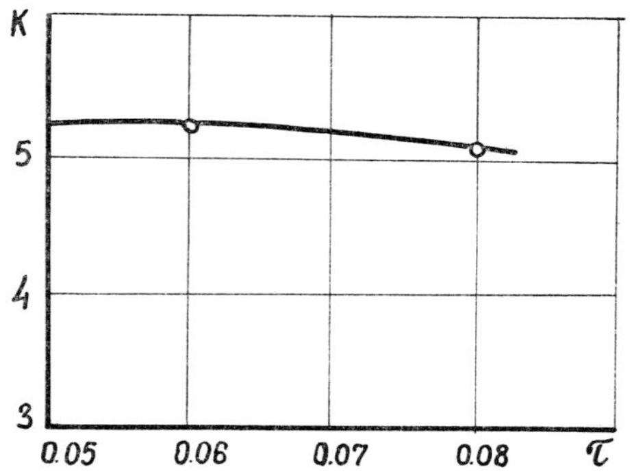

FIG. 40. Lift-to-drag ratio for optimum body.

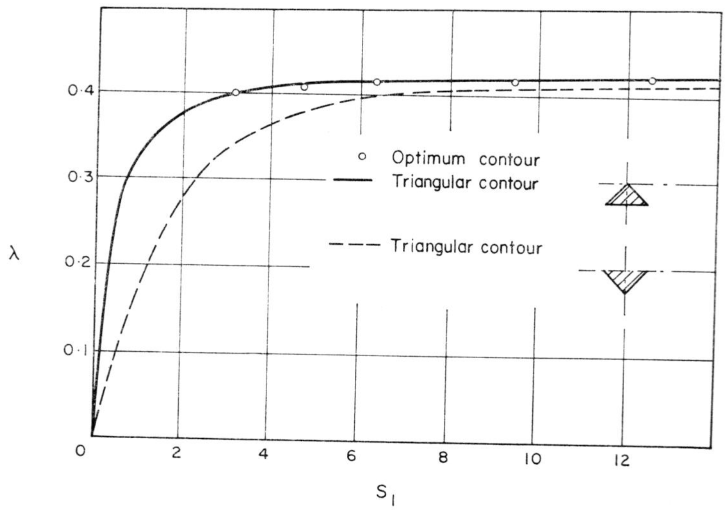

FIG. 41. Lift-to-drag ratio coefficient versus base area.

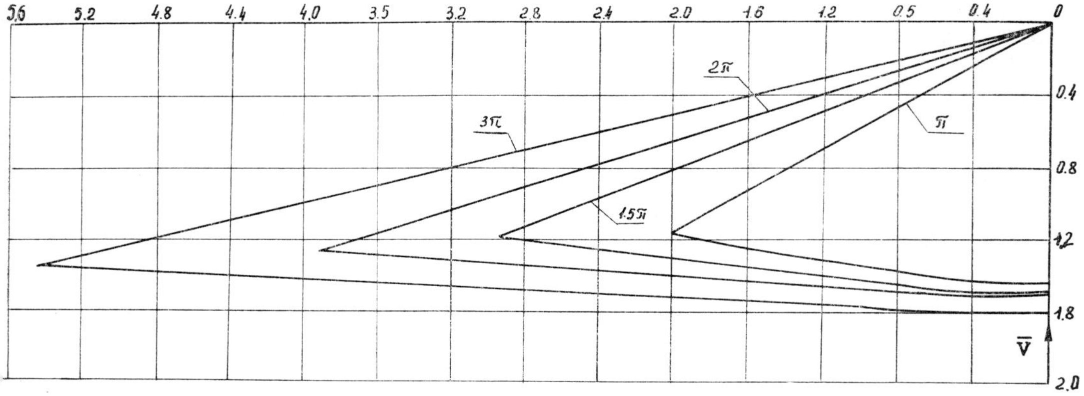

FIG. 42. Cross-sections of optimum bodies.

The bottom surface of the wing being the windward side has a slight convex to the incoming flow. The obtained numerical results confirmed with high accuracy the hypothesis that the top surfaces of the optimum wing are planes set along the flow. On the whole the shape of the optimum contour is close to the triangular shape with a windward base. So it is clear why the lift-to-drag ratio of the wing with the optimum triangular cross-section almost coincides with the lift-to-drag ratio of the optimum conical wing. For several values of the parameters $S = \pi \times 10^{-2}$; $2\pi \times 10^{-2}$; $4\pi \times 10^{-2}$ at $a = 10^{-3}$ the calculation carried out using formulae (7.3)–(7.4) did not assume the body to be thin. In this case the results obtained were identical with the ones above. The shape of the body and the lift-to-drag ratio in particular for the same initial parameters did not differ from the preceding ones. In conclusion we shall evaluate the convergence of the solution. The numerical checking of the fulfilment of Euler's equation and boundary conditions was carried out for the case considered.

The calculated errors proved to be $\sim 7\%$ and coincided with the accuracy of the method having a mesh cell of the order of $\tau = 0.07$.

REFERENCES

1. A. L. GONOR, Hypersonic gas flow around conical bodies, *Izv. Akad. Nauk SSSR, Otd. Nauk, Mekh. i Mashinostr.*, no. 1, 1959. (M. D. Friedman Transl. no. 6-163.)
2. A. F. MESSITER, Lift of slender delta wings according to Newtonian theory, *AIAA J.*, vol 1, no. 4, pp. 794–802, 1963.
3. W. D. HAYES and R. E. PROBSTEIN, *Hypersonic Flow Theory*, 2nd edition, vol. 1. Academic Press, N.Y., 1966.
4. D. A. BABAEV, Numerical solution of the problem of supersonic flow past the lower surface of a delta wing (translated from Russian), *AIAA J.*, vol. 1, no. 9, pp. 2224–2231, 1963.
5. D. A. BABAEV, Flow about a triangular wing for large values of *M*, *Zh. Vych Mat. i Mat. Fiz.*, vol. 3, no. 2, pp. 528–532, 1963.
6. G. P. VOSKRENSENSKII, Numerical solution of the problem of a supersonic gas flow past an arbitrary surface of a delta wing in the compression region, *Izv. Akad. Nauk SSSR, Mekh. Zhidk. i Gaza*, no. 4, pp. 134–142, 1968.
7. P. KUTLER and H. LOMAX, A systematic development of the supersonic flow fields over and behind wings and wing-body configurations using a shock-capturing finite-difference approach, AIAA Paper, no. 71-99, 1971.
8. K. HIDA, Thickness effect on the force of slender delta wings in hypersonic flow, *AIAA J.*, vol. 3, no. 3, pp. 3427–3433, 1965.

9. A. I. GOLUBINSKII, Hypersonic flow past some triangular wings with attached shock wave, *Izv. Akad. Nauk SSSR, Mekh. Zhidk. i Gaza*, no. 5, 1966.
10. D. A. MELNIKOV, Supersonic flow past a triangular wing, *Izv. Akad. Nauk SSSR, Mekh. i Mashinostr.*, no. 6, 1962.
11. L. C. SQUIRE, Calculated pressure distributions and shock shapes on conical wings with attached shock waves, *Aeronaut. Quart.*, vol. xix, pp. 31–50, 1968.
12. B. M. BULAKH, Some problems of the non-linear conical flows. *Prikl. Mat. i Mekh.*, vol. 25, no. 2, 1961.
13. H. KENNET, The inviscid hypersonic flow on the windward side of a delta wing. JAS Paper, no. 63-55, 1963.
14. A. P. BAZZHIN, On the calculation flow past the lower surface of a triangular wing at high angle of attack, *Inzhen. Zh., Moscow*, vol. 4, no. 2, 1964.
15. A. P. BAZZHIN, On the calculation flow past a delta wing at a high angle of attack, *Izv. Akad. Nauk SSSR, Mekh. Zhidk. i Gaza*, no. 5, 1966.
16. G. G. CHERNYI, Wings in hypersonic flow, *Prikl. Mat. i Mekh.*, vol. 29, no. 4, 1965.
17. E. A. AKINRELERE, The calculation of invisced hypersonic flow past the lower surface of a delta wing, *J. Fluid Mech.*, vol. 44, part 1, 1970.
18. A. L. GONOR, Flow past delta wings in a hypersonic stream, *Prikl. Mat. i Mekh.*, vol. 34, no. 3, 1970.
19. A. L. GONOR, V. I. LAPYGIN and N. A. OSTAPENKO, The conical wing in hypersonic flow, Paper presented at the Second International Conference on Numerical Methods in Fluid Dynamics, Berkeley, U.S.A., 1970 (*Lecture Notes in Physics*, vol. 8, 1971).
20. V. Z. LAPYGIN, The computation of the flow past V-shaped wings by the time-dependent methods, *Izv. Akad. Nauk SSSR, Mekh. Zhidk. i Gaza*, no. 3, 1971.
21. T. STRAND, Wings and bodies of revolution of minimum drag in Newtonian flow, Convair Report N ZA-303, 1958.
22. D. G. HULL and A. MIELE, Three-dimensional shapes of minimum total drag in Newtonian flow, *J. Astron. Sci.*, vol. 12, no. 2, 1965.
23. G. I. MAIKAPAR, The wing with a maximum lift-to-drag ratio at hypersonic speeds, *Prikl. Mat. i Mekh.*, vol. 30, no. 1, 1966.
24. A. MIELE, Maximum lift-to-drag ratio of a slender wing at hypersonic speeds, *Z. Flugwiss.*, vol. 15, no. 7, 1967.
25. A. L. GONOR and A. I. SHVETZ, Supersonic flow past V-shaped wings, Report no. 613, Institute of Mechanics, Moscow State University, 1966.
26. A. H. LUSTY JR. and A. MIELE, Bodies of maximum lift-to-drag ratio in hypersonic flow, *AIAA J.*, vol. 4, no. 12, 1966.
27. A. MIELE, Lift-to-drag ratio of a slender wing at hypersonic speeds, *J. Astron. Sci.*, vol. 13, no. 6, 1966.
28. A. L. GONOR, N. A. OSTAPENKO and V. I. LAPYGIN, The flow past a conical wing and optimal form at hypersonic speeds, Report No. 981, Institute of Mechanics, Moscow State University, 1969.
29. V. I. LAPYGIN, Conical bodies of maximum lift-to-drag ratio in hypersonic flow, Transactions No. II, Institute of Mechanics, Moscow State University, 1971.
30. A. MIELE, I. G. HEIDEMAN and R. E. PRITCHARD, Conical bodies of given length and volume having maximum lift-to-drag ratio at hypersonic speeds, Part I. *J. Astron. Sci.*, vol. xv, no. 2, 1968.
31. HO-YI-HUANG, Conical bodies of given length and volume having maximum lift-to-drag ratio at hypersonic speeds, Part 2, *J. Astron. Sci.*, vol. xv, no. 3, 1968.
32. K. FERRARI, Aerodynamic problems of re-entry, VIIth Congress of the International Council of the Aeronautical Sciences, Meccanica, March 1971.
33. V. D. PERMINOV, Wings with optimum characteristics in hypersonic flow, *Izv. Akad. Nauk SSSR, Mekh. Zhidk. i Gaza*, no. 6, 1969.
34. L. F. CRABTREE, and D. A. TREADGOLD, Experiments on hypersonic lifting bodies, ICAS Paper, no. 66-24, 1966.
35. R. RANDALL, P. BELL and I. BURK, Pressure distribution tests of several sharp leading wings, bodies and body-wing combinations at Mach 5 and 8, AEDS/TN, 173, 1966.
36. H. K. CHENG, Hypersonic shock layer theory of a yawed-cone and other three-dimensional pointed bodies, WADC-TN, 59–335, 1959.
37. R. E. MELNIK and R. A. SCHEUING, Shock layer structure and entropy layers in hypersonic conical flows, *Progress Astronaut. and Rocketry*, vol. 7, Academic Press, N.Y., 1962.
38. YA. G. SAPUNKOV, Hypersonic flow past a circular cone at an angle of attack, *Prikl. Mat. i Mekh.*, vol. 27, no. 1, 1963.
39. YA. G. SAPUNKOV, Hypersonic flow past a circular cone at an angle of attack, *Prikl. Mat. i Mekh.*, vol. 27, no. 5, 1963.
40. YA. G. SAPUNKOV, Hypersonic flow past conical bodies, *Izv. Akad. Nauk SSSR, Mekh. Zhidk. i Gaza*, no. 1, 1966.
41. A. G. MUNSON, The vortical layer on an inclined cone, *J. Fluid Mech.*, pp. 625–643, 1964.

42. R. F. Melnik, Newtonian entropy layer in the vicinity of a conical symmetry plane, *AIAA J.*, vol. 3, no. 3, 1965.
43. H. S. Tsien, The Poincare–Lighthill–Kuo method, *Advances in Appl. Mech.*, vol. 4, 1956.
44. D. Küchemann, Hypersonic aircraft and their aerodynamic problems, *Progress in Aeron. Sci.*, vol. 6, Pergamon, 1965.
45. A. L. Gonor and A. I. Shvetz, Investigation of pressure distribution on certain star-shaped bodies for Mach number $M = 4$, *Prikl. Mekh. i Tekh. Fiz.*, no. 6, 1965.
46. A. L. Gonor and A. I. Shvetz, Investigation of the shocks at supersonic flow past a star-shaped body, *Izv. Akad. Nauk SSSR, Mekh. Zhidk. i Gaza*, no. 3, 1966.
47. A. L. Gonor and A. I. Shvetz, Flow past V-shaped wings in a supersonic stream for Mach number $M = 3.9$, *Izv. Akad. Nauk SSSR, Mekh. Zhidk. i Gaza*, no. 6, 1967.
48. G. I. Maikapar, On the wave drag of axisymmetric bodies at supersonic speeds, *Prikl. Mat. i Mekh.*, vol. 23, no. 2, 1959.
49. T. Nonweiler, Aerodynamic problems of manned space vehicles, *J. Royal Aeronaut. Soc.*, vol. 63, no. 5850, 1959.
50. A. L. Gonor, Exact solution of the problem of supersonic flow of a gas past some three-dimensional bodies, *Prikl. Mat. i Mekh.*, vol. 28, no. 5, 1964.
51. A. L. Gonor, Certain three-dimensional flows with Mach-type shock wave interactions, *Izv. Akad. Nauk SSSR, Mekh. Zhidk. i Gaza*, no. 6, 1966.
52. V. V. Rusanov, Calculation of the interaction of shock waves with an obstacle, *Zh. Vych. Mat. i Mat. Fiz.*, vol. 2, 1961.
53. Yu. I. Zaitsev and V. V. Keldysh, Particular cases of flow near supersonic edges and lines of intersection of shock waves, Lecture Notes, *TsAGI*, vol. I, no. 1, 1970.
54. Yu. A. Panov, Interaction of an incident three-dimensional shock wave with a turbulent boundary layer, *Izv. Akad. Nauk SSSR, Mekh. Zhidk. i Gaza*, no. 3, 1968.
55. V. V. Keldysh, Intersection of the two plane shocks in three-dimensional space, *Prikl. Mat. i Mekh.*, vol. 30, no. 1, 1966.
56. F. L. Chernousko, The method of local variations for the numerical solution of the optimum problems, *Zh. Vych. Mat. i Mat. Fiz.*, vol. 5, no. 4, 1965.
57. I. A. Krylov and F. L. Chernousko, The solution of the optimum problems of control by the local variations method, *Zh. Vych. Mat. i Mat. Fiz.*, vol. 6, no. 2, 1966.
58. N. V. Banichuk, V. M. Petrov and F. L. Chernousko, The method of local variations for the optimum problems with nonadditive functionals, *Zh. Vych. Mat. i Mat. Fiz.*, vol. 9, no. 3, 1969.
59. V. A. Bashkin, Exploratory wind-tunnel investigation of delta wings at $M = 5$ and angles of attack 0–70°, *Izv. Akad. Nauk SSSR, Mekh. Zhidk i Gaza*, no. 3, 1967.

4

Theoretical Prediction of Base Pressure for Steady Base Flow

M. TANNER
DFVLR-AVA, Göttingen

SUMMARY

The present paper gives a review of various theories developed for the prediction of base pressure in two-dimensional steady base flow. Theories based on the Chapman–Korst flow model are compared with one another and with experimental results. The new theory of Tanner, based on the concept of mass outflow from the dead-air region, is explained. The various achievements in this field will be pointed out and so also those problems which still should be investigated.

NOTATION

x	coordinate in flow direction
y	coordinate perpendicular to flow direction
d	body thickness at flow separation
p	static pressure
p_1	static pressure in undisturbed flow
p_B	base pressure
p_R	static pressure at reattachment point
c_p	pressure coefficient $= (p - p_1)/q_1$
c_{pB}	base pressure coefficient $= (p_B - p_1)/q_1$
c_{pR}	pressure coefficient at reattachment point $= (p_R - p_1)/q_1$
U	velocity
U_1	velocity in the undisturbed flow
U_2	velocity at the outer edge of the mixing region, corresponding to the base pressure
U_D	velocity on the dividing streamline
U_m	velocity in the plane of symmetry of the wake
Φ	dimensionless velocity $= U/U_2$
Φ_D	dimensionless velocity on the dividing streamline $= U_D/U_2$
ρ	density of air
ρ_1	density of air in the undisturbed flow

ρ_2	density of air at the outer edge of the mixing region
q_1	dynamic pressure in the undisturbed flow $= \frac{1}{2}\rho_1 U_1^2$
δ	boundary layer thickness
δ_1	boundary layer displacement thickness
δ_2	boundary layer momentum thickness
ϑ	boundary layer momentum thickness after expansion at the separation corner
δ_T	thickness of thermal layer
y_r	dimension in the dead-air region (see Fig. 27)
y_t	dimension in the dead-air region (see Fig. 27)
y_g	dimension in the dead-air region (see Fig. 27)
h	shear layer thickness at the end of the outflow region (see Fig. 27)
h_a	dimension in the dead-air region (see Fig. 27)
H_1	wake thickness downstream of the reattachment point, Fig. 36
M	Mach number
M_1	Mach number in the undisturbed flow
M_2	Mach number at the outer edge of the mixing region
M_4	Mach number downstream of reattachment point
M_R	Mach number corresponding to the pressure p_R at reattachment
ζ	dimensionless y-coordinate
ψ	dimensionless x-coordinate
ξ	transformed dimensionless x-coordinate
η	dimensionless coordinate $= \sigma \frac{y}{x}$
T	absolute temperature
T_2	absolute temperature at the outer edge of the mixing region
T_0	total temperature
T_{02}	total temperature at the outer edge of the mixing region
D	drag
c_D	drag coefficient
m_a	maximum rate of mass outflow from the dead-air region, eq. (55)
c_{Qa}	dimensionless mass outflow per unit span $= m_a/(\rho_1 V_1 d)$
γ	ratio of specific heats
Θ	Prandtl–Meyer expansion angle
μ	coefficient of viscosity
ϵ	turbulent viscosity
σ	similarity parameter of jet mixing
C	Crocco number
c_Q	bleed number
Pr_t	turbulent Prandtl number
r	recovery factor
H	boundary layer shape parameter $= \delta_1/\delta_2$
H^*	transformed shape parameter
φ	wedge angle

b_3	coefficient in eq. (78)
N	Nash's reattachment parameter
R	Roberts' reattachment parameter

1. INTRODUCTION AND GENERAL REMARKS

The flows over bluff bodies have for a long time been an interesting problem not only for experimenters but also for theorists. While the classical potential-flow theory gives zero drag for every body, regardless of its shape, experiments show that especially bluff bodies have a great drag, which cannot be explained by effects of friction forces on its surface. It was recognized that an essential feature of the flow past bluff bodies is flow separation which causes a great change in the pressure distribution on the body. Behind the separation points there exists a remarkably low pressure as compared with the pressure in the undisturbed stream, and this low pressure is the origin of a great part of the pressure drag.

The first theory for predicting pressure drag of blunt bodies was developed by Kirchhoff.[1] In this theory the free streamlines which originate at the separation points extend to infinity downstream. Between the free streamlines behind the body there is a quiescent fluid the pressure of which is equal to the static pressure in the undisturbed flow. The base pressure is thus $p_B = p_1$. The theoretical drag coefficient for the flow past a flat plate perpendicular to the air-stream is $c_D = 0.88$ compared with the much higher value $c_D = 2$ as given by measurement (see e.g. refs. 2 and 3). The disagreement between theory and experiment depends on the fact that the base pressure actually is not equal to the undisturbed static pressure but in general remarkably lower.

Various improvements of the theory by Kirchhoff made it possible to account for base pressures which are lower than the pressure in the undisturbed flow. Of such improved theories in this connection may be mentioned those of Riabouchinsky,[4] Eppler,[5] and Roshko.[6] A further theory which does not use free streamlines but replaces the dead-air region by a finite body was developed by the present author.[7] All these theories give a pressure distribution on the front part of the body which is in good agreement with measurements, if the experimental base pressure is used in the calculations.

Therefore, at incompressible speeds there exist many theories which enable us to predict the pressure drag of a blunt body in dependence of its base pressure, in good agreement with experiments. The base pressure, however, cannot be determined by these theories. In spite of this shortcoming, the afore-mentioned theories, which give a connection between the base pressure and the pressure drag, are very important at incompressible flow, and it would be desirable to extend them for compressible subsonic flow too. A detailed treatment of these theories may be found, e.g. in refs. 4 and 8.

In order to obtain a theory for predicting the base pressure, the actual viscous flow in the separated region behind the body must be considered. This has been done in the theories which will be discussed in the following sections.

In this review we will confine ourselves to steady two-dimensional base flows, i.e. to two-dimensional base flows without periodic vortex shedding. Therefore, in general there must be a splitter plate in the near wake of the body, the length of which should be at least 4 to 5 times the body thickness at separation.[13] For such flows careful surveys have failed to reveal any evidence of periodicity.[28]

At Mach numbers above 1, the periodic vortex shedding is very much reduced, i.e. the strength of the vortex sheet is then much smaller than at subsonic velocities. Then the base pressures with and without splitter plate are nearly equal as shown by some measurements.[20,28] Therefore it can often be assumed that at supersonic speeds the base flow is always steady, an approximation which is likely to be very good at Mach numbers above 1.5.[20]

2. THE THEORY OF CHAPMAN

We will start with the description of the theory of Chapman,[9,10] and will do it in detail, because this theory, together with that of Korst, forms a basis for most theoretical works done later in this field.

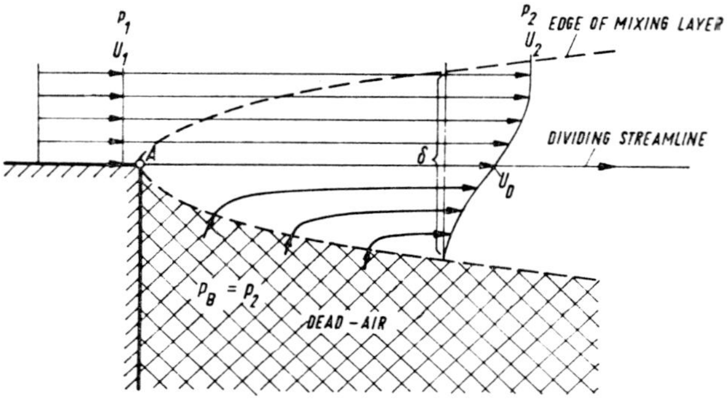

FIG. 1. The laminar mixing region. After Chapman *et al.*, ref. 10. A = separation point.

Before developing the basic idea for calculating dead-air pressure, Chapman outlines the results of the laminar mixing-layer theory which forms the basis for these calculations. Typical streamlines in the viscous mixing region and a representative velocity profile are shown in Fig. 1. A uniform stream of velocity U_1, Mach number M_1, and pressure p_1 approaches the corner A. At the corner it comes in contact with a dead-air region having large dimensions compared to the thickness δ of the mixing-layer developing downstream of point A. Because, in general, the static pressure at the outer edge of the mixing region differs from that upstream of separation, we will denote the pressure (and velocity) with different symbols in both cases. For an "ideal" free jet boundary, with very small curvature of streamlines at separation, the pressures p_2 and p_1 are, however, almost identical.

The thickness of the laminar mixing-layer grows parabolically with distance from the origin of mixing. The velocity profiles at different streamwise stations are similar; hence, the velocity ratio $\Phi_D = U_D/U_2$ along the dividing streamline (see Fig. 1) does not change with Reynolds number or with distance from separation. This velocity ratio changes only slightly with variation in Mach number and in temperature–viscosity relationship. Some computed values of Φ_D from ref. 11 are tabulated (Table 1) for two different temperature–viscosity relationships.

It will be remarked that the tabulated values of Φ_D involve no empirical constants and are

TABLE 1. COMPUTED VALUES OF $\Phi_D = U_D/U_2$ (ref. 11).

M_2	$\mu \sim T$	$\mu \sim T^{0.76}$
0	0.587	0.587
1	0.587	0.588
2	0.587	0.591
3	0.587	0.593
5	0.587	0.597

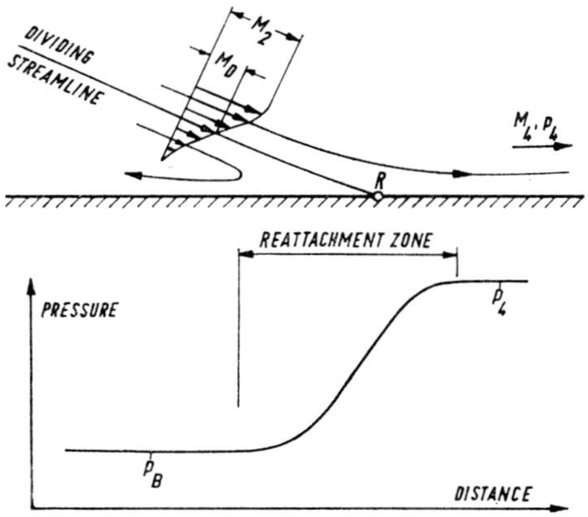

FIG. 2. Streamlines and pressure distribution in the reattachment zone, schematically. R = reattachment point.

exact within the framework of the boundary layer equations. In the calculation of "dead-air" pressure (= base pressure), the essential mechanism is considered to be a balance between mass flow scavenged from the dead-air region by the mixing layer and mass flow reversed back into the dead-air region by the pressure rise through the reattachment zone. For steady flow without bleeding into the dead-air region the dividing streamline at separation as calculated from mixing-layer theory must also be a dividing streamline at reattachment. In Fig. 2 one can see the reattachment zone with corresponding pressure distribution.

The base pressure is determined by requiring that the total pressure along the dividing streamline

$$p_{DO} = p_B \left[1 + \frac{\gamma - 1}{2} M_D^2\right]^{\gamma/(\gamma-1)} \tag{1}$$

must be equal to the terminal static pressure p_4 downstream of reattachment (Fig. 1). Thus the flow is divided into two regions: a viscous layer where the pressure is assumed to be constant, and a reattachment zone where the compression is assumed to be such that not much total pressure is lost along the dividing streamline. For isentropic compression

$$p_B = \frac{p_4}{\left[1 + \frac{\gamma - 1}{2} M_D^2\right]^{\gamma(\gamma-1)}}. \tag{2}$$

If the dead-air temperature T_d is assumed as equal to the total temperature of the outer stream

$$T_d = T_{02} = T_2 \left[1 + \frac{\gamma - 1}{2} M_2^2\right] \tag{3}$$

and if the Prandtl number is assumed to be unity the Busemann isoenergetic integral for a perfect gas gives for the Mach number

$$M_D^2 = \frac{\Phi_D^2 M_2^2}{1 + \frac{\gamma - 1}{2} M_2^2 (1 - \Phi_D^2)}. \tag{4}$$

By substituting in eq. (2) with $\Phi_D = U_D/U_2$ one gets:

$$\frac{p_B}{p_4} = \left[\frac{1 + (1 - \Phi_D^2)\frac{\gamma - 1}{2} M_2^2}{1 + \frac{\gamma - 1}{2} M_2^2}\right]^{\gamma/(\gamma-1)}. \tag{5}$$

Since Φ_D is independent of Reynolds number, the dead-air pressure p_B also is independent of Reynolds number. Body shape should therefore affect p_B only through its effect on p_4, the reference pressure.

It is more convenient, however, to express the ratio p_B/p_4 in terms of the Mach number M_4 which exists just downstream of the reattachment zone. Because the outer edge of the laminar viscous layer curves smoothly, the trailing shock wave does not form within or near the viscous layer, and the flow along this outer edge is isentropic. Hence the values of M_4 and p_4 for two-dimensional flow are, in the terminology of ref. 9, the same as the "equivalent freestream conditions" approaching separation. For isentropic flow along the outer edge of the viscous layer

$$\frac{p_4}{p_B} = \frac{p_4}{p_2} = \left[\frac{1 + \frac{\gamma - 1}{2} M_2^2}{1 + \frac{\gamma - 1}{2} M_4^2}\right]^{\gamma/(\gamma-1)}. \tag{6}$$

By combining with eq. (5), there results

$$M_4^2 = (1 - \Phi_D^2) M_2^2. \tag{7}$$

Equations (5) and (7) provide an explicit equation for dead-air pressure

$$\frac{p_B}{p_4} = \left[\frac{1 + \frac{\gamma - 1}{2} M_4^2}{1 + \frac{\gamma - 1}{2} \frac{M_4^2}{1 - \Phi_D^2}}\right]^{\gamma/(\gamma-1)}. \tag{8}$$

The theory applies also for low-speed flow. By taking the limit of eq. (8) as $M_4 \to 0$, there results

$$c_{pB} = \frac{p_B - p_4}{q_4} = \frac{p_B - p_4}{\frac{\gamma}{2} p_4 M_4^2} = -\frac{\Phi_D^2}{1 - \Phi_D^2} = -0.526 \tag{9}$$

with $\Phi_D = 0.587$ (Table 1).

Equation (9) for incompressible flow, just like eq. (8) for compressible flow, should be independent of the Reynolds number or the shape of the dead-air region.

The chief approximations and restricting assumptions made in the foregoing analysis should be noted. One essential approximation is that the compression is isentropic along the dividing streamline through the reattachment zone. Another approximation is that the dividing streamline terminates at a point where the pressure is p_4 and not at the reattachment point where the pressure is p_R. It should also be remembered that the substitution $\Phi_D = 0.587$ in eqs. (8) and (9) is restricted to steady, two-dimensional, pure laminar, separated flows having zero boundary layer thickness at the separation point. If the boundary layer thickness at separation were sizable, eq. (8) would still apply, but the velocity profiles at different stations along the mixing layer would not be similar and Φ_D would not be 0.587. Then the value of Φ_D would have to be calculated by solving the partial differential equations of viscous flow for each case.

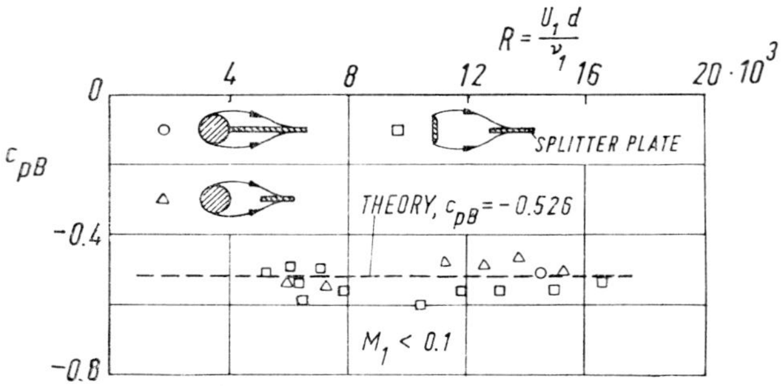

FIG. 3. Comparison of theory of Chapman with experiments of Roshko conducted at low speed, $M_1 < 0.1$. After Chapman *et al.*, ref. 10.

We will now compare the theoretical results with some experimental data. In Fig. 3 the theoretical value for incompressible flow $c_{pB} = -0.526$ is shown together with some values measured by Roshko.[12] The measurements show no dependence on the Reynolds number, which is in agreement with the theory. Otherwise, the scatter of the experimental values is relatively large, which makes the apparent independence of Reynolds number less convincing.

Furthermore, the results obtained by Roshko show no, or only a small influence of the shape of the body. This is in strong disagreement with experimental results by the present author,[13] which are shown in Fig. 4. According to these, the coefficient of base pressure depends strongly on the shape of the body, i.e. in this case on the wedge angle φ. The highest value for c_{pB} ($c_{pB} = -0.20$) corresponds to the flow past a flat plate parallel to the undisturbed flow ($\varphi = 0°$), and the lowest value ($c_{pB} = -0.65$) to the flow past a flat plate per-

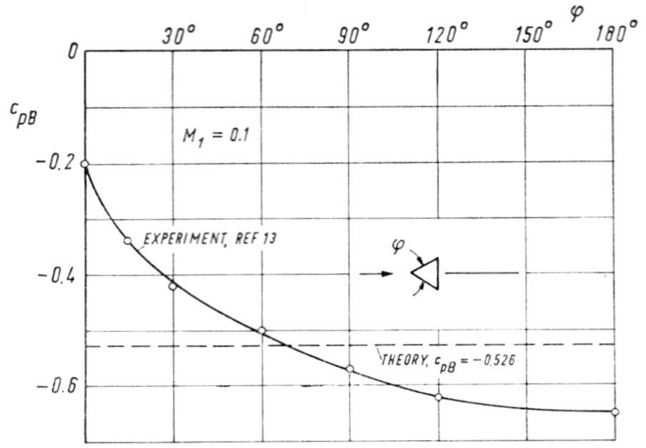

FIG. 4. Comparison of theory of Chapman with experiments of Tanner conducted at the Mach number $M_1 = 0.1$.

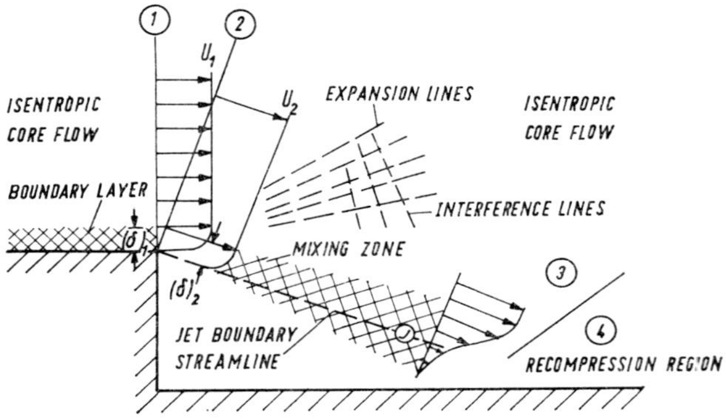

FIG. 5. Flow model for two-dimensional base pressure problem. After Korst, ref. 14.

pendicular to the flow direction ($\varphi = 180°$). Since c_{pB}-values of the order of -0.20 have been measured also by other authors (see e.g. refs. 28 and 61) for bodies corresponding to a flat plate parallel to the air stream, the validity of the theoretical value $c_{pB} = -0.526$ seems to be questionable.* A further comparison of the theory of Chapman with experiments will follow at the end of section 3.

3. THE THEORY OF KORST

In the theory of Korst[14] for base pressure in transonic and supersonic turbulent flow, the flow model is nearly the same as in the theory of Chapman for laminar flow. The various

* It must be assumed, however, that the shear layer was laminar only over a part of its length in some of these tests. Therefore the base pressure coefficient for $\varphi = 0°$ possibly is $|c_{pB}| < 0.20$ for pure laminar separation, i.e. the difference between theory and experiment is for this case perhaps still greater than shown in Fig. 4.

flow regions will, however, be more precisely defined in the present context (Fig. 5). They are:

(a) Flow approaching the trailing edge, in front of cross-section 1.
(b) Flow expanding around the trailing edge, between 1 and 2.
(c) Flow along the dead-air region, with constant pressure mixing in the compressible jet boundary, between 2 and 3.
(d) Recompression at the end of the dead-air region; e.g., by a plane shock between 3 and 4.

The flow mechanism is formulated by Korst in the following five points:

(a) The flow in the dissipative shear-flow domain has its static pressure impressed by the adjacent, nearly uniform free stream, i.e.

$$\frac{p_i}{p_{i2}} = 1, \tag{10}$$

where p_i is the static pressure in the mixing region and p_{i2} the static pressure at the outer boundary of the mixing region.

(b) The expansion of the free stream between cross-sections 1 and 2, follows the Prandtl–Meyer solution for the simple wave, functionally expressed by

$$\frac{p_2}{p_1} = \frac{p_2}{p_1}(M_1, \Theta_2 - \Theta_1) \tag{11}$$

$$M_2 = M_2 (M_1, \Theta_2 - \Theta_1). \tag{12}$$

In the dissipative domain, between 1 and 2, one may formally express

$$\Phi_2(\zeta) = \Phi_2 \left[\Phi_1(\zeta_1), M_1, \frac{p_2}{p_1}\right] \tag{13}$$

$$\frac{(\delta)_2}{(\delta)_1} = \frac{(\delta)_2}{(\delta)_1}\left[\Phi_1(\zeta_1), M_1, \frac{p_2}{p_1}\right] \tag{14}$$

where Φ is the dimensionless velocity and ζ_1 the dimensionless y-coordinate $= y/(\delta)_1$. $\Phi_1 = U/U_1$, $\Phi_2 = U/U_2$.

(c) The mixing process in the jet boundary, between cross-sections 2 and 3, takes place at constant pressure. Therefore

$$\frac{p_3}{p_2} = 1 \tag{15}$$

which generally will require that the thickness of the dissipative region remains small compared to the step height or radius of curvature related to the jet boundary. Under this condition, and with restriction to the isoenergetic turbulent jet boundary, the results of a mixing theory by Korst,[15,16] become applicable, the main points of which are briefly summarized below.

The mixing theory introduces a reference system of coordinates (X, Y) as an orthogonal and generally curvilinear system of coordinates which follows the boundary of the corresponding inviscid jet (Fig. 6). The inviscid jet is defined as the hypothetical frictionless jet which has the same approach Mach number M_1, expands through the same pressure ratio

p_2/p_1 and is influenced by the same geometry of physical boundaries as the actual viscous jet.

Also specified is an intrinsic system of coordinates (x, y) which is displaced from the reference system of coordinates in the Y-direction, so that, as the angle between the two coordinate systems will be small, the following relations are used:

$$X \approx x$$
$$Y = y - y_m(x), \text{ with } y_m(0) = 0. \tag{16}$$

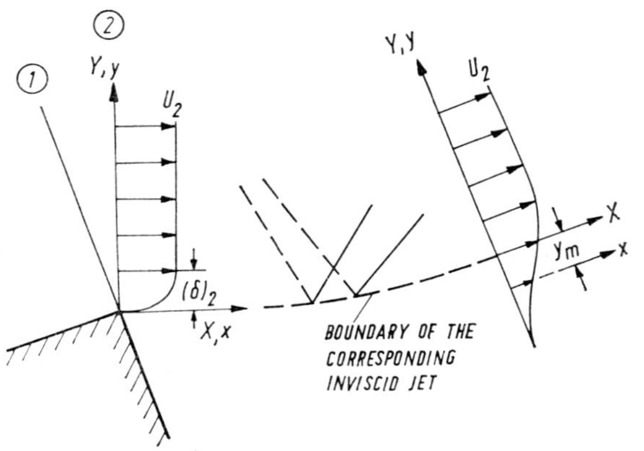

FIG. 6. System of coordinates employed for jet mixing in two-dimensional base pressure model. After Korst, ref. 14.

Essentially, the mixing theory is a momentum-integral method which utilizes a highly simplified equation of motion

$$\frac{\partial U}{\partial x} = \frac{\epsilon}{U_2}\frac{\partial^2 U}{\partial y^2} \tag{17}$$

with Prandtl's exchange coefficient ϵ as turbulent viscosity. After introducing the dimensionless variables

$$\Phi = \frac{U}{U_2}, \ \psi = \frac{x}{(\delta)_2}, \ \zeta = \frac{y}{(\delta)_2}$$
$$\epsilon = \frac{1}{2\sigma^2}\psi(\delta)_2 \, U_2 \, f(\psi) \tag{18}$$

where σ is a similarity parameter and $(\delta)_2$ is the boundary layer thickness after the expansion at the corner (see Fig. 5), and with $f(\psi) \to 1$ as $\psi \to \infty$ and by using the transformation

$$\xi = \frac{1}{2\sigma^2}\int_0^{\psi} \psi f(\psi) \, d\psi \tag{19}$$

for the ψ-coordinate, the following formulation of the mixing problem, with Φ from eq. (18), is obtained from eq. (17):

$$\frac{\partial \Phi}{\partial \xi} = \frac{\partial^2 \Phi}{\partial \zeta^2}. \tag{20}$$

The following initial and boundary conditions are valid:

$$\begin{aligned} &\Phi\,(0,\,\zeta) \equiv 0 && \text{for } -\infty < \zeta < 0 \\ &\Phi\,(0,\,\zeta) = \Phi_2\,(\zeta) && \text{for } 0 < \zeta < 1 \\ &\Phi\,(0,\,\zeta) \equiv 1 && \text{for } 1 < \zeta < +\infty \\ &\Phi\,(\xi,\,-\infty) \to 0 && \text{for } \xi > 0 \\ &\Phi\,(\xi,\,+\infty) \to 1 && \text{for } \xi > 0. \end{aligned} \tag{21}$$

The general solution within the intrinsic system is

$$\Phi = \Phi\,[\Phi_2(\zeta),\,\eta_P,\,\eta] = \tfrac{1}{2}\,[1 + \operatorname{erf}\,(\eta - \eta_P)] + \frac{1}{\sqrt{\pi}} \int_{\eta-\eta_P}^{\eta} \Phi_2\left(\frac{\eta - \beta}{\eta_P}\right) e^{-\beta^2}\,d\beta \tag{22}$$

where η_P is a position parameter given by

$$\eta_P = \frac{1}{2\sqrt{\xi}} \tag{23}$$

and

$$\eta = \zeta \eta_P. \tag{24}$$

(d) The pressure increase in the recompression region at the end of the dead-air region is determined by the flow conditions in the adjacent free stream, according to point (a), and the free-stream compression has to be specified, e.g. by a plane angle shock between cross-sections 3 and 4. Thus the static pressure in the dissipative region behind the shock may be found if approach conditions of the free stream are given and the deflection angle $(\Theta_3 - \Theta_4)$ is defined:

$$\frac{p_4}{p_3} = \frac{p_4}{p_3}\,(M_2,\,\Theta_3 - \Theta_4). \tag{25}$$

(e) The requirement of conservation of mass in the dead-air region must be satisfied. A discriminating (= reattachment) streamline at location η_D within the mixing region, as shown in Fig. 7, is defined such that it has a level of mechanical energy, expressed by the stagnation pressure at cross-section 3, p_{03D}, so that recompression to the static pressure p_4 in cross-section 4 is possible by the complete conversion of the kinetic energy, i.e.

$$\frac{p_{03D}}{p_3} = \frac{p_4}{p_3} \tag{26}$$

where, using the reversible adiabatic relation,

$$\frac{p_{03D}}{p_3} = \left[1 + \frac{\gamma - 1}{2}\,M_{3D}^2\right]^{\gamma/(\gamma-1)}. \tag{27}$$

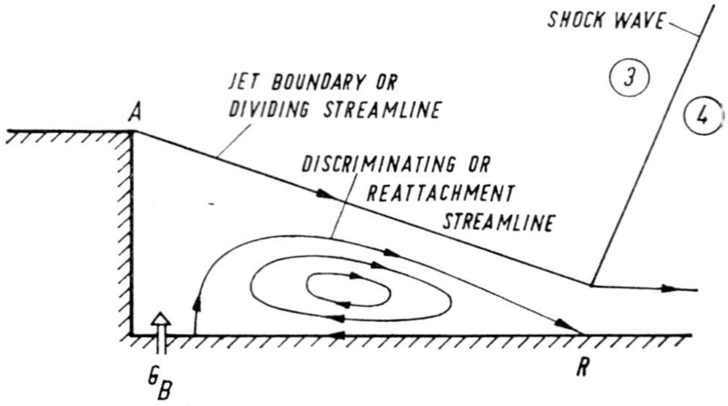

FIG. 7. Definition of jet boundary and discriminating streamlines. A = separation point, R = reattachment point, G_B = rate of mass flow bled into the dead-air region.

This expresses the escape criterion satisfied by the reattachment streamline. Fluid along streamlines having $p_{03}/p_4 < 1$ will not be able to penetrate into the region where the static pressure is equal to p_4, while fluid along streamlines with $p_{03}/p_4 > 1$ will pass through the recompression zone.

An "open wake" is defined as a wake where mass is added to the dead-air region at the rate G_B by direct mass injection. The conservation of mass in the dead-air region requires that

$$G_B + G_D = 0 \tag{28}$$

where

$$G_D = \int_{y_J}^{y_D} \rho U \, dy. \tag{29}$$

Equation (28) expresses the fact, that at the end of the dead-air region, i.e. at the reattachment point R, the rate of mass flow injected in the dead-air region ($= G_B$) must flow out between the reattachment streamline (index D) and the separation streamline (index J) and this mass flow is equal to G_D.

If there is no bleed into the wake, G_B is equal to zero. Because then the reattachment streamline (index D) and the jet boundary or dividing streamline (index J) are identical (Fig. 7), also G_D is equal to zero, as can be seen from eq. (29). A wake without external bleeding into the wake is called by Korst a "closed wake". Then

$$\frac{p_{03J}}{p_{03D}} = \frac{p_{03J}}{p_4} = 1 \tag{30}$$

where the subscript J refers to conditions along the jet boundary ($=$ dividing) streamline.

The system of equations listed under points (a) to (e) is principally sufficient to determine the solution of the base-pressure problem. However, exact theoretical solutions have not yet been obtained for eqs. (13) and (14). But fortunately the case of thin approaching boundary layers ($\eta_P \rightarrow 0$ in eq. (22)) eliminates these two relationships entirely and, at the same time, will not require any empirical information concerning the mixing component in the case of a "closed wake", while for an "open wake" only one empirical coefficient, namely σ, has to be known. It shall be noted that as $\Phi\ (\eta_P > 0;\ \eta_J) < \Phi\ (\eta_P = 0;\ \eta_J)$, where the subscript J refers to the dividing streamline, the "restricted base pressure theory" ($\eta_P = 0$) establishes the lower limit for base pressures obtained with a finite approaching boundary layer thickness.

As $\eta_P \to 0$, the equation for the velocity distribution in the shear layer equation (22) reduces to

$$\Phi = \tfrac{1}{2}\,(1 + \operatorname{erf} \eta) \tag{31}$$

where

$$\eta = \sigma \frac{y}{x}. \tag{32}$$

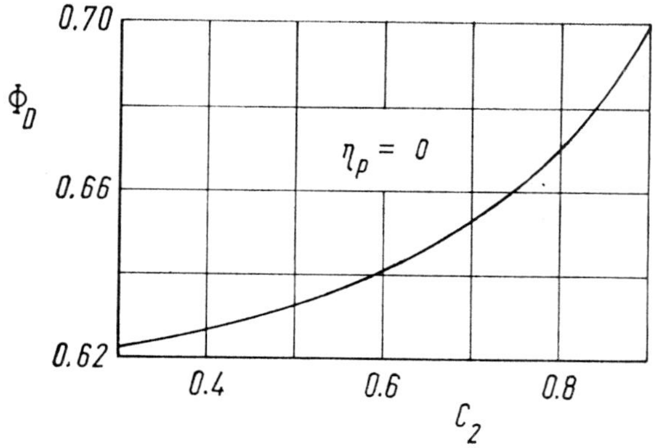

FIG. 8. Dimensionless velocity along jet-boundary (= dividing) streamline as function of the Crocco number C_2 at the edge of the mixing region. After Korst, ref. 14. Zero boundary layer thickness ($\eta_p = 0$), C_2 as given by eq. (33).

Without presenting the mathematical equations of the solution, the velocity on the dividing streamline as calculated by Korst on an electronic digital computer, is presented in Fig. 8 in dependence of the Crocco number C_2 defined by

$$C_2 = \left[1 + \frac{2}{(\gamma - 1)\, M_2^2}\right]^{-1/2}. \tag{33}$$

It may be noted that $C_2 = 0$ corresponds to $M_2 = 0$, $C_2 = 0.4$ to $M_2 = 1.82$ and $C_2 = 0.9$ to $M_2 = 6.7$. Therefore the velocity on the dividing streamline increases with increasing Mach number.

A comparison between theory and experiment for the case without base bleed can be seen in Fig. 9 (reproduced from ref. 14). The theoretical base pressure ratio p_B/p_1 seems to agree well with the measured values of Chapman *et al.*[17] and Eggink.[18] A further comparison between the theories of Korst[14] and Chapman[10] and experimental results of Rogers *et al.*[19] and Tanner[20] is shown in Fig. 10. At transonic Mach numbers the measured values of the base pressure coefficient c_{pB} are much lower than those predicted by the theories. At higher Mach numbers there is a better agreement between theory and measurements. According to the measurements of the present author (ref. 20) the base pressure also strongly depends

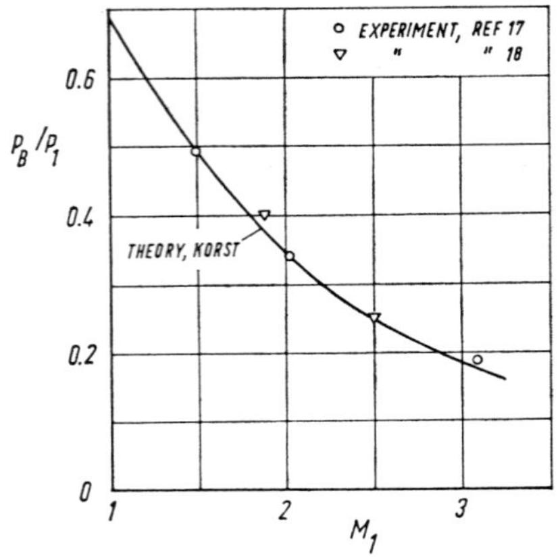

FIG. 9. Comparison of theory of Korst with experiment for the flow past a back step. After Korst, ref. 14.

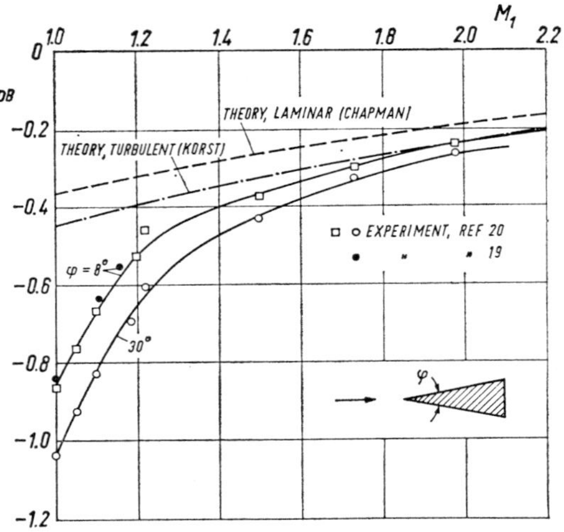

FIG. 10. Comparison of theories of Chapman and Korst with experiments of Rogers *et al.*, ref. 19, and Tanner, ref. 20.

on the wedge angle, i.e. on the body shape, whereas the shown theoretical values are independent of the geometry of the body.*

As will be discussed in the next section, there exists a shortcoming in the theories of Chapman and Korst, namely in the assumption about the pressure at the reattachment point. Obviously the great difference between theory and experiment, as shown in Fig. 10, depends at least partly on this fact. In Fig. 11 results for an "open wake", i.e. for a supersonic flow past a back step with bleeding into the dead-air region, are shown. The bleeding

* Korst[14] also introduces the concept of a "reduced Mach number" for a "generalized back step", i.e. to account for the effects of upstream and downstream wall inclinations. This will not be explained in this connection. It will be remarked, however, that for wedges, as shown in Fig. 10, the reduced Mach number is (nearly) identical with the Mach number of the undisturbed flow, M_1.

air has the same stagnation temperature as the main stream and is added slowly into the wake at the rate G_B, and the dimensionless bleed number c_Q is given by

$$c_Q = \frac{G_B \sqrt{T_{02}}}{d \cdot p_{02}} \sqrt{\frac{R}{g\gamma}} \tag{34}$$

where R is the gas constant and g the gravitational constant. The theoretical calculations are based on a value of $\sigma = 12$. Negative bleed number means suction. The experimental values for $M_1 = 2.0$ obtained at the University of Illinois agree well with the corresponding theoretical curve.

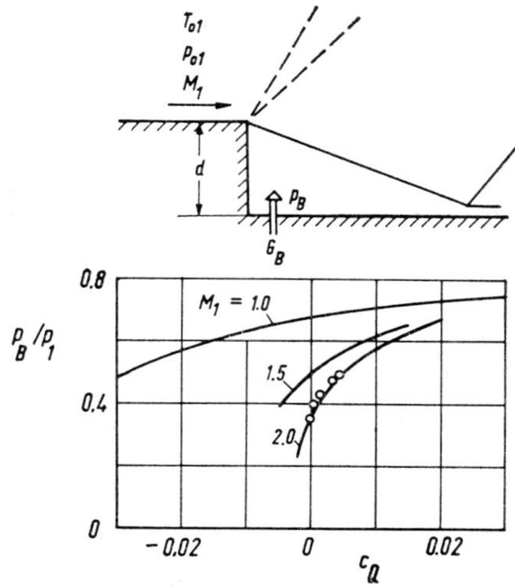

FIG. 11. Base pressure in supersonic flow past a back step with bleeding into the wake. After Korst, ref. 14. ○ Experimental values obtained at the University of Illinois (ref. 14), G_B = rate of mass flow bled into the dead-air region.

It should be noted that independently of the thoughts of Chapman and Korst, a base pressure theory with a similar flow model was formulated by Kirk.[21] To account for the effects of a finite boundary-layer thickness at separation, Kirk assumes that the shear layer or mixing region behaves as if it started some distance ahead of the base instead of starting at the separation point. It was found that, qualitatively, the effect of a boundary layer is to increase the base pressure over its value for vanishing boundary layer thickness.

4. THE THEORY OF NASH

In the theory of Nash[22] for turbulent base flows the flow model is essentially the same as used by Chapman and Korst. The most significant difference in relation to the both older theories lies in the fact that Nash takes into consideration that the pressure at the real reattachment point differs from the final recovery pressure far downstream. With this theory it is also possible to predict the influence of boundary layer thickness on the base pressure. In the following we will explain the main new thoughts of the theory without presenting all mathematical details.

(a) The velocity profile in the free shear layer is approximated by the error function

$$\Phi = \tfrac{1}{2}\left[1 + \operatorname{erf}\frac{\sigma y}{x}\right]. \tag{35}$$

The velocity distribution is thus the same as in the "restricted theory" by Korst for zero boundary layer thickness.

By using this velocity profile, the velocity on the reattachment streamline at incompressible flow is $\Phi_D = 0.615$, provided that the boundary layer thickness is zero. This value of Φ_D is greater than that computed by the more precise methods of Tollmien[23] and Görtler,[24] which amounts to $\Phi_D = 0.585$. Nash therefore uses the value

$$\Phi_D = 0.58 \tag{36}$$

at subsonic speeds. For supersonic velocities he assumes that Φ_D follows the law

$$\Phi_D^2 = 0.348 + 0.018\, M_2 \tag{37}$$

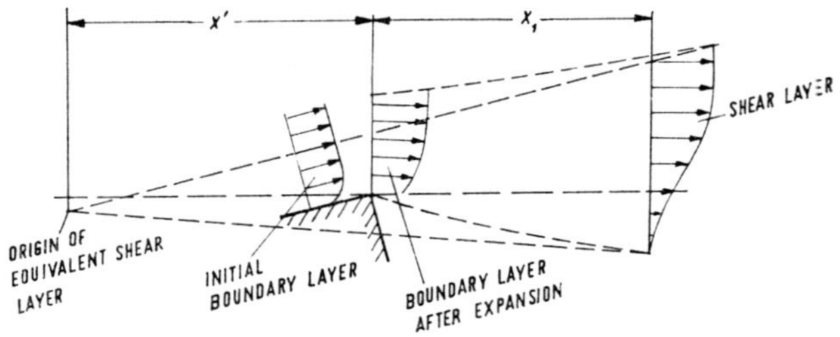

FIG. 12. Model for accounting for the effects of initial boundary layer on free shear layer. After Nash, ref. 22. x' = distance between origin of equivalent shear layer and separation point, x_1 = distance from separation point to actual shear layer.

which gives a smaller increase of Φ_D with Mach number than calculated from the error function velocity profile. After the definition of Φ_D through eqs. (36) and (37) Nash, however, uses the error function velocity profile, eq. (35), in deriving the final equations for the prediction of base pressure.

(b) To account for the effects of a finite boundary layer thickness at separation the simple method proposed by Kirk[21] is used. Here the real shear layer developing from an initial boundary layer is replaced by an equivalent asymptotic shear layer growing over a greater distance from zero thickness. The distance between the origin of the equivalent shear layer and the separation point is equated to a simple multiple of the boundary layer momentum thickness at separation. This method is shown in Fig. 12, schematically.

At supersonic Mach numbers the effect of the rapid expansion at the separation corner on the boundary layer momentum thickness will be considered by the equation

$$\frac{\rho_2 U_2 \vartheta}{\rho_1 U_1 \delta_2} = \frac{M_1^2}{M_2^2} \tag{38}$$

where the subscript 1 refers to the flow approaching the separation point and the subscript 2 to the flow after full expansion at the corner, i.e. at the edge of the dead-air region. δ_2 is the momentum thickness of the boundary layer approaching the base and ϑ the momentum thickness of the boundary layer immediately downstream of the centred expansion fan at the base. At subsonic speeds, if $M_2 < 1$, it is assumed that $\vartheta = \delta_2$. Equation (38) should give good results, if the boundary layer velocity profile is "full".

(c) As most important improvement over the older theories Nash takes account of the fact that the pressure at the real reattachment point differs from the final recovery pressure far downstream. Actually reattachment takes place at a point of strong positive pressure gradient (Fig. 13) as is shown by various measurements at subsonic[13] and supersonic

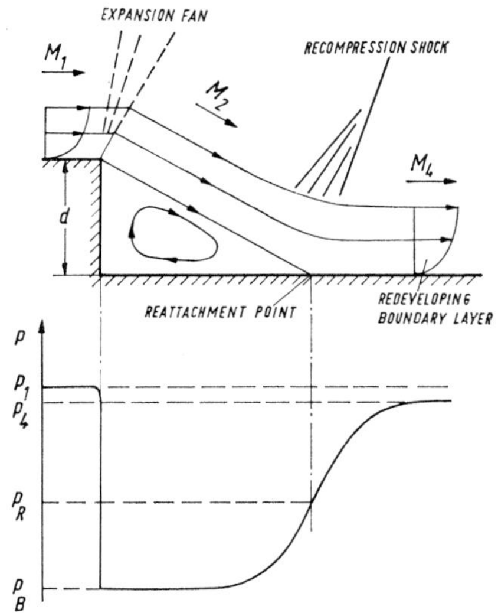

FIG. 13. Supersonic flow over a back step with pressure distribution, schematically.

flow,[25,26,27] which means that a substantial residual pressure recovery is achieved downstream of the reattachment point. Nash therefore introduces a new parameter N, which is defined by

$$N = \frac{p_R - p_B}{p_1 - p_B} \tag{39}$$

where p_R is the pressure at the reattachment point, p_B the base pressure and p_1 the undisturbed static pressure in the approaching flow. The compression from the dead-air pressure p_B to p_R is assumed to be isentropic. The value of N must be determined from measurements.

If there is no bleed into the dead-air region,* the solution for subsonic flow is given by

$$\frac{\delta_2}{l} = C_1 \frac{A(A - B)}{B} \tag{40}$$

* By this theory it is also possible to predict the influence of base bleed. In ref. 22 there are, however, no calculated results given for this case.

where

$$\lambda_b = \frac{1 + \frac{\gamma - 1}{2} M_2^2}{1 + \frac{\gamma - 1}{2} M_2^2 (1 - \Phi_D^2)} \tag{41}$$

$$A = \log \lambda_b \tag{42}$$

$$B = \log \left[\frac{p_R}{p_1}\right]^{(\gamma - 1)/\gamma} \tag{43}$$

$$C_1 = \frac{\sqrt{\pi}}{(\gamma - 1)\, \sigma M_2^2} \tag{44}$$

and l is the length of the free shear layer (from separation to reattachment), which must be taken from experiment. From eq. (40) the value of M_2 and thus the ratio p_B/p_1 can be estimated if M_1 and δ_2/l are given.

For supersonic flow the solution is given by

$$\frac{\delta_2}{d} = C_2 \frac{A(A - B)}{B} \tag{45}$$

where

$$C_2 = \left[\frac{M_2\, T_2}{M_1\, T_1}\right]^3 \frac{C_1}{\sin\,(\Theta_2 - \Theta_1)} \tag{46}$$

and A, B, C_1 have the same meaning as for subsonic flow, d is the height of the step and Θ is the Prandtl–Meyer angle corresponding to Mach number M.

For intermediate Mach numbers as M_2 exceeds unity, although M_1 is below unity, the solution is

$$\frac{\delta_2}{l} = C_3 \frac{A(A - B)}{B} \tag{47}$$

where

$$C_3 = \left[\frac{M_2\, T_2}{M_1\, T_1}\right]^3 C_1, \text{ for } \gamma = 1.4. \tag{48}$$

Nash points out that in the subsonic case, eqs. (40) to (44) and (47) to (48), the solution contains two parameters, which are as yet unspecified, the length l of the free shear layer and the reattachment parameter N. These must be determined from experiments. In the supersonic case only N has to be taken from measurements.

As the thickness of the boundary layer approaching the step decreases to zero, the base pressure tends to its limiting value. Then

$$\frac{p_B}{p_1} = \left(\frac{p_B}{p_1}\right)_{\text{lim}} = \frac{1}{1 + \frac{1}{N}\,[(\lambda_b)^{\gamma/(\gamma - 1)} - 1]}. \tag{49}$$

In Fig. 14 some data on the reattachment parameter N as collected by Nash[22] from results of various measurements are shown in dependence of the Mach number. Most of the experimental values were obtained in tests on back steps or airfoils, i.e. on bodies, which approximate to a wedge with a wedge angle $\varphi = 0°$. The scatter is considerable but there seems to be a definite trend of N with free-stream Mach number.

The results at subsonic Mach numbers were achieved on a model which approximates to a simple back step.[28] Calculations based on measurements, at the Mach number 0.1, of the present author[13] seem to indicate that at low subsonic speeds N also depends on the shape of the body. For a 30° wedge N amounts to 0.98 and for a 120° wedge to 0.71, compared with values $N = 1.5$ to 1.6 as predicted by Nash *et al.*[22] for the back step ($\varphi \approx 0°$) and for the reattachment of a two-dimensional jet on a flat plate by Bourque and Newman.[32]

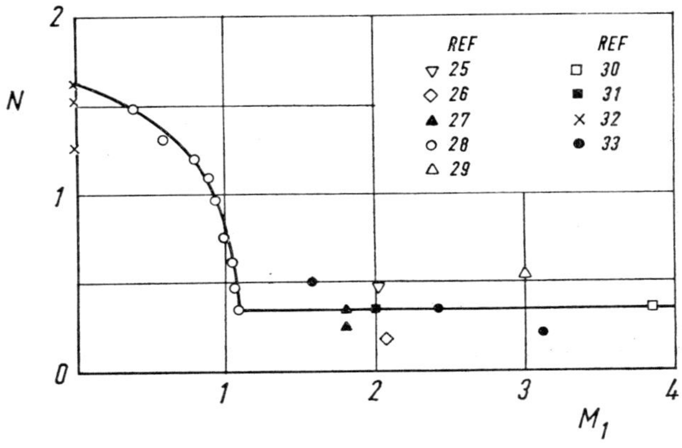

FIG. 14. Variation of the reattachment parameter N with Mach number. After Nash, ref. 22.

According to Nash it should be possible that for the flow past a body with splitter plate N has a different value than for the flow past the same body without splitter plate, even if the flow is steady in both cases. Therefore the base pressure is possibly different in both cases. This is obviously a problem which should be cleared up by suitable experiments.

The variation of base pressure with boundary layer momentum thickness at the Mach number $M = 2.0$ is shown in Fig. 15. The theoretical curve is computed from eq. (45) with $N = 0.35$ and assuming Φ_D to vary according to eq. (37). The agreement between theory and measurement is very good. At higher Mach numbers ($M = 2.3$ and $M = 3.0$) the agreement between theory and measurement is less good than for $M = 2.0$.

Figure 16 shows the coefficient of base pressure on a step at subsonic and transonic Mach numbers. The rapid decrease of c_{pB} at Mach numbers above $M_1 = 0.9$ and the minimum value at about $M_1 = 1.1$ is given by the theory in qualitative agreement with experimental values. The values valid for incompressible flow for σ and Φ_D were used in the calculations ($\sigma = 12$, $\Phi_D = 0.58$). It must be remembered, however, that the length of the free shear layer l and the reattachment parameter N have been taken from measurements. In spite of this, the introduction of the parameter N is a great improvement over the assumption $N = 1$, as implicitly used in the older theories by Chapman, Korst and Kirk.

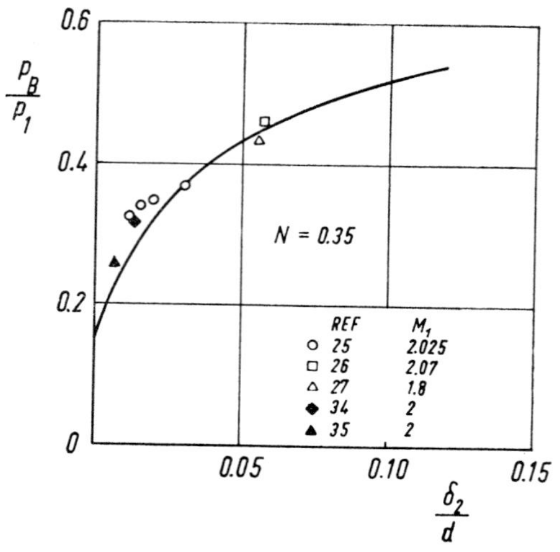

FIG. 15. Variation of base pressure with boundary layer momentum thickness at the Mach number $M_1 \approx 2$. After Nash, ref. 22. d = body thickness at separation.

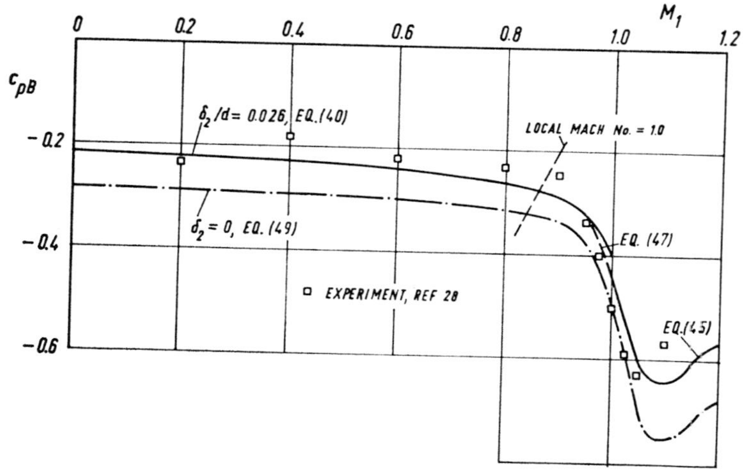

FIG. 16. The base pressure for the flow past a back step at subsonic and transonic speeds. After Nash, ref. 22.

5. LATER DEVELOPMENTS IN THE THEORY BASED ON THE CHAPMAN–KORST FLOW MODEL

5.1. *The Effect of Rapid Expansion at Separation on the Boundary Layer Momentum Thickness*

The assumptions leading to eq. (38) as given by Nash to account for the effect of the rapid expansion, occurring at separation at supersonic speeds, on the boundary layer momentum thickness has been criticized by Wazzan.[36] He points out that especially if the velocity profile of the approaching boundary layer is not "full", deviations from the results given by Nash are likely to occur. Furthermore, viscous effects should not be unimportant

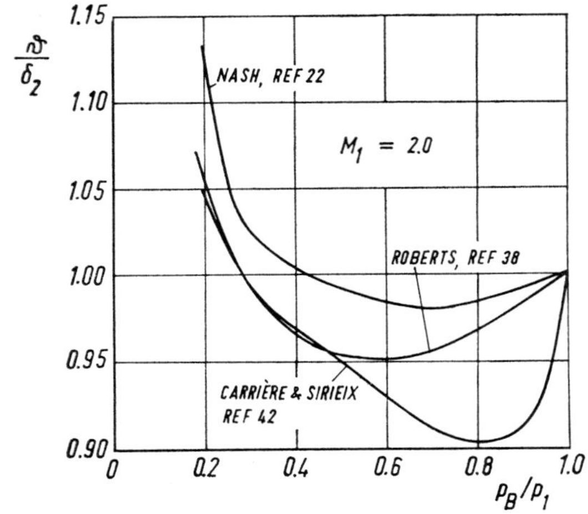

FIG. 17. Effect of abrupt expansion on boundary layer momentum thickness at the Mach number $M_1 = 2$. After Roberts, ref. 38.

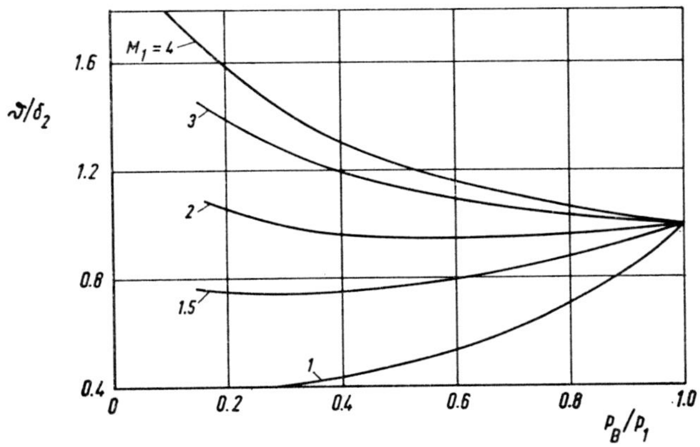

FIG. 18. Effect of abrupt expansion on boundary layer momentum thickness at various Mach numbers. After Roberts, ref. 38.

contrary to the assumption by Nash. Therefore the value of ϑ/δ_2 should be smaller than given by eq. (38).

This problem has also been tackled by Carrière and Sirieix[37] and by Roberts.[38] The results of these investigations are shown in Fig. 17, together with those by Nash. It can be seen that the more accurate calculations of refs. 37 and 38 give a smaller value for the ratio ϑ/δ_2 than that predicted by Nash. This is in qualitative agreement with the before-mentioned criticism by Wazzan.[36]

The quantity ϑ/δ_2 as calculated by Roberts[38] is given in Fig. 18.

5.2. *The Reattachment Criterion*

The reattachment criterion used by Chapman and Korst was simply that the reattachment pressure is equal (or nearly equal) to the static pressure in the undisturbed flow. The first improvement on this criterion was due to Nash who introduced a parameter N, which is given by eq. (39), to account for the fact that the pressure at the reattachment point actually is smaller than the maximum static pressure downstream of the reattachment point. The value $N = 0.35$ was predicted from measured base pressures at supersonic speeds. There is, however, a considerable scatter in the values, as can be seen from Fig. 14. Furthermore, later unpublished work by Nash indicated that N varied with Mach number and with the ratio of boundary layer momentum thickness to step height.[39] McDonald[39] therefore tries to improve the reattachment criterion. He divides the region of pressure rise at the end of the dead-air region into an isentropic region of flow reattachment followed by a region of rehabilitation, i.e. a region of development of the newly attached boundary layer in the downstream direction. The theoretical investigations of McDonald lead to the formulation of the reattachment criterion in terms of the final shape parameter of the reattached boundary layer $H = \delta_1/\delta_2$. This parameter should be of the flat plate type. Using a proper value of H, the base pressure can be determined. In the equivalent incompressible boundary layer the shape parameter has a value of $H = 1.40$, approximately.

Roberts' contribution to the reattachment problem[38] uses as starting point an analogy between the pressure rise from reattachment pressure p_R to the pressure p_4 downstream of reattachment in the base flow model, and the pressure rise from the static pressure in the oncoming flow p_1 to the pressure at separation point p_s for shock induced separation of a boundary layer on a flat plate. Roberts employs the method of Reshotko and Tucker[40] according to which, if friction can be neglected, across a discontinuity the following relationship is valid:

$$\frac{M_2}{M_1} = \frac{f(H_2^*)}{f(H_1^*)} \tag{50}$$

where the subscript 1 refers to conditions ahead of the discontinuity and the subscript 2 to conditions behind the discontinuity. M is the free-stream Mach number and H^* is the transformed shape parameter, which for $\gamma = 1.40$ is related to the compressible shape parameter H by the relation

$$H = H^* \,(1 + 0.2\, M^2) + 0.2\, M^2. \tag{51}$$

The function $f(H^*)$, eq. (50), is given by

$$f(H^*) = \frac{H^{*2}}{(H^{*2} - 1)^{1/2}\,(H^* + 1)}\, e^{1/(H^*+1)}. \tag{52}$$

Roberts introduces the reattachment parameter R, which is defined by

$$R = \frac{M_4}{M_R} \tag{53}$$

where M_4 is the Mach number corresponding to the pressure p_4 downstream of reattachment and M_R the Mach number corresponding to the reattachment pressure p_R.

Figure 19 shows the parameter N, as introduced by Nash, and Fig. 20 the parameter R as introduced by Roberts. Experimental results from refs. 25, 27, 34, 35, 37 and 41 have been used. N is seen to vary between 0.1 and 0.5 and the scatter of the points is considerable. Nash's reattachment criterion of $N = 0.35$ provides therefore no good correlation of the data, as pointed out by Roberts.

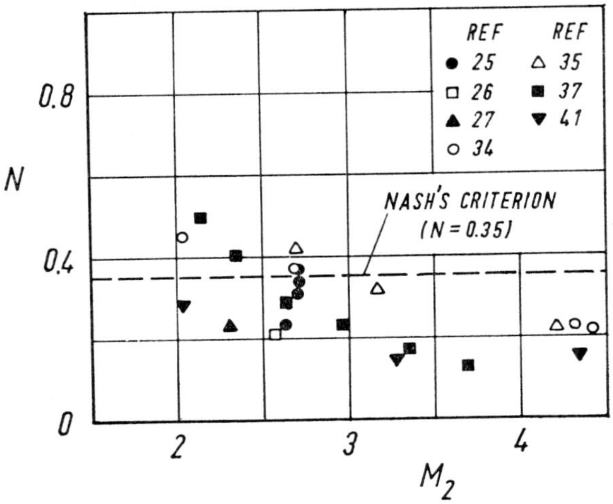

FIG. 19. Variation of reattachment parameter N with the Mach number M_2 on the edge of the mixing layer. Filled symbols denote tests on back steps, unfilled symbols on aerofoils and blunt-trailing-edge sections. After Roberts, ref. 38.

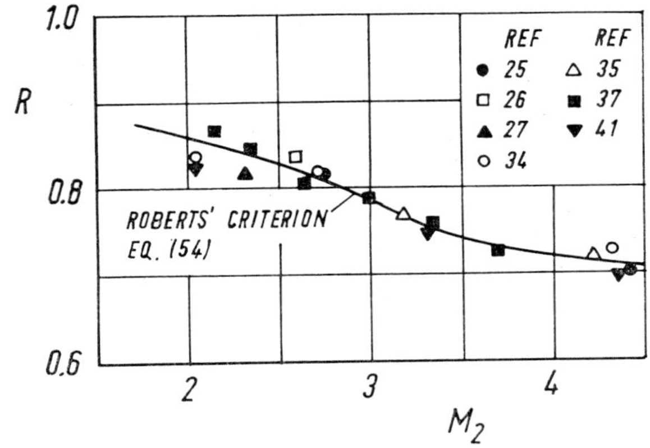

FIG. 20. Variation of reattachment parameter R with the Mach number M_2 on the edge of the mixing layer. Filled symbols denote tests on back steps, unfilled symbols on aerofoils and blunt-trailing-edge sections. After Roberts, ref. 38.

A much better correlation is given by the parameter R. Assuming the recompression to occur through an oblique shock wave, the empirical relation by Roberts

$$R = 0.799 + 0.1560\, M_2 - 0.08237\, M_2^2 + 0.009564\, M_2^3 \tag{54}$$

gives a good fit to the data.

The values of H_4^* in Fig. 21 show that H_4^* varies with Mach number contrary to a constant value of $H_4^* = 1.40$ as suggested by McDonald. It seems therefore that the use of the parameter R as introduced by Roberts gives a better correlation of the conditions at reattachment than the parameters N by Nash and H_4^* by McDonald.

It should perhaps be remarked that the values for R as shown in Fig. 20 are all estimated for Mach numbers greater than $M_2 = 2$. Since the theory of Roberts is valid for supersonic Mach numbers only the parameter R is, contrary to the parameter N of Nash, applicable for supersonic flow only.

A recompression criterion based on the angle of declination of the external streamlines, at the reattachment point, relative to the wall to which the flow is reattaching, has been suggested by Carrière and Sirieix.[42] If it is assumed that the wall can be extended downstream in the same plane, their criterion relates the pressure at the reattachment point to the final recovery pressure. Some tests by the same authors have shown that the base pressure is not affected by the length of the downstream wall, so long as it extends a short distance beyond the actual reattachment point. A more recent analysis using this reattachment criterion is given in a paper by Jacques and Gailly.[43]

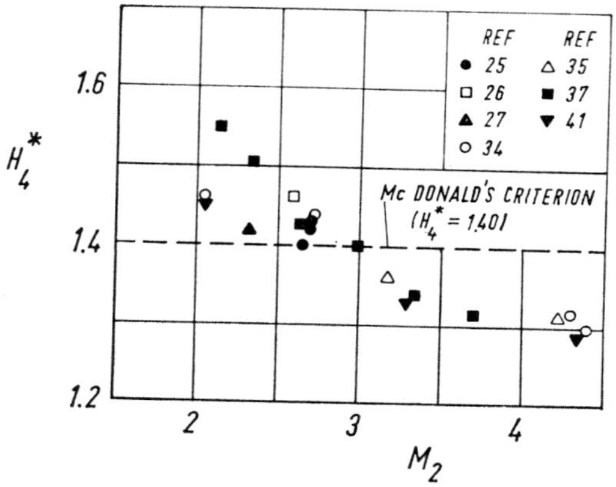

FIG. 21. Variation of H_4^* with the Mach number M_2 on the edge of the mixing layer. Filled symbols denote tests on back steps, unfilled symbols on aerofoils and blunt-trailing-edge sections. After Roberts, ref. 38.

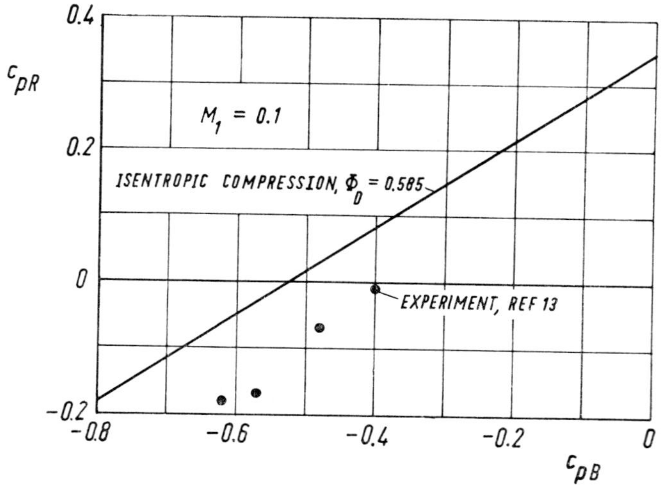

FIG. 22. Comparison of measured reattachment pressure coefficients with calculated values. In the calculations the compression along the dividing streamline is assumed to be isentropic.

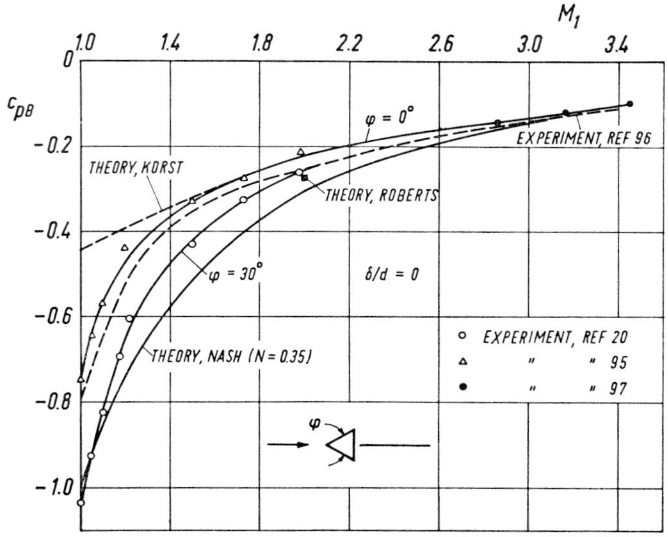

FIG. 23. Comparison of theoretical base pressures for zero boundary layer thickness with measured values. $\varphi = 0°$ means a simple back step.

In a very recent paper Batham[44] takes another approach to the reattachment problem. According to his analysis the pressure coefficient at reattachment is a function of the Mach number and decreases with increasing Mach number. Further discussion of the paper by Batham is given in refs. 45 and 46. A full treatment of the analysis of Batham is given in ref. 47.

In general it is assumed that the compression along the reattachment streamline is isentropic. Some results of the present author seem to indicate, however, that this assumption possibly must be regarded only as an approximation. In Fig. 22 the measured reattachment pressure is shown as a function of the base pressure for the incompressible flow past wedges.[13] It can be seen that the measured c_{pR}-values are all lower than those according to the assumption of an isentropic compression with a velocity on the dividing streamline of $\Phi_D = 0.585$.

6. A COMPARISON OF THE VARIOUS THEORIES BASED ON THE CHAPMAN–KORST FLOW MODEL

In the first sections we have given a description of the theories of Chapman, Korst and Nash, which all are based on the Chapman–Korst flow model. Subsequently later improvements on the theory were explained. We will now compare the results given by these theories with one another and with experimental results.

In Fig. 23 the coefficient of base pressure is shown as a function of the Mach number at transonic and supersonic velocities, for the case of zero boundary layer thickness at separation. The experimental values of the present author have been taken from refs. 20 and 95 and the base pressure coefficients collected by Nash from ref. 96.

It can be seen that the theory of Nash[22] (with $N = 0.35 = \text{const}$) gives the trend of variation of c_{pB} with Mach number in better agreement with experiment than the theory of Korst.[14] This improvement is due to the more realistic reattachment conditions used by

Nash as discussed in section 4. It seems, however, that the theoretical base pressures predicted by Nash for zero boundary layer thickness are too low. The reason for this result is that the theory of Nash[22] evidently overestimates the influence of the boundary layer thickness on the base pressure as will be seen later in this section. It is also possible that the value $N = 0.35$ is not a good value for the reattachment parameter. N should perhaps also be varied with Mach number as mentioned in section 5. Furthermore, the parameter N seems to be a function of the body shape too, as already mentioned in section 4, since the measured c_{pB}-values for the 30° wedge are lower than those for the flow past a simple back step ($\varphi = 0°$). The criticism of Roberts[38] that the value $N = 0.35$ provides a poor reattachment criterion seems therefore to be justified.

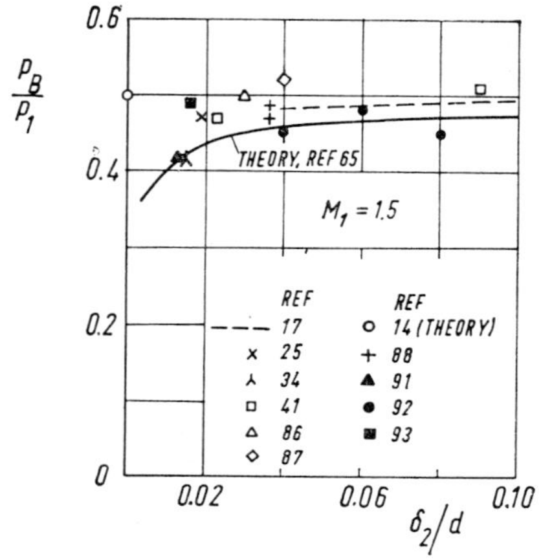

FIG. 24. Comparison of theoretical influence of boundary layer momentum thickness on base pressure with experiment at $M_1 = 1.5$. After McDonald, ref. 65.

The one theoretical base pressure coefficient taken from the work of Roberts[38] seems also to be too low in spite of the apparently better reattachment criterion used by Roberts (see section 5.2). It is possible also that the theory of Roberts overestimates the influence of the boundary layer thickness on the base pressure. We will return to this question later in this section.

The theory of Korst gives too high values for the base pressure at low supersonic Mach numbers, as already shown in section 3. It also does not account in all cases for the effect of body shape on the base pressure as was explained in section 3. However, it can be seen that the agreement between this theory and experimental results obtained for $\varphi = 0°$ (which is the flow past a simple back step, approximately) is very good for Mach numbers above 1.6. Therefore it can be assumed also that the base pressure of profiles with a blunt trailing edge can be predicted at Mach numbers higher than $M_1 = 1.6$ by the theory of Korst in reasonable agreement with experiment. This is almost a paradox: the theory which uses a "wrong" reattachment criterion gives the best results! Merely due to this fact it is obvious that there is a need for more investigations in this field. One possible reason for this state of affairs is, among others, that according to some new results (see Fig. 22) the recompression along the dividing streamline is perhaps not isentropic, contrary to the common assumption.

We will now compare the theoretical influence of boundary layer thickness on base

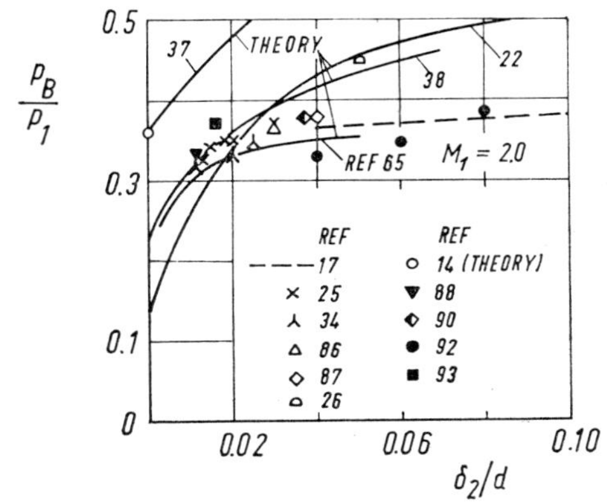

FIG. 25. Comparison of theoretical influence of boundary layer momentum thickness on base pressure with experiment at $M_1 = 2.0$. After McDonald, ref. 65.

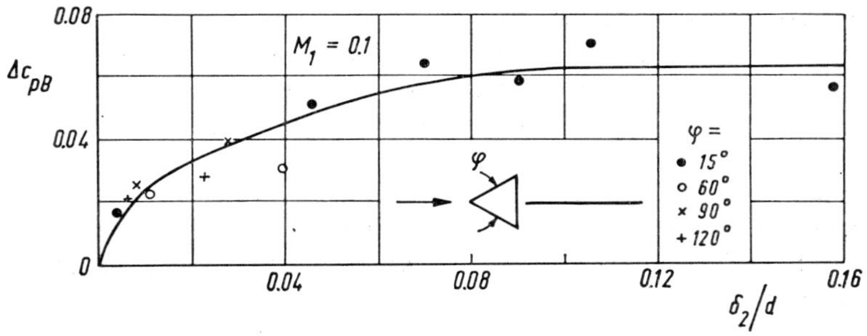

FIG. 26. Increase of base pressure coefficient with boundary layer momentum thickness at $M_1 = 0.1$.

pressure with experiment. In Fig. 24 the base pressure ratio p_B/p_1 is plotted against the dimensionless boundary layer momentum thickness at separation δ_2/d. The Mach number of the oncoming flow is $M_1 = 1.5$. The base pressure at first increases with increasing boundary layer momentum thickness, but then seems to approach a nearly constant value at greater momentum thicknesses. The theory of McDonald[65] agrees, at least qualitatively, with experiment.

Figure 25 shows similar results obtained at the Mach number $M_1 = 2.0$. It is obvious that the theory of Nash[22] overestimates the effect of the boundary layer thickness on the base pressure. This is also true for the theory of Carrière and Sirieix, ref. 37. The theory of Roberts[38] gives a smaller dependence of the base pressure on the boundary layer thickness than both before-mentioned theories. It seems, however, that the best agreement with measurements is given by the theory of McDonald[65] which at the same time gives the smallest increase of base pressure with boundary layer momentum thickness of all theories shown in this figure. Some experimental results by the present author,[94] which are shown in Fig. 26, indicate also that at low Mach numbers the base pressure approaches a constant value as the boundary layer momentum thickness increases over a certain value.

Remark

Before continuing with the description of the theory of the present author it may be mentioned that all theories described before dissect the base flow into a number of discrete parts and analyse them separately. This way of solution is called by Nash[96] "the analytic approach". There are, however, additionally a different group of theories which are called by Nash[96] "integral methods". The first-mentioned theories are in the view of the present author more important and have therefore been reviewed here. The integral methods are, however, important too especially for laminar base flows. The first work in this direction was that of Crocco and Lees, ref. 98. Later contributions are, for example, due to Rom.[99,100,101] An extensive survey of laminar wakes is given in the very recent book of Berger.[102] A more general representation of separated flows may be found in the book of Chang.[103]

7. THE THEORY OF TANNER

7.1. *Introduction*

It is characteristic for all theories explained before that the base pressure is assumed to depend on the pressure recovery, which can be sustained by the flow in the free shear layer. The pressure at the reattachment point depends on the reattachment criterion and can generally be assumed to be known. The base pressure can then be found by subtracting from this known pressure the computed pressure rise occurring between base and the reattachment point.

The present author has used another approach to the base pressure problem.[48] In this theoretical investigation (preceded by an earlier, in which the same basic principle is used, ref. 49) attention is given to the concept of outflow from the dead-air region. This basic concept will be explained by referring to Fig. 27.

In the region from the separation point to the section I there is assumed to occur a mixing process which corresponds to a constant pressure mixing between a uniform external stream and a fluid at rest. Due to this mixing fluid is withdrawn from the dead-air region. This phenomena will be called "outflow from the dead-air region". The streamline ③, originating at the separation point, separates the fluid from the outer flow from that withdrawn from the dead-air region.

At section I, the position of which need not be quantitatively fixed, the mass flow between the boundary ② and the dividing streamline ③ has its maximum value because then the backflow into the dead-air region—which exists at greater distances from the separation points—begins. This maximum value is given by

$$m_a = \int_{y_r}^{y_t} \rho U \, dy. \tag{55}$$

It is the assumption of Tanner[48] that there is a connection between m_a as given by eq. (55) and the pressure drag, and therefore also between m_a and the base pressure.

The pressure drag in a steady separated flow will be generated through the energy losses of the main flow, which are produced by supporting the out- and inflow in the dead-air region. At the end of the dead-air region, i.e. at the reattachment point, the total energy loss,

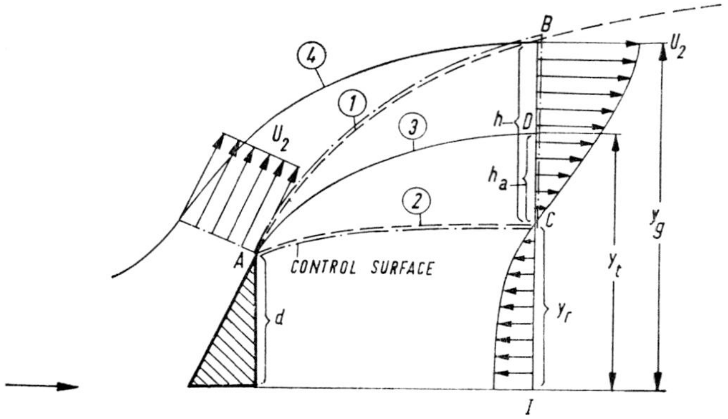

FIG. 27. Flow model for derivation of the second outflow function. (1) = outer boundary of mixing region, (2) = boundary of dead-air region, (3) = dividing streamline, (4) = particular streamline cutting boundary (1) at section I.

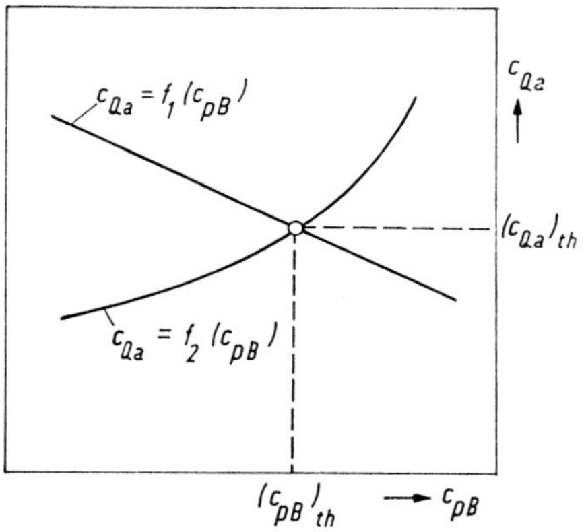

FIG. 28. Theoretical prediction of c_{pB} and c_{Qa}, schematically.

generated by the body, can be found in the flow downstream of reattachment, i.e. in the wake of the body. Therefore, there must be a connection between the drag of the body* and the maximum rate of outflow from the dead-air region. As will be shown later it is possible to derive two formulae for the dependence between the base pressure and the rate of mass outflow from the wake. In dimensionless form one may write

$$c_{Qa} = f_1(c_{pB}, c_D) \tag{56}$$

$$c_{Qa} = f_2(c_{pB}, c_D). \tag{57}$$

It is characteristic for the solution that both equations must be satisfied simultaneously. As the final solution one, therefore, becomes

$$c_{pB} = (c_{pB})_{\text{th}}$$

$$c_{Qa} = (c_{Qa})_{\text{th}}$$

as shown schematically in Fig. 28.

* This is true at subsonic speeds. If the shock wave is attached to the nose of the body, there probably is a connection between the *base* drag and the maximum rate of outflow from the dead-air region.

Before deriving eqs. (56) and (57), a formula for the distribution of temperature and density in the free shear layer will be developed.

7.2. *Distribution of Temperature in the Mixing Region*

According to Korst and Chow[50] the static temperature ratio in a mixing region, for fluids of unity turbulent Prandtl number, is given by (pressure = const):

$$\frac{T}{T_2} = \frac{\rho_2}{\rho} = \frac{t_0 - C_2^2 \Phi^2}{1 - C_2^2} \tag{58}$$

where T is the temperature in the mixing region and T_2 the temperature at the outer boundary of the mixing region. $\Phi = U/U_2$ is the local dimensionless velocity, with U_2 as velocity on the edge of the mixing region, and C_2 is the Crocco number of the free stream

$$C_2 = \frac{U_2}{\sqrt{2c_p T_{02}}} \tag{59}$$

where c_p means the specific heat at constant pressure. Further, $t_0 = T_0/T_{02}$ is the ratio of the total temperature in the mixing region to the total temperature on the edge of the mixing region.

For a constant value of the total temperature in the mixing layer, t_0 is equal to one. By substituting this value for t_0 in eq. (58) and, additionally, the value for C_2 from eq. (59) one obtains

$$\frac{T}{T_2} = \frac{1 - \dfrac{U_2^2}{2c_p T_{02}} \left(\dfrac{U}{U_2}\right)^2}{1 - \dfrac{U_2^2}{2c_p T_{02}}}. \tag{60}$$

After some calculations it follows:

$$\frac{T}{T_2} = 1 + \frac{\gamma - 1}{2} M_2^2 \left[1 - \left(\frac{U}{U_2}\right)^2\right] \tag{61}$$

where M_2 is the Mach number corresponding to the velocity U_2 and γ is the ratio of the specific heats. For turbulent Prandtl numbers not equal to unity one can write[51]

$$\frac{T}{T_2} = 1 + r\frac{\gamma - 1}{2} M_2^2 \left[1 - \left(\frac{U}{U_2}\right)^2\right] \tag{62}$$

where r is the recovery factor. Assuming a constant pressure, the density distribution wil be given by

$$\frac{\rho_2}{\rho} = \frac{T}{T_2} = 1 + r\frac{\gamma - 1}{2} M_2^2 \left[1 - \left(\frac{U}{U_2}\right)^2\right] \tag{63}$$

which will be used in the following calculations. The measurements of Seban *et al.*[52] sugges a value of $r = 0.80$ to be valid in the near wake.

Because the turbulent Prandtl number in the free shear layer is less than unity ($Pr_t \approx$ 0.50), the thickness of the thermal layer δ_T is likely to be greater than the thickness of th shear layer δ. This fact has been found in various measurements.[53,54] Seban and Back[54] have measured the value $\delta_T/\delta = 1.39$ at a point far from the origin of the boundary layer. Therefore

$$\frac{\delta_T}{\delta} = \frac{1}{\sqrt{Pr_t}} = 1.41 \tag{64}$$

seems to be a useful approximation for the wake downstream of reattachment.

As shown by the measurements of Seban and Back[54] the ratio δ_T/δ seems to vary with the distance from the origin of the boundary layer. A value of $\delta_T/\delta = 1.20$ was measured at a point closer to the origin of the boundary layer. Therefore it is plausible that at the end of the outflow region of the steady wake, $\delta_T/\delta < 1.41$. The real value cannot be determined theoretically but must be found from measurements.

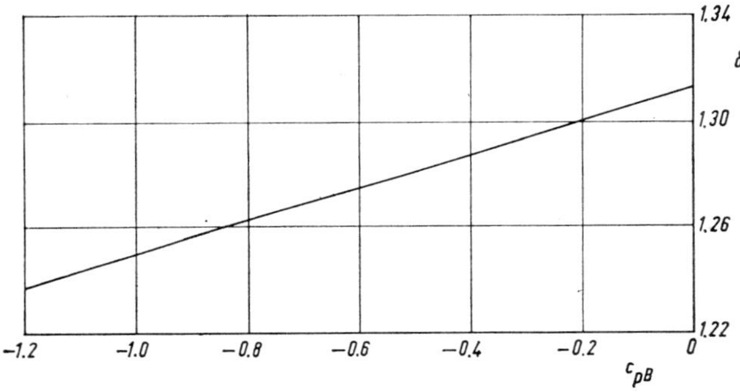

FIG. 29. Ratio of thermal layer thickness δ_T to shear layer thickness δ in dependence of the base pressure coefficient.

By using the measured value $c_{pB} = -0.20$ for the flow past a back step ($\varphi = 0°$), the value $\delta_T/\delta = 1.30$ could be determined by using eqs. (78), (79), and (80). By using the measured values $c_{pB} = -0.65$ and $c_D = 1.48$ for the flow past a perpendicular flat plate ($\varphi = 180°$) the value $\delta_T/\delta = 1.266$ was found by using the above-mentioned equations. It was then assumed that δ_T/δ is a function of the base pressure only. Figure 29 shows the assumed linear relationship between c_{pB} and δ_T/δ. In following this will be used for all Mach numbers. Because δ_T/δ varies not strongly with c_{pB}, this assumption should be a good first approximation for δ_T/δ at the end of the outflow region.

7.3. *The Momentum and Continuity Equations*

The steady subsonic flow past a blunt body is shown schematically in Fig. 30. We will define section I of Fig. 27 and Fig. 30 as that section where the rate of mass outflow from the dead-air region ceases. By the scheme of Fig. 27 the four important boundaries are:

(a) The boundary ① separates the external flow from the mixing region.
(b) The mass outflow from the dead-air region takes place through the boundary ②.
(c) The boundary ③ is the dividing or reattachment streamline.
(d) The boundary ④ is that particular streamline in the external flow, which cuts the boundary of the mixing region at section I.

As is shown by the measurements of Fage and Johansen[55] the velocity and therefore the pressure at boundary ① of the mixing region tends to remain constant and to have a value corresponding to the pressure at the separation point, i.e. the base pressure. Other measurements by Fail *et al.*[56] verify this. Their results show that the pressure at the boundary ① remains constant about up to the half distance from separation point to the reattachment point. The section I where the outflow from the dead-air region ceases and the inflow into it begins is at the midlength of the dead-air region, approximately, as can be assumed from Fig. 30. Therefore, the static pressure along the outer edge of the mixing region—along the boundary ①—should be practically constant up to section I.

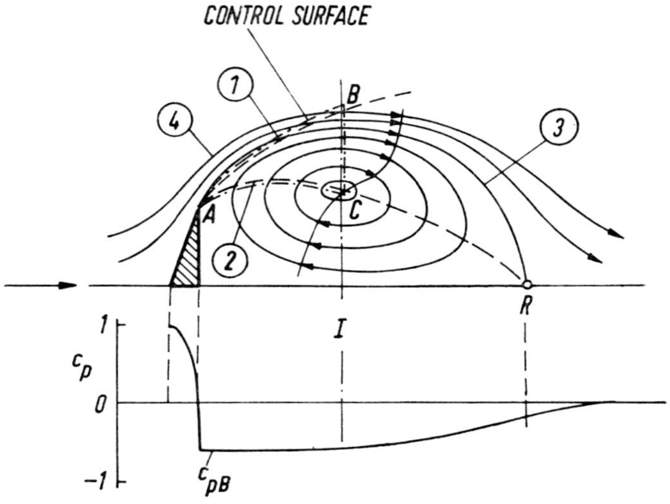

FIG. 30. The subsonic steady base flow past a blunt body. Mean streamlines and pressure distribution, schematically. R = reattachment point, I = section at which the outflow from the dead-air region ceases, ① = outer boundary of mixing region, ② = boundary of dead-air region, ③ = dividing streamline, ④ = particular streamline cutting boundary ① at section I.

We will now choose a control volume with the boundaries *AB*, *BC* and *CA* and apply the momentum equation for steady flow on this region. The pressure along *AB* is practically constant as said before. Since the flow within the area *ABCA* approximates the flow in a free jet boundary, the static pressure in the whole control volume can be assumed to be constant. In the momentum equation

$$\bar{P} = \int_A \rho \bar{V} (\bar{V} \cdot \bar{n}) \, dA \tag{65}$$

therefore the vector of the net force $\bar{P}$, which acts on the boundary A, must be equal to zero. Therefore it follows for the momentum equation:

$$\int_A \rho \bar{V} (\bar{V} \cdot \bar{n}) \, dA = 0 \tag{66}$$

where $\bar{V}$ is the vector of velocity, dA an element of the surface of the control volume and $\bar{n}$ the unit vector perpendicular to the surface of the control volume.

The scalar value of the velocity of the flow into the control volume through the boundary AB has the value $U = U_2 =$ const. at this boundary and the corresponding density is $\rho = \rho_2 =$ const. Therefore

$$\int_{AB} \rho \bar{V} (\bar{V} \cdot \bar{n})\, dA = \int_{AB} \rho_2\, \bar{U}_2 (\bar{U}_2 \cdot \bar{n})\, dA = \bar{U}_2 \int_{AB} \rho_2\, \bar{U}_2 \cdot \bar{n}\, dA. \tag{67}$$

Due to continuity

$$\int_{AB} \rho_2\, \bar{U}_2 \cdot \bar{n}\, dA = \int_{y_t}^{y_g} \rho U\, dy. \tag{68}$$

Therefore the absolute value of the vector*

$$\left| \int_{AB} \rho \bar{V} (\bar{V} \cdot \bar{n})\, dA \right| = U_2 \int_{y_t}^{y_g} \rho U\, dy. \tag{69}$$

For the section BC

$$\int_{BC} \rho \bar{V} (\bar{V} \cdot \bar{n})\, dA = \int_{y_r}^{y_g} \rho U^2\, dy. \tag{70}$$

Because the net momentum flux vanishes, it follows the equation

$$\int_{y_r}^{y_g} \rho U^2\, dy - U_2 \int_{y_r}^{y_g} \rho U\, dy = 0. \tag{71}$$

This is the momentum equation in its final form. This equation together with eq. (55) forms the basis for the derivation of the second outflow function.

7.4. *The Velocity Profile*

We will start from the velocity profile as given by Görtler[24] considering the four first terms in the infinite series, i.e.:

$$\frac{U}{U_2} = \tfrac{1}{2}\,[1 + F_1' + F_2' + F_3' + F_4'] \tag{72}$$

with

$$F_1' = \frac{2}{\sqrt{\pi}} \int_0^{\eta} e^{-t^2}\, dt \tag{73}$$

and

$$\eta = \sigma \frac{y}{x} \tag{74}$$

where σ is a similarity parameter and with values of F_2' to F_4' as tabulated by Görtler.[24]

* Since the sum of the momentum flux through the control volume is equal to zero, the vectors of momentum flowing into and out from the control volume must lie on the same line. Therefore, we can restrict ourselves to the absolute values of the vectors.

This velocity profile agrees quite well at positive and small negative η-values with measurements and its shape is essentially unchanged by compressibility as shown in refs. 38, 57, and 82.

It will be assumed that the velocity profile lies between the η-values from $\eta = -1.4$ to $\eta = +1.4$. At $\eta = -1.4$, which corresponds to $y = y_r$, the velocity calculated from eq. (72) still has the value $U/U_2 = 0.056$, whereas at this point the velocity actually must be equal to zero. We therefore replace the Görtler profile, eq. (72), by the sinus profile

$$\frac{U}{U_2} = \tfrac{1}{2}\left[1 + \sin\left(\frac{\pi(y - y_r)}{h} - \frac{\pi}{2}\right)\right]. \tag{75}$$

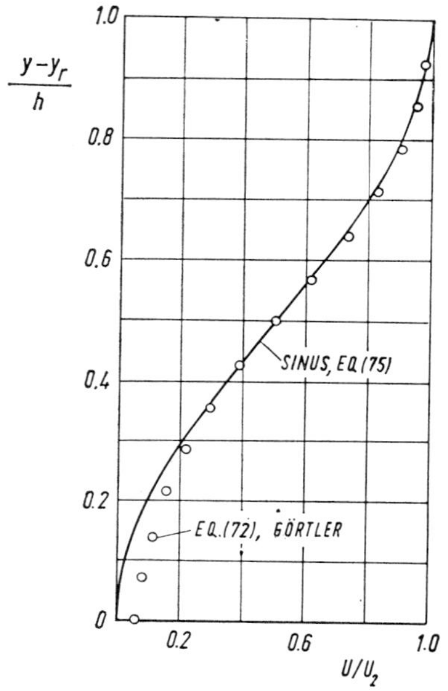

FIG. 31. Velocity profile in the mixing region at the end of the outflow region.

As is shown in Fig. 31, this is really a good approximation for the Görtler profile for $(y - y_r)/h > 0.3$. For smaller y-values, $(y - y_r)/h < 0.3$, the velocity for the sinus profile is smaller than for the Görtler profile, and as $(y - y_r) \to 0$, the velocity for the sinus profile approaches to zero as required.

7.5. *The Second Outflow Function*

Now, we will remember our concept of section 7.1 and look first to the second outflow function. The rate of mass outflow per unit span may be given by

$$m_a = b_3\, \rho_2\, U_2 h \tag{76}$$

where b_3 is a dimensionless coefficient. By introducing a dimensionless rate of mass outflow

$$c_{Qa} = \frac{m_a}{\rho_1\, U_1\, d} \tag{77}$$

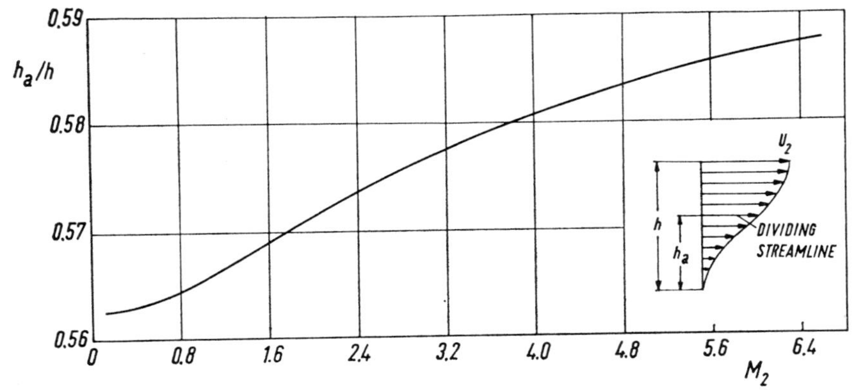

FIG. 32. The position of the dividing streamline, h_a/h, as function of the Mach number M_2 at the edge of the mixing region.

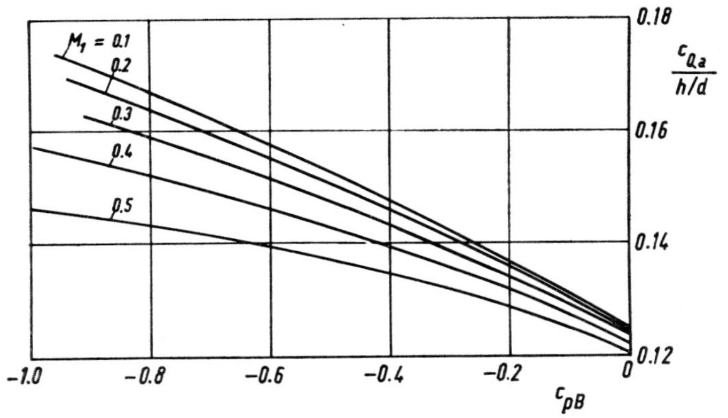

FIG. 33. The second outflow function for Mach numbers from $M_1 = 0.1$ to 0.5.

and the velocity of sound a, eq. (76) can be written

$$c_{Qa} = b_3 \frac{\rho_2}{\rho_1} \frac{a_2}{a_1} \frac{M_2}{M_1} \frac{h}{d}. \tag{78}$$

The quantities b_3 and h_a/h can be determined by solution of eqs. (55) and (71).*

The position of the dividing streamline, h_a/h, is shown in Fig. 32 as a function of the Mach number at the edge of the dead-air region M_2. In incompressible flow, $M_2 \approx 0.1$, h_a/h has a value of 0.562. The value of h_a/h increases somewhat with increasing Mach number.

The second outflow function $c_{Qa}/(h/d)$ is shown in Figs. 33, 34, and 35 in dependence of the base pressure coefficient c_{pB}. It will be remembered that h/d, which is proportional to the jet spread parameter σ, cannot be determined theoretically so far.

* $h_a = y_t - y_r$ and $h = y_g - y_r$.

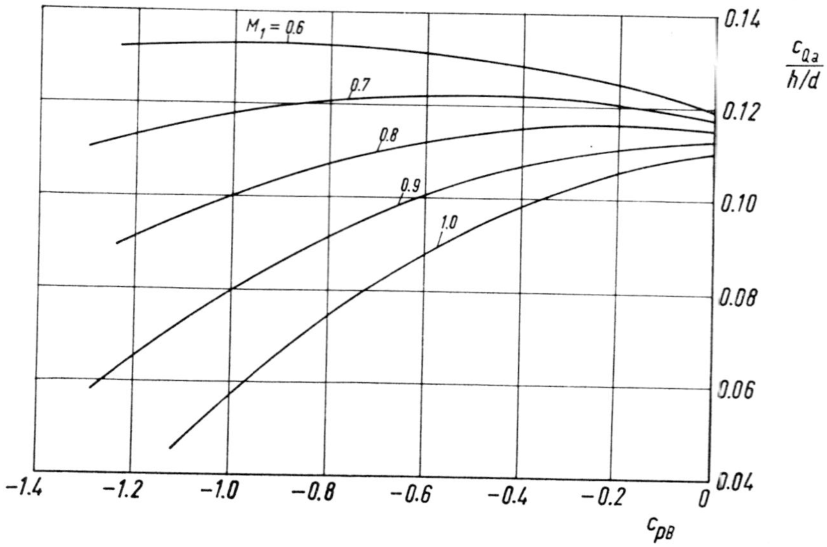

FIG. 34. The second outflow function for Mach numbers from $M_1 = 0.6$ to 1.0.

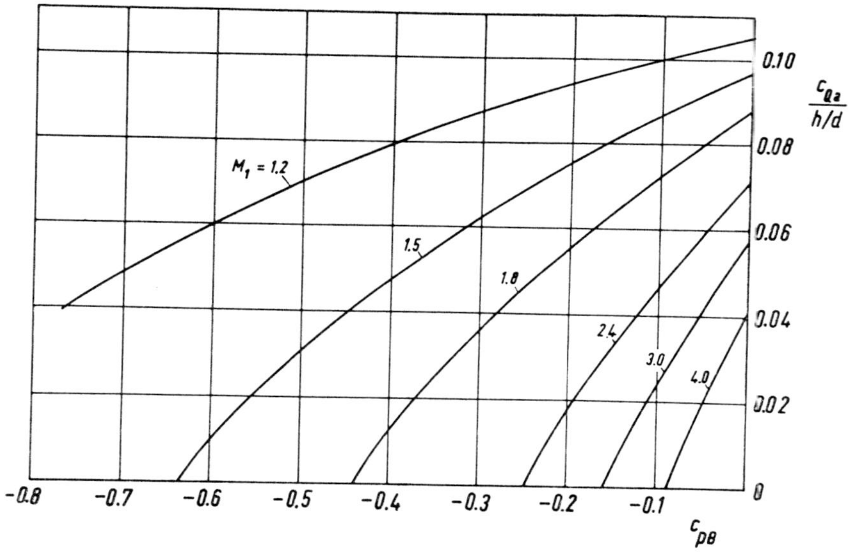

FIG. 35. The second outflow function for Mach numbers from $M_1 = 1.2$ to 4.0.

7.6. *The First Outflow Function*

For deriving the first outflow function the equations of continuity, energy and momentum are applied on another control volume as shown in Fig. 36. Section I lies at the end of the outflow region. Section II is chosen to be downstream of the reattachment point, at the smallest distance from the body at which the static pressure is already with sufficient accuracy equal to the static pressure in the undisturbed flow, $p = p_1$. The equations for continuity, energy and momentum may be written as follows:

$$\int_{y_t}^{y_g+h_1} \rho U \, dy = \int_0^{H_1} \rho U \, dy \tag{79}$$

$$\int_{y_t}^{y_g+h_1} (c_p T + \tfrac{1}{2}U^2) \rho U \, dy = \int_0^{H_1} (c_p T + \tfrac{1}{2}U^2) \rho U \, dy \tag{80}$$

$$c_D = \frac{2}{d} \int_0^{H_1} \frac{\rho}{\rho_1} \frac{U}{U_1} \left(1 - \frac{U}{U_1}\right) dy. \tag{81}$$

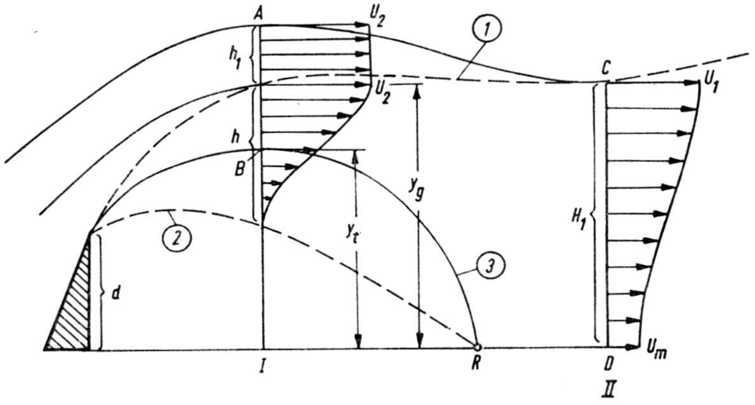

FIG. 36. Flow model for derivation of the first outflow function. (1) = outer boundary of mixing region, (2) = boundary of dead-air region, (3) = dividing streamline, R = reattachment point, I = section at which the outflow from the dead-air region ceases, II = section in the wake at which already $p = p_1$ = const.

These three equations contain the quantities h_1/d, h/d and H_1/d*. It is therefore possible to express all of these, e.g. in drag coefficient c_D. By expressing h/d in c_D and using an additional equation connecting c_D and c_{pB} from any classical potential flow solution (see section 1) it is possible to get the dependence

$$\frac{h}{d} = f(c_{pB}) \tag{82}$$

between c_{pB} and h/d. By using eq. (78) one at least gets the first outflow function

$$c_{Qa} = f_1(c_{pB}, c_D). \tag{83}$$

It must be stressed that the (h/d)-values in eqs. (78) and (83) generally are different. In eq. (78) there is a h/d satisfying the mixing requirements, whereas in eq. (83) h/d must satisfy the

* Since $y_g - y_t = h_a$ and h_a/h is given in Fig. 32.

drag requirements. Only at the point of intersection of both the functions both (h/d)-values are of equal magnitude.

7.7. *The Velocity Profile in Section II*

The velocity profile downstream of the reattachment point, in section II, is approximated by

$$\frac{U}{U_1} = \frac{U_m}{U_1} + \tfrac{1}{2}\left(1 - \frac{U_m}{U_1}\right)\left[1 + \sin\left(\frac{\pi y}{H} - \frac{\pi}{2}\right)\right]. \tag{84}$$

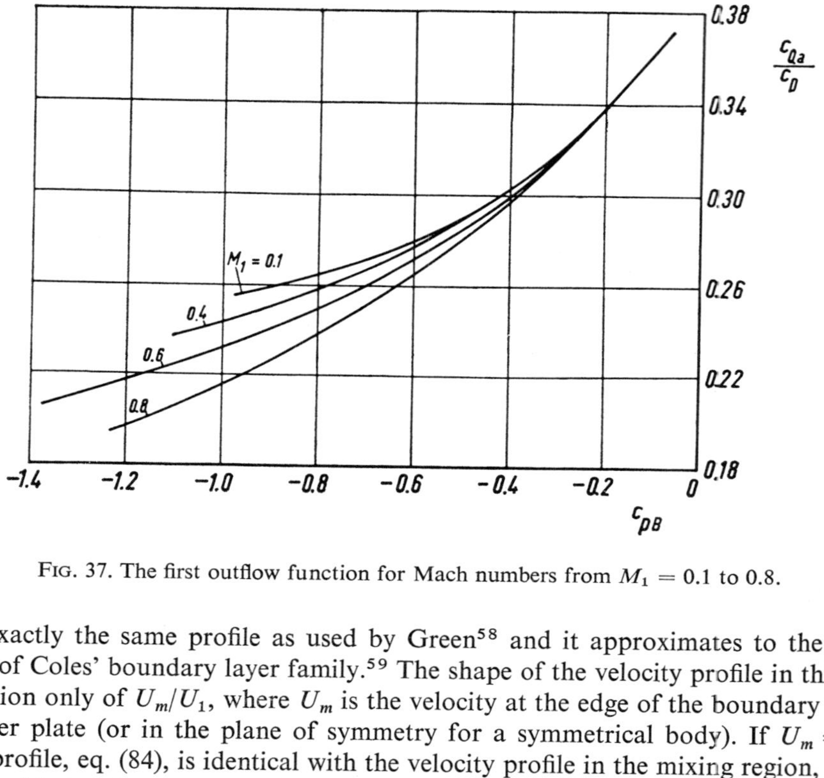

FIG. 37. The first outflow function for Mach numbers from $M_1 = 0.1$ to 0.8.

This is exactly the same profile as used by Green[58] and it approximates to the "wake" function of Coles' boundary layer family.[59] The shape of the velocity profile in this family is a function only of U_m/U_1, where U_m is the velocity at the edge of the boundary layer on the splitter plate (or in the plane of symmetry for a symmetrical body). If $U_m = 0$, the velocity profile, eq. (84), is identical with the velocity profile in the mixing region, eq. (75).

By considering the entropy and mass flow condition in the wake, especially in section II (entropy cannot decrease, the mass flow in the wake cannot decrease), it was found that the smallest possible value for U_m/U_1 at subsonic speeds is equal to 0.16. This value was then used in section II. It may be stressed that the value of U_m/U_1 is not very critical.

Figure 37 shows the first outflow function as calculated from eqs. (79), (80) and (81) with the computer IBM 7040.

7.8. *Theoretical Results*

7.8.1. *Incompressible flow*

In incompressible flow one can use the semi-empirical relation for h/d, derived in ref. 49:

$$\frac{h}{d} = 1.13 \sqrt{|c_{pB}|} + 1.32 \frac{|c_{pm}|}{|c_{pB}|} \tag{85}$$

which is valid for negative base pressure coefficients. This equation makes it possible to predict the base pressure coefficient theoretically. In Fig. 38 such a prediction is shown for a 60° wedge. Figure 39 shows a comparison between the theoretical and experimental dimensionless rate of mass outflow from the dead-air region, and Fig. 40 the theoretical and experimental base pressure coefficients in dependence of the wedge angle. The agreement between theory and experiment is good. In Fig. 41 one can see the theoretical prediction of the base pressure of a circular cylinder at laminar separation. The theoretical value $c_{pB} = -0.45$ is in agreement with the experimental, which are $c_{pB} = -0.47$ as measured by Roshko[12] and $c_{pB} = -0.43$ as measured by Tanner.[49]

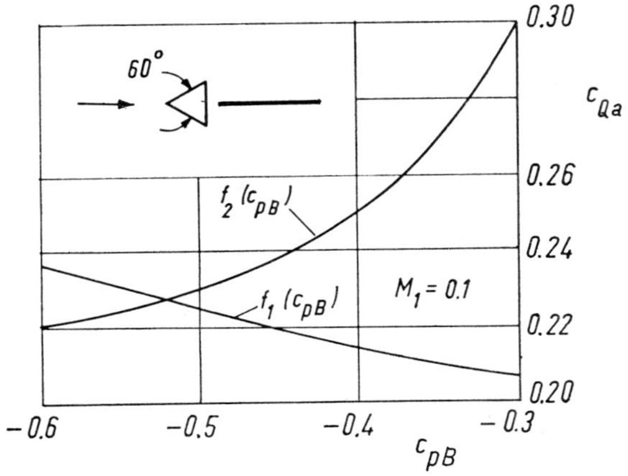

FIG. 38. Theoretical prediction of base pressure coefficient and dimensionless rate of mass outflow from the dead-air region for the flow past a 60° wedge.

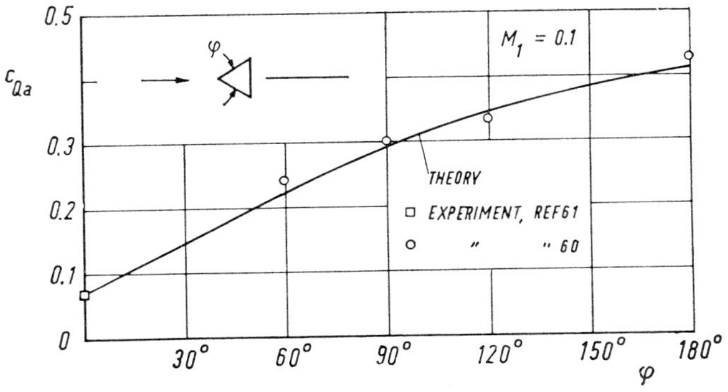

FIG. 39. Comparison of theoretical rate of mass outflow from the dead-air region with experiment.

7.8.2. *Compressible flow*

Using measured values of the drag and base pressure for a 30° wedge with a long splitter

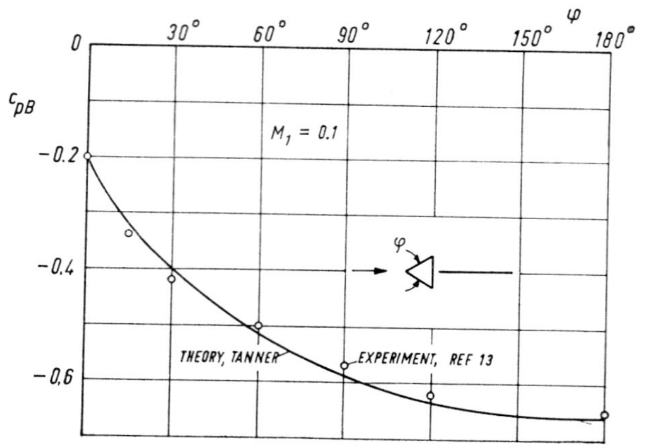

FIG. 40. Comparison of theoretical base pressure coefficients with experimental values.

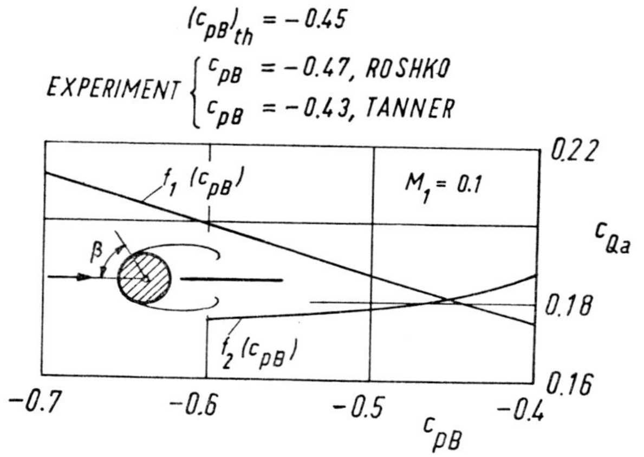

FIG. 41. Theoretical prediction of base pressure coefficient for the flow past a circular cylinder. Laminar separation with $\beta = 81°$.

plate,[20] the theoretical value of h/d was determined (Fig. 42). It can be seen that h/d increases with increasing Mach number.

The theoretical prediction of the influence of the Mach number on the dimensionless rate of mass outflow from the dead-air region and on the base pressure coefficient, for a flow past a back step, is shown in Fig. 43. Here the value of h/d is assumed to be constant, $h/d = 0.50$. The theoretical dependence of base pressure on Mach number is shown in Fig. 44, together with some experimental values by Nash *et al.*[28] A better agreement with experiment than for a constant h/d is obtained if it is assumed that h/d increases with increasing Mach number, which is plausible according to Fig. 42.

In Fig. 45 the theoretical c_{Qa}-value is shown as a function of the Mach number for a flow past a back step. Here again, h/d is assumed to be constant with a value of 0.50. It can be seen that c_{Qa} decreases with increasing Mach number.

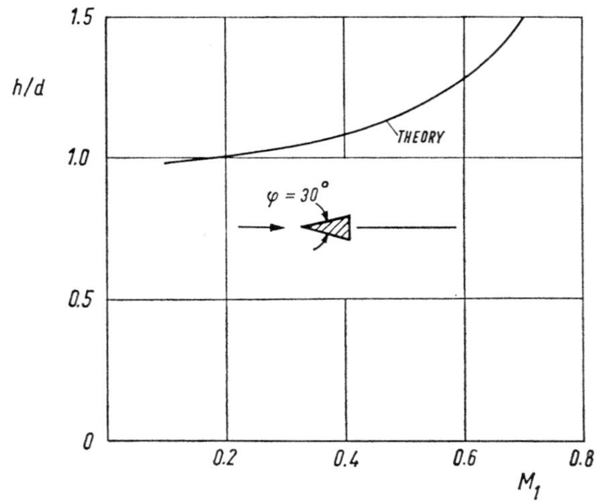

FIG. 42. Theoretical value of h/d as function of the Mach number for the flow past a 30° wedge.

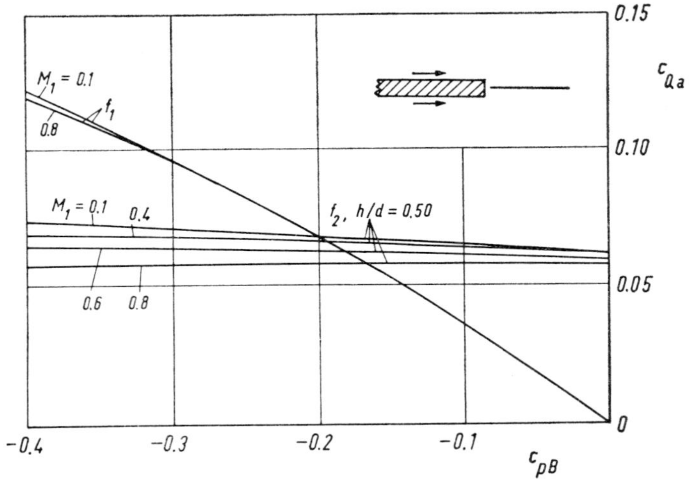

FIG. 43. Theoretical prediction of base pressure coefficient and dimensionless rate of mass outflow from the dead-air region for the subsonic flow past a back step.

7.9. *Concluding Remarks*

In its present state the theory of Tanner is valid for subsonic flows only. This is due to the fact that the first outflow function is valid only for Mach numbers at the edge of the mixing region, which are below unity. By developing a first outflow function valid for supersonic Mach numbers it is, in principle, possible to extend the theory for prediction of base pressures at supersonic velocities, too.

It should be noted that the ratio of the temperature layer thickness to the shear layer thickness, at the end of the outflow region, has been taken from experiments, for the wedge angles $\varphi = 0°$ and $\varphi = 180°$. It was thereafter assumed that this ratio was a function of the base pressure only. Further experimental results for this problem would be welcome.

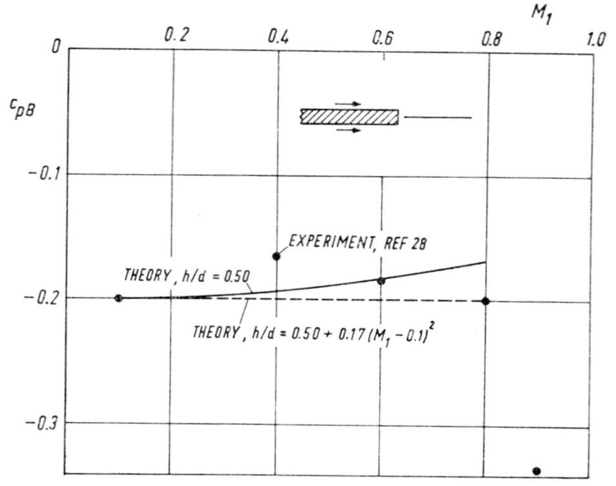

FIG. 44. Comparison of theoretical base pressure coefficients with experiment for the subsonic flow past a back step.

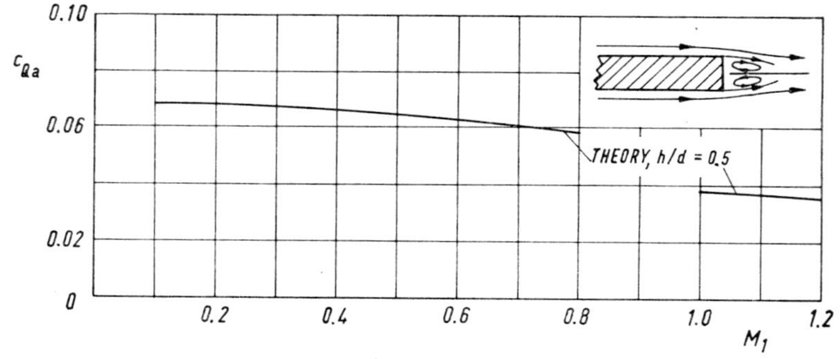

FIG. 45. Theoretical rate of mass outflow from the dead-air region for the subsonic and transonic flow past a back step.

Another parameter, which at this state of the theory cannot be determined theoretically, is h/d. This quantity is closely related to the jet spread parameter σ as used in other theories. Because the jet spread parameter is of great importance, we will give a short survey of works concerning this subject in the next section.

8. THE SIMILARITY PARAMETER OF THE MIXING REGION

8.1. *The Influence of the Shape of the Body at Incompressible Speeds*

In a mixing layer, the dimensionless coordinate in direction perpendicular to the mean flow is usually given by

$$\eta = \sigma \frac{y}{x} \tag{86}$$

where σ is a similarity parameter. For turbulent mixing, the mixing zone width varies linearly with the x-coordinate, which means that at constant Mach number σ also is constant.

At incompressible flow the value $\sigma = 12$ generally is used and this seems to be a good value for conditions, which correspond to the mixing between a uniform stream and a quiescent fluid.* Such a flow is, for example, the flow past a back step, if the walls upstream and downstream of the step are parallel. Some calculations of the present author,[62] which are based on experimental values,[49] show, however, that the angle between the wall from which the flow separates and the wall on which the flow reattaches, has a great influence on the value of σ. Figure 46 shows the similarity parameter as function of the wedge angle φ. For the flow past a flat plate perpendicular to the air stream ($\varphi = 180°$), σ has a value of about 6.

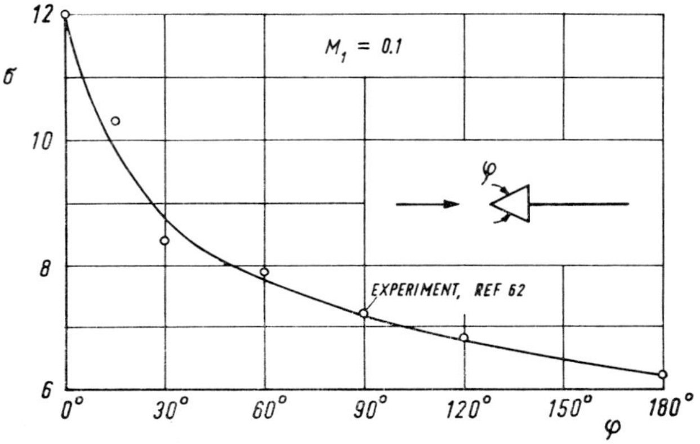

FIG. 46. Influence of wedge angle on the similarity parameter of the mixing region for incompressible flow.

Therefore the spread angle of the mixing layer in this case is about twice the value for a flow past a back step, i.e. for $\varphi = 0°$.

8.2. *The Influence of Compressibility*

Various experiments have shown that the similarity parameter σ is a function of the Mach number. In the light of a relatively limited number of experiments, Korst and Tripp[63] suggested the linear relationship

$$\frac{\sigma}{\sigma_0} = 1 + 0.23\, M_2 \tag{87}$$

for the relation between σ and Mach number M_2 at the outer edge of the mixing region. σ_0 is the value of the similarity parameter in incompressible flow. This empirical relation has been used in many base pressure analyses (e.g. in refs. 14, 22 and 37).

McDonald[39] takes account for the experimental results obtained in ref. 64 and develops the equation

$$\frac{\sigma}{\sigma_0} = \left[1 + \frac{\gamma - 1}{2} M_2^2\right] [1 + 0.035\, M_2^2] \tag{88}$$

* It might be mentioned that after careful measurements of Reichardt[104] σ has the value of 13.5.

for the influence of the Mach number. In a later paper[65] he, in view of remarks made in refs. 66 and 67, suggests a different relation between σ and Mach number M_2, which is given by

$$\frac{\sigma}{\sigma_0} = \frac{\left(1 + \frac{\gamma - 1}{2} M_2^2\right)(1 + 0.35\, M_2^2)}{1 + 0.004\, M_2^4}. \tag{89}$$

In an analysis given by Channapragada[68] the shear stress is based on Prandtl's mixing length hypothesis. Using the Howarth[84] transformation the shear stress for an incompressible flow is obtained. Following the assumption of Mager[85] that the shear stress is unaffected by the transformation, Channapragada develops the formula

$$\frac{\sigma}{\sigma_0} = \frac{\rho}{\rho_r} \tag{90}$$

where ρ_r is the density at some reference condition. Defining a compressible divergence factor R_1, which is a function of the Mach number as given in ref. 68, the final equation is

$$\frac{\sigma}{\sigma_0} = R_1 \left[1 + \frac{\beta_1}{Z}\right]^{-1} \tag{91}$$

with

$$Z = 1 + \frac{\gamma - 1}{2} M_1^2 \tag{92}$$

and

$$\beta_1 = \frac{T_{01}}{T_{02}} \tag{93}$$

where T_{01} is the total temperature in the outer flow and T_{02} that in the dead-air region. It will be remarked that when $M_1 \to \infty$, the value of $\sigma/\sigma_0 \to 4$, which therefore is the upper limit for this ratio.

Further contributions to the jet spread problem have been given, e.g. by Bauer in refs. 69 and 70.

In Fig. 47 various methods for predicting the effect of the Mach number on the jet spreading parameter σ are shown together with a great number of experimental results collected from refs. 50, 59 and 64 to 74. The results show a considerable scatter, but nevertheless there is a definite trend for the ratio σ/σ_0 to increase with increasing Mach number. Because of lacking experimental results at Mach numbers over 3.5 it is rather difficult to say which of the semiempirical methods gives the best agreement with measurements. It seems, however, that the method of ref. 39 gives too large σ/σ_0-values at Mach numbers above 3.5, approximately. Further measurements, in the whole Mach number range, would be welcome.

9. CONCLUSION

In the preceding sections we have given a review of existing theories for prediction of base pressure at steady base flows. The main points originating from this investigation may be summarized as follows:

A comparison of the theories based on the Chapman–Korst flow model with experiment shows that at supersonic velocities the theory of Korst gives the most satisfactory base

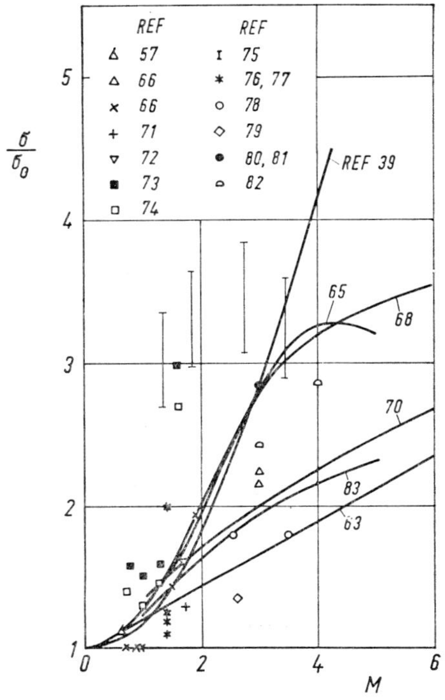

FIG. 47. Influence of Mach number on the similarity parameter of the mixing region.

pressure values for zero boundary layer thickness, provided that the Mach number is not too low, $M_1 > 1.6$. Then the agreement with experimental values for bodies approximating to a simple back step is good. In the transonic Mach number range the theory of Korst, however, gives base pressures which are much higher than those predicted by measurements. Also the influence of the body shape is not in all cases accounted for by the theory. Further, in this theory the reattachment criterion obviously is not in accordance with real physical facts.

Other theorists have later improved the reattachment criterion used by Korst. Especially Nash was then able to get, at least qualitatively, a better agreement with experiment in the transonic Mach number range. The reattachment parameter introduced by Nash and those introduced by Roberts and McDonald must, however, be taken from measurements and it seems that they vary with Mach number and body shape in a manner which at present is difficult to account for. Furthermore, the assumption of an isentropic compression along the dividing streamline, as used in all of these theories, is possibly not justified.

The influence of the boundary layer will be given by various theories in qualitative agreement with measurements. It seems, however, that the effect of the boundary layer thickness on the base pressure is overestimated by most theories. The best agreement with experiment is obviously given by the theory of McDonald. The effect of base bleed can also be calculated by some of the theories.

A different approach to the base pressure problem is given by the theory of Tanner, which is based on the concept of outflow from the dead-air region. This theory is at present valid for subsonic flow only. At incompressible flow the base pressure is given in good agreement with experiment for various body shapes. For the flow past a back step, the influence of compressibility seems to be properly accounted for. The method gives also the maximum rate of mass outflow from the dead-air region. It remains to see how successful the theory is in predicting the base pressure at supersonic Mach numbers. For this theory a quantity, which is proportional to the similarity parameter of the jet mixing region, must be taken from experiment and additionally also the ratio of temperature layer thickness to shear layer thickness at the end of the outflow region.

It can thus be seen that up to now many encouraging results have been achieved by theoretical investigations concerning the base pressure problem. But this review shows also that in spite of it the need for further research in this field is urgent.

10. REFERENCES

1. G. Kirchhoff, Zur Theorie freier Flüssigkeitsstrahlen, *Crelles Journal für Mathematik* **70** (1869), pp. 289–298.
2. L. Prandtl, A. Betz and C. Wieselsberger, *Ergebnisse der Aerodynamischen Versuchsanstalt zu Göttingen*, II. Lfg. Oldenbourg, München-Berlin 1923, p. 33; IV. Lfg. 1932, p. 99.
3. K. Kraemer, Die Druckverteilung am Keil bei inkompressibler Strömung, ein Beitrag zum Totwasserproblem, *Mitteilungen aus dem Max-Planck-Institut für Strömungsforschung und der Aerodynamischen Versuchsanstalt*, No. 30 (1964), pp. 1–86.
4. G. Birkhoff and E. H. Zarantonello, *Jets, Wakes and Cavities.* Academic Press, New York, 1957.
5. R. Eppler, Beiträge zur Theorie und Anwendung der unstetigen Strömungen, *J. Rat. Mech. Anal.* **3** (1954), pp. 591–644.
6. A. Roshko, A new hodograph for free-streamline theory, NACA TN 3168 (1954).
7. M. Tanner, Zur Bestimmung des Totwasserwiderstandes mit Anwendung auf Totwasser hinter Keilen, *Mitteilungen aus dem Max-Planck-Institut für Strömungsforschung und der Aerodynamischen Versuchsanstalt*, No. 31 (1964), pp. 1–58.
8. D. Gilbarg, Jets and cavities, *Handb. Phys.* IX, Springer-Verlag, 1960, pp. 311–445.
9. D. R. Chapman, An analysis of base pressure at supersonic velocities and comparison with experiment, NACA Rep. 1051 (1951).
10. D. R. Chapman, D. M. Kuehn and H. K. Larson, Investigation of separated flows in supersonic and subsonic streams with emphasis on the effect of transition, NACA Rep. 1356 (1958).
11. D. R. Chapman, Laminar mixing of a compressible fluid, NACA Rep. 958 (1950).
12. A. Roshko, On the drag and shedding frequency of two-dimensional bluff bodies, NACA TN 3169 (1954).
13. M. Tanner, Totwasserbeeinflussung bei Keilströmungen, *Deutsche Luft- und Raumfahrt Forschungsbericht* 64–39 (1964).
14. H. H. Korst, A theory for base pressures in transonic and supersonic flow, *J. Appl. Mech.* **23** (1956), pp. 593–600.

14a. H. H. Korst, Zur theoretischen Bestimmung des Dellendruckes bei abgelöster Strömung, *Oesterreichisches Ingenieur-Archiv* **10** (1958), pp. 227–238.

15. H. H. Korst, Auflösung eines ebenen Freistrahlrandes bei Berücksichtigung der ursprünglichen Grenzschichtströmung, *Oesterreichisches Ingenieur-Archiv* **7**, No. 2 (1954).
16. H. H. Korst, R. H. Page and M. E. Childs, Compressible two-dimensional jet mixing at constant pressure, University of Illinois, ME-TN 392-1 (1954).
17. D. R. Chapman, W. R. Wimbrow and R. H. Kester, Experimental investigations of base pressures on blunt-trailing-edge wings at supersonic velocities, NACA TN 2611 (1952).
18. H. Eggink, The improvement in pressure recovery in supersonic wind tunnels, ARC R&M 2703 (1952).
19. E. W. E. Rogers, C. J. Berry and V. G. Quincey, Tests at transonic speeds on wings with wedge sections and sweep varying between 0° and 60°, ARC R&M 3348 (1963).
20. M. Tanner, Druckverteilungsmessungen an Keilen bei kompressibler Strömung, *Zeitschrift für Flugwissenschaften* **18** (1970), pp. 202–208.
21. F. N. Kirk, An approximate theory of base pressure in two-dimensional flow at supersonic speeds, RAE TN Aero 2377 (1959).
22. J. F. Nash, An analysis of two-dimensional turbulent base flow, including the effect of the approaching boundary layer, ARC R&M 3344 (1963).
23. W. Tollmien, Berechnung turbulenter Ausbreitungsvorgänge, *Zeitschrift für angewandte Mathematik und Mechanik* **6** (1926), pp. 468–478. (Translated as NACA Tech. Memo. 1085 (1945).)

24. H. GOERTLER, Berechnung von Aufgaben der freien Turbulenz auf Grund eines neuen Näherungsansatzes, *Zeitschrift für angewandte Mathematik und Mechanik* **22** (1942), pp. 244–254.
25. M. SIRIEIX, Pression de culot et processus de mélange turbulent en écoulement supersonique plan, *La Recherche Aéronautique* No. 78 (1960), p. 13.
26. M. A. BADRINARAYANAN, An experimental investigation of base flows at supersonic speeds, *J. Roy. Aer. Soc.* **65** (1961), p. 475.
27. H. THOMANN, Measurements of heat transfer and recovery temperature in regions of separated flow at a Mach number of 1.8, FFA Rep. 82, Sweden (1959).
28. J. F. NASH, V. G. QUINCEY and J. CALLINAN, Experiments on two-dimensional base flow at supersonic and transonic speeds, NPL Aero. Rep. 1070 (1963).
29. S. M. BOGDONOFF, C. E. KEPLER and E. SANLORENZO, A study of shock wave turbulent boundary layer interaction at $M = 3$, Princeton University, Dept. of Aero. Eng. Rep. 222 (1953).
30. I. E. VAS and S. M. BOGDONOFF, Interaction of a shock wave with a turbulent boundary layer at $M = 3.85$, Princeton University, Dept. of Aero. Eng. Rep. 294; AFOSR TN 55-199 (1955).
31. R. J. HAKKINEN, I. GREBER, L. TRILLING and S. S. ABARBANEL, The interaction of an oblique shock wave with a laminar boundary layer, M.I.T. Fluid Dynamic Research Group TR 57–1 (1957).
32. C. BOURQUE and B. G. NEWMAN, Reattachment of a two-dimensional jet to an adjacent flat plate, *Aero. Quart.* **11** (1960), pp. 201–232.
33. R. C. HASTINGS, Unpublished work, mentioned in ref. 22 (possibly identical with ref. 41).
34. G. E. GADD, D. W. HOLDER and J. D. REGAN, Base pressure in supersonic flow, ARC CP 271 (1955).
35. V. VAN HISE, Investigation of variation in base pressure over the Reynolds number range in which wake transition occurs for two-dimensional bodies at Mach numbers from 1.95 to 2.92, NASA TN D-167 (1959).
36. A. R. WAZZAN, Review of recent developments in turbulent supersonic base flow, *AIAA Journal* **3** (1965), pp. 1135–1138.
37. P. CARRIÈRE and M. SIRIEIX, Facteurs d'influence du recollement d'un écoulement supersonique, *Proc. 10th Int. Congr. of Appl. Mech.*, Stresa, Italy, 1960 (1962), p. 205.
38. J. B. ROBERTS, On the prediction of base pressure in two-dimensional supersonic turbulent flow, ARC R&M 3434 (1966).
39. H. MCDONALD, Turbulent shear layer reattachment with special emphasis on the base pressure problem, *Aero. Quart.* **15** (1964), pp. 247–280.
40. E. RESHOTKO and M. TUCKER, Effect of a discontinuity on turbulent boundary-layer-thickness parameters with application to shock-induced separation, NACA TN 3454 (1955).
41. R. C. HASTINGS, Turbulent flow past two-dimensional bases in supersonic streams, ARC R&M 3401 (1963).
42. P. CARRIÈRE and M. SIRIEIX, Résultats récents dans l'étude des problèmes de mélange et de recollement, *Proc. 11th Int. Congr. of Appl. Mech.*, Munich, 1964 (1966).
43. R. JACQUES and A. GAILLY, Mélange supersonique turbulent et application aux problèmes de recollement, AGARD Conference Proceedings No. 4, Separated Flows, Part 1 (1966), pp. 272–301.
44. J. P. BATHAM, A reattachment criterion for turbulent supersonic separated flows, *AIAA Journal* **7** (1969), pp. 154–156.
45. R. H. PAGE, Comments on "A reattachment criterion for turbulent supersonic flows", *AIAA Journal* **7** (1969), pp. 1659–1660.
46. J. P. BATHAM, Reply by author to R. H. Page, *AIAA Journal* **7** (1969), pp. 1660–1661.
47. J. P. BATHAM, Analysis of turbulent supersonic separated flows, ARC FM 3964 (1968).
48. M. TANNER, Ein Beitrag zur Theorie der kompressiblen abgelösten Strömungen um Keile, *AVA Bericht* 71 A 29 (1971).
49. M. TANNER, Ein Verfahren zur Berechnung des Totwasserdruckes und Widerstandes von stumpfen Körpern bei inkompressibler, nichtperiodischer Totwasserströmung, *Mitteilungen aus dem Max-Planck-Institut für Strömungsforschung und der Aerodynamischen Versuchsanstalt*, No. 39 (1967), pp. 1–79.
50. H. H. KORST and W. L. CHOW, Non-isoenergetic turbulent ($Pr_t = 1$) jet mixing between two compressible streams at constant pressure, NASA CR-419 (1966).
51. H. SCHLICHTING, *Boundary Layer Theory*, 6th edition, Verlag G. Braun, Karlsruhe; McGraw-Hill, New York (1968).
52. R. A. SEBAN, A. EMERY and A. LEVY, Heat transfer to separated and reattachment subsonic turbulent flows obtained downstream of a surface step, *J. Aerospace Sci.* **26** (1959), pp. 809–814.
53. P. S. KLEBANOFF and R. W. DIEHL, Some features of artifically thickened fully developed turbulent boundary layers with zero pressure gradient, NACA TN 2475 (1951).
54. R. A. SEBAN and L. H. BACK, Velocity and temperature profiles in turbulent boundary layers with tangential injection, *J. Heat Trans.* **84,** (1962), pp. 45–54.
55. A. FAGE and F. C. JOHANSEN, The structure of vortex sheets, *Phil. Mag.* **5,** Ser. 7 (1928), pp. 417–441; ARC R&M 1143 (1927).

56. R. Fail, J. A. Lawford and R. C. W. Eyre, Low-speed experiments on the wake characteristics of flat plates normal to an airstream, ARC R&M 3120 (1959).
57. L. C. Crane, The laminar and turbulent mixing of jets of compressible fluids, *J. Fluid Mech.* **3** (1957), pp. 81–92.
58. J. E. Green, Two-dimensional turbulent reattachment as a boundary-layer problem, AGARD Conference Proceedings No. 4, Separated Flows, Part 1 (1966), pp. 393–428.
59. D. Coles, The law of the wake in the turbulent boundary-layer, *J. Fluid Mech.* **1** (1956), pp. 191–226.
60. M. Tanner, Messungen im Rückströmgebiet hinter verschiedenen Keilen, *AVA-Bericht* 66 A 22 (1966).
61. I. Tani, Experimental investigation of flow separation over a step, Grenzschichtforschung IUTAM-Symposium, Freiburg/Br. 1957, Springer-Verlag (1958), pp. 377–386.
62. M. Tanner, Einfluß der Körperform auf den Aehnlichkeitsparameter der turbulenten Ausbreitungszone bei inkompressibler Strömung, *AVA-Bericht* IB 187-72 A 05 (1972).
63. H. H. Korst and W. Tripp, The pressure on a blunt trailing edge separating two supersonic two-dimensional air streams of different Mach numbers and stagnation pressures, but identical stagnation temperatures, Paper presented at the Mid-West conference on solid and fluid mechanics, University of Michigan, 1957.
64. R. C. Maydew and J. F. Reed, Turbulent mixing of compressible free jets, *AIAA Journal* (1963), pp. 1443–1444.
65. H. McDonald, The turbulent supersonic base pressure problem: A comparison between a theory and some experimental evidence, *Aero. Quart.* **17** (1966), pp. 105–126.
66. R. C. Maydew and J. F. Reed, Turbulent mixing of axi-symmetric compressible jets (in the half-jet region) with quiescent air, Sandia Corporation, Alburquerque, New Mexico, Report SC-4764 (1963).
67. G. W. Zumwalt and H. H. Tang, Transient base pressure study, Oklahama State University Research Dept. SBW-6 (1964).
68. R. S. Channapragada, Compressible jet spread parameter for mixing zone analyses, *AIAA Journal* **1** (1963), pp. 2188–2190.
69. R. C. Bauer, An analysis of two-dimensional laminar and turbulent compressible mixing, *AIAA Journal* **4** (1966), pp. 392–395.
70. R. C. Bauer, Another estimate of the similarity parameter for turbulent mixing, *AIAA Journal* **6** (1968), pp. 925–927.
71. P. B. Gooderum, G. P. Wood and M. J. Brevoort, Investigation with an interferometer of the turbulent mixing of a free supersonic jet, NACA Rep. 963 (1950).
72. D. Bershader and S. I. Pai, On turbulent mixing in two-dimensional supersonic flow, *J. Appl. Phys.* **21** (1950), p. 616.
73. B. B. Carey, An optical study of two-dimensional jet mixing, Ph.D. Thesis, Dept. of Physics, University of Maryland (1954).
74. S. I. Pai, *Fluid Dynamics of Jets*, chapter 5, p. 101, D. van Nostrand Co., Inc., New York (1954).
75. A. R. Anderson and F. R. Johns, Characteristics of free supersonic jets, exhausting into quiescent air, *Jet Propulsion* **25** (1955), pp. 13–15.
76. N. H. Johannesen, The mixing of free axially-symmetrical jets of Mach number 1.40, ARC R&M 3291 (1957).
77. N. H. Johannesen, Further results on the mixing of free axially symmetric jets of Mach number 1.40, ARC R&M 3292 (1959).
78. A. F. Charwat and J. K. Yakura, An investigation of two-dimensional supersonic base pressure, *J. Aero. Sci.* **25** (1958), pp. 122–128.
79. E. T. Pitkin and I. Glassman, Experimental mixing profiles of a Mach 2.6 free jet, *J. Aero Sci.* **25** (1958), pp. 791–793.
80. R. S. Channapragada, A compressible jet spread parameter for mixing zone analyses. Part I, United Technology Center, Tech. Memo. 14-63-U25 (1963).
81. R. S. Channapragada, A compressible jet spread parameter for mixing zone analyses. Part II, United Technology Center, Tech. Memo. 14-63-U31 (1963).
82. M. Sirieix and J. L. Solignac, Contribution à l'étude experimentale de la couche de mélange turbulent isobare d'un écoulement supersonique, AGARD Conference Proceedings No. 4, Separated Flows, Part 1 (1966), pp. 241–270.
83. R. S. Channapragada and J. P. Woolley, Turbulent mixing of parallel compressible free jets, AIAA Paper No. 65.006, June 14–18 (1965).
84. L. Howarth, Concerning the effect of compressibility on laminar boundary layers and their separation, *Proc. Roy. Soc.* **194** (1948), pp. 16–42.
85. A. Mager, Transformation of the compressible turbulent boundary layer, *J. Aero. Sci.* **25** (1958), pp. 305–311.
86. D. Beastall and H. Eggink, Some experiments on breakaway in supersonic flow. Part II, RAE TN Aero. 2061 (1951).

87. W. R. Wimbrow, Effects of base bleed on the base pressure of blunt trailing edge airfoils at supersonic speeds, NACA RM A54A07 (1954) (British TIL 4148).
88. J. D. Morrow and E. Katz, Flight investigation at Mach numbers from 0.6 to 1.7 to determine drag and base pressures on a blunt trailing edge airfoil and drag of diamond and circular arc air-foils at zero lift, NACA TN 3548 (1955).
89. R. Rebuffet, Effects de supports sur l'écoulement à l'arrière d'un corps, AGARD Report 302 (1959).
90. A. Roshko and G. J. Thomke, Flow separation and re-attachment behind a downstream facing step, Douglas Aircraft Company Report SM-43056-1 (1964).
91. R. A. White, Turbulent boundary layer separation from smooth convex surfaces in supersonic two-dimensional flow, Ph.D. Thesis, University of Illinois (1963).
92. K. L. Goin, Effects of plan form, airfoil section and angle of attack on the pressure along the base of blunt trailing edge wings at Mach numbers of 1.41, 1.62 and 1.96, NACA RM L52D21 (Ministry of Aviation TIB 3324) (1952).
93. E. J. Saltzman, Preliminary base pressures obtained from the X-15 airplane at Mach numbers from 1.1 to 3.2, NACA TN D-1056 (1961).
94. M. Tanner, Einfluß der Reynoldsazhl und der Grenzschichtdicke auf den Totwasserdruck bei der Umströmung von Keilen, *Deutsche Luft- und Raumfahrt Forschungsbericht* 65–18 (1965).
95. M. Tanner, Der Einfluß der Hinterkantenform auf den Widerstand eines Rechteckflügels im Machzahlbereich von $Ma_\infty = 0{,}5$ bis 2,2, *Deutsche Luft- und Raumfahrt Forschungsbericht* 71–85 (1971).
96. J. F. Nash, A discussion of two-dimensional turbulent base flows, ARC R&M 3468 (1967).
97. M. Tanner, Basisdruckmessungen an einer Platte und an einem Keil in dem Machzahlbereich von $Ma = 2{,}8$ bis 6,8, *AVA-Bericht* 70 A 21 (1970).
98. L. Crocco and L. Lees, A mixing theory for the interaction between dissipative flows and nearly isentropic streams, *J. Aero. Sci.* **19** (1952), pp. 649–676.
99. J. Rom, Supersonic two-dimensional base type flows, Technion (Israel) TAE Rep. 27 (1963).
100. J. Rom, Analysis of the near-wake pressure in supersonic flow using the momentum integral method, Technion (Israel) TAE Rep. 35 (1964).
101. J. Rom, Theory for supersonic two-dimensional laminar base-type flows using the Crocco–Lees mixing concepts, *J. Aero. Space Sci.* **29** (1962), pp. 963–968.
102. Stanley A. Berger, *Laminar Wakes*, American Elsevier Publishing Company, Inc., New York (1971).
103. Paul K. Chang, *Separation of Flow*, Pergamon Press (1970).
104. H. Reichardt, Gesetzmäßigkeiten der freien Turbulenz, *VDI-Forschungsheft* 414 (1942), 2. Aufl. (1951).

5

INTRODUCTION TO AERODYNAMIC NOISE THEORY

H. V. FUCHS and A. MICHALKE

Deutsche Forschungs- und Versuchsanstalt für Luft- und Raumfahrt, Institut für Turbulenzforschung, Berlin

CONTENTS

5

INTRODUCTION TO AERODYNAMIC NOISE THEORY

H. V. Fuchs and A. Michalke

Deutsche Forschungs- und Versuchsanstalt für Luft- und Raumfahrt, Institut für Turbulenzforschung, Berlin

ABSTRACT

Aerodynamic noise theory is reviewed in a somewhat unified form. The analysis is mainly based on Lighthill's wave-equation approach, although some derivations were kept more general in view of flow acoustic problems other than those directly associated with free turbulent jets. In Part I a general integral solution to the inhomogeneous wave equation is discussed including the effect of boundaries. Lighthill's acoustic analogy is described in terms of equivalent simple sources. In Part II certain approximations are given for the far-field sound intensity. Dimensional analysis is performed neglecting the effect of retarded time. Jet noise is treated particularly in Part III on the assumption of a specific structure of the turbulence convected by the flow. Some approaches alternative to Lighthill's theory are indicated. Part IV treats spectral methods in jet noise theory. Individual frequency components of sound are derived by means of a Fourier transform with respect to time. Their directivities are discussed on the basis of two differing methods.

GENERAL INFORMATION

This contribution is a review rather than a progress report. It may well serve as a thorough introduction to theoretical problems of aerodynamic sound generation. The material presented was originally compiled in 1970. It was first used as complementary lecture notes for students attending six-day short courses on aircraft noise in March 1971 which were arranged by the University of Tennessee, Space Institute, Tullahoma, and the Technische Universität, Institut für Luft- und Raumfahrt, Aachen. The presentation in four parts corresponding to four separate lectures was retained here, although all four parts together provide the background for theoretical approaches to aerodynamic noise in general and jet noise in particular.

The main purpose of the following notes will be to outline the starting-points and the general background for aerodynamic noise theories. It is noted that the framework for this has evolved from the pioneer work of Lighthill[1,2] and has been contributed to by a substantial number of workers like Curle, Ffowcs-Williams, Lilley, Phillips, Powell, Proudman and Ribner, to mention only a few. The corresponding literature is cited only where it is felt to be auxiliary to a proper comprehension of a subject. The list of references is therefore rather fragmentary and should not be taken as representative in any respect.

During the past twenty years a great deal of research activities has gone into aerodynamic noise theory. It cannot possibly be the aim here to review all the material on hand. Rather is it intended to make flow noise theory understandable to those who are not yet familiar but wish to make themselves thoroughly acquainted with this subject of still increasing importance.

The derivation of the governing equations will be kept as general as possible, whereas specifications feasible for special configurations and boundaries will be treated less comprehensively. More practical aspects as the reduction of noise from distinct sources can not be dealt with either.

NOTATION

a, a_0	speed of sound
A	surface
A_i	vector defined in section 5 of Part III
B	skalar defined in section 5 of Part III
c_i	particle velocity (vector)
$c_{ph} = \omega/\alpha$	phase velocity of an azimuth-frequency component
C	constant defined in eq. (53) Part III
D	nozzle diameter of circular jet
E	energy density in eq. (79) Part III
f	frequency
f_{peak}	frequency of peak in pressure spectrum
f_i	force density (vector), eq. (2) Part I.
F_i	vector defined in eq. (4) Part I
$F_{m\omega}$	phase of the complex $Q_{m\omega}$
g	function defined in eq. (42) Part I
	argument of delta function
	axial amplitude distribution in Part IV
h	displacement of two monopoles
I	sound intensity in far-field, eq. (19) Part II
$I_{m\omega}$	double integral defined by eq. (33) Part IV
$\lvert\hat{I}_r\rvert$	jet thickness factor
$\lvert\hat{I}_x\rvert$	convection factor
J	intensity per unit volume of turbulence as defined in section 6.3 Part II
$k = \omega/a_0$	wave-number of sound
k_i	wave-number vector
K_i	vector defined in eq. (37) Part I
l	displacement of two dipoles
L	linear length scale
	length of source region in Part IV
L_c	typical correlation length, see sketch on p. 249
m	rate of mass injection density, eq. (1) Part I
	integer azimuthal wave-number in Part IV
M	scalar denoted by eq. (25) Part I
	Mach number
$M_c = c_{ph}/a_0$	convection Mach number
n_i	outward normal to a surface
N_i	energy-flux density (vector) in eq. (79) Part III
O	origin of coordinates

p	pressure
p_m	complex *m*th azimuthal component of pressure
P	field point
	total sound power radiated
	covariance in eq. (18) Part III
	energy-source density in eq. (79) Part III
P_ω	frequency component of pressure
$P_{m\omega}$	azimuth-frequency component of pressure
q, q'	source functions, eqs. (11) and (10) Part I
Q	source point
$\hat{Q} \equiv \hat{Q}_\omega$	wave-number–frequency spectral function of source term
$\tilde{Q}$	radial amplitude distribution
$Q_{m\omega}$	azimuth-frequency component of source term
$\tilde{Q}_{m\omega}$	amplitude distribution of $Q_{m\omega}$ defined in eq. (35) Part IV
r	distance from Q to P
	radial coordinate of a cylindrical coordinate in system in Part IV
$\tilde{r}$	radial coordinate of a polar coordinate system
r_0	distance between source point and measuring point in Part IV
R	correlation coefficient
	radius of source region in Part IV
s	entropy
S	covariance in eq. (36) Part III
	Strouhal number
t	time
t^+	retarded time, see eq. (13) Part I
T_{ij}	tensor defined in eq. (25) Part I
U_0	jet exit velocity
U_c, U_i	eddy convection speed
V	volume
V_c	correlation volume
W	covariance in eq. (43) Part III
x_i	Cartesian coordinates of field point
y_i	Cartesian coordinates of source point
z_i	displacement in source region, see Fig. 7
α	axial wave number
β	anisotropy factor
γ	ratio of specific heats
δ	Dirac delta function
δ_{ij}	Kronecker delta
ϵ	mean rate of dissipation of energy per unit mass, number smaller than unity in section 5 of Part III
ζ_i	displacement in far-field, see Fig. 7
η	efficiency of aerodynamic sound production
η_i	displacement in convected system of coordinates
ϑ, θ	angle of emission
λ	wavelength of radiated sound
	integration variable proportional to τ'

Λ	length scale characteristic of flow or flow element
Π	covariance in eq. (20) Part III
ρ	fluid density
τ	time delay
τ'	integration variable in delta function
τ_{ij}	viscous stress tensor
χ	angle defined in eq. (21) Part IV
φ	azimuthal coordinate of a cylindrical coordinate system
ω	radian frequency

PART I: GENERAL ASPECTS OF AERODYNAMIC NOISE THEORY

1. GOVERNING EQUATIONS

We postulate a general fluid motion without any restrictions concerning the specific features of the continuous medium carried with the flow. The conservation equation for the mass in a volume element may then be written, in differential form and Cartesian coordinates x_i, as

$$\frac{\partial \rho}{\partial t} + \frac{\partial \rho\, c_i}{\partial x_i} = m\,(x_i, t). \tag{1}$$

A list of the symbols used here and in the subsequent sections may be found in the Notation above. The somewhat uncommon rate of mass injection density m is not given any specific meaning here. It is kept nonzero because this term on the right of the continuity equation does not complicate the following developments. It could strictly be justified and given real significance only for a flow in which the fluid consists of a mixture of two or more phases, e.g. gas bubbles in water or dust particles in a gas.[3]

The conservation equation for the momentum usually reads:

$$\rho\,\frac{\partial c_i}{\partial t} + \rho c_j\,\frac{\partial c_i}{\partial x_j} + \frac{\partial p}{\partial x_i} - \frac{\partial \tau_{ij}}{\partial x_j} = f_i(x_i, t). \tag{2}$$

The introduction of an externally applied force density f_i on the right of the momentum equation may be accepted more readily, although again it is not specified in our analysis. The inclusion of finite m and f_i here is merely to demonstrate their principal impact as inhomogeneities of a wave equation to be derived in the next section.

Finally an equation of state is postulated as a fixed functional relation connecting density ρ, pressure p and entropy s:

$$\rho = \rho(p, s); \quad d\rho = \left(\frac{\partial \rho}{\partial p}\right)_s dp + \left(\frac{\partial \rho}{\partial s}\right)_p ds, \tag{3}$$

where the reciprocal of $(\partial\rho/\partial p)_s$ equals the speed of sound squared, a^2. An alternative form of the momentum equation (2) may be obtained by subtracting c_i times the equation of continuity (1):

$$\frac{\partial \rho\, c_i}{\partial t} + \frac{\partial \rho\, c_i\, c_j}{\partial x_j} + \frac{\partial p}{\partial x_i} - \frac{\partial \tau_{ij}}{\partial x_j} = f_i + m c_i = F_i(x_i, t). \tag{4}$$

Equations (1) and (4) will serve as the stepping-stone to a theoretical model of sound generation by unsteady flows of utmost generality.

2. DERIVATION OF AN INHOMOGENEOUS WAVE EQUATION

It was Lighthill's[1] original idea to combine two equations similar to (1) and (4) in order to derive a wave equation for the fluid density ρ. Ribner[4] formulated an equivalent equation with the pressure p replacing ρ as the dependent variable. Along these lines, we differentiate eq. (1) by t:

$$\frac{\partial^2 \rho}{\partial t^2} + \frac{\partial^2 \rho\, c_i}{\partial x_i\, \partial t} = \frac{\partial m}{\partial t} \tag{5}$$

and eq. (4) by x_i:

$$\frac{\partial^2 \rho\, c_i}{\partial x_i\, \partial t} + \frac{\partial^2 \rho\, c_i\, c_j}{\partial x_i\, \partial x_j} + \frac{\partial^2 p}{\partial x_i^2} - \frac{\partial^2 \tau_{ij}}{\partial x_i\, \partial x_j} = \frac{\partial F_i}{\partial x_i}. \tag{6}$$

Subtracting (6) from (5) yields:

$$\frac{\partial^2 \rho}{\partial t^2} - \frac{\partial^2 p}{\partial x_i^2} = \frac{\partial^2}{\partial x_i\, \partial x_j}(\rho c_i\, c_j - \tau_{ij}) - \frac{\partial F_i}{\partial x_i} + \frac{\partial m}{\partial t}. \tag{7}$$

By adding $-a_0^2\, \partial^2 \rho / \partial x_i^2$ on both sides of eq. (7), one may formally derive a kind of wave equation, where a_0 is an arbitrary constant with the dimensions of a velocity:

$$\frac{\partial^2 \rho}{\partial t^2} - a_0^2 \frac{\partial^2 \rho}{\partial x_i^2} = \frac{\partial^2}{\partial x_i\, \partial x_j}(\rho c_i\, c_j - \tau_{ij}) + \frac{\partial^2}{\partial x_i^2}(p - a_0^2 \rho) - \frac{\partial F_i}{\partial x_i} + \frac{\partial m}{\partial t}. \tag{8}$$

In an alternative procedure one may add $a_0^{-2}\, \partial^2 p / \partial t^2$ on both sides of eq. (7), with the result:

$$\frac{1}{a_0^2}\frac{\partial^2 p}{\partial t^2} - \frac{\partial^2 p}{\partial x_i^2} = \frac{\partial^2}{\partial x_i\, \partial x_j}(\rho c_i\, c_j - \tau_{ij}) - \frac{\partial F_i}{\partial x_i} + \frac{\partial m}{\partial t} + \frac{1}{a_0^2}\frac{\partial^2}{\partial t^2}(p - a_0^2 \rho). \tag{9}$$

We note that neither equation involves the restrictive assumption that the fluid obeys an equation of state. Being no more than a statement of mass and momentum conservation, in fact all continuous flows satisfy a wave equation like (8) or (9).

It also lies within the validity of these equations to choose the value of a_0 such that it equals the adiabatic speed of sound in the ambient uniform medium at rest in which the unsteady flow may be embedded. It is then obvious that far from the flow region itself the right-hand sides of eqns. (8) or (9) must vanish identically, leaving the well-known homogeneous wave equations which, under isentropic conditions, govern linear acoustics in a uniform medium at rest. One may now just as well imagine the medium as being at rest at any point in space and, to be consistent, interpret all the additional effects caused by the flow as the result of inhomogeneities, which are continuously distributed throughout a limited part of the medium. This idea seems a very logical one from the viewpoint of classical acoustics. For the region surrounding the flow it may help to illuminate what types of virtual sound sources are active in the noise-producing areas.

Analytically, the action of flow-induced sources upon a uniform medium at rest may be expressed by a series of forcing terms on the right of eqs. (8) or (9). For compactness, we write these source terms q' and q respectively:

$$\frac{\partial^2 \rho}{\partial t^2} - a_0^2 \frac{\partial^2 \rho}{\partial x_i^2} = q'(x_i, t) \tag{10}$$

and

$$\frac{1}{a_0^2}\frac{\partial^2 p}{\partial t^2} - \frac{\partial^2 p}{\partial x_i^2} = q(x_i, t). \tag{11}$$

It is emphasized that in q' or q not only are taken into account all sound-generating mechanisms but also are incorporated all such effects as convection and refraction of sound in the flow. The speed of sound in particular, may locally deviate very much from a_0, the value far from the flow.

If one were able to somehow fully identify q or q' such that it could be treated analytically in all details, then the problem of sound generation would in fact reduce to solving the inhomogeneous wave equation for a given source function. Of course, q or q' cannot in general be regarded as specified, since both source functions involve the quantities which are to be solved for, and to suppose that q or q' are *a priori* known is strictly to suppose the problem previously solved.

3. GENERAL INTEGRAL OF THE WAVE EQUATION

Suppose the source function q in eq. (11) completely known as explicit function in time and space. Then extremely general solutions of the wave equation are available from classical theories. For the formal transformation of a differential equation like (11) into an integral equation for the pressure, refer to a textbook on electromagnetics by Stratton (*Electromagnetic Theory*, New York, 1941), where this is rigorously treated in subsection 8.1.

In our notation, the volume V encloses both the noise-producing source region, where $q(t)$ is supposed to be known at any locus $Q(y_i)$, and the field point $P(x_i)$, where the fluctuating pressure $p(t) - p_0$ is to be calculated. The closed surface A envelopes the whole volume V as indicated in Fig. 1. The result, well known as Kirchoff integral, then reads for the pressure perturbation field:

$$p(x_i, t) - p_0 = \frac{1}{4\pi}\int_V \frac{1}{r}[q]\, dV(y_i) + \frac{1}{4\pi}\int_A \left\{\frac{1}{r}\left[\frac{\partial p}{\partial n}\right] + \frac{1}{r^2}\frac{\partial r}{\partial n}[p] + \frac{1}{a_0 r}\frac{\partial r}{\partial n}\left[\frac{\partial p}{\partial t}\right]\right\} dA(y_i). \tag{12}$$

Here and in the following the square brackets will, if not otherwise stated, always indicate evaluation at the so-called retarded time t^+:

$$[q] = q\,|_{t\,\mathrm{repl.by}\,t^+} = q(y_i, t^+) = q\left(y_i, t - \frac{r}{a_0}\right) \tag{13}$$

where r denotes the distance between the field point $P(x_i)$ and the point of emission $Q(y_i)$,

$$r = |x_i - y_i| = \sqrt{(x_1 - y_1)^2 + (x_2 - y_2)^2 + (x_3 - y_3)^2}. \tag{14}$$

n in eq. (12) is in the direction of the outward normal n_i to the surface A, which may be in one or several closed parts.

The physical meaning of the retarded time effects is that it takes a finite time for the sound emitted to travel the distance r from a source point to the distant point $P(x_i)$. Hence, it is essential to take values of the integrands in (12) at times varying with the integration variable y_i in order to get the correct superposition of all sound rays at a fixed point x_i at any instant t.

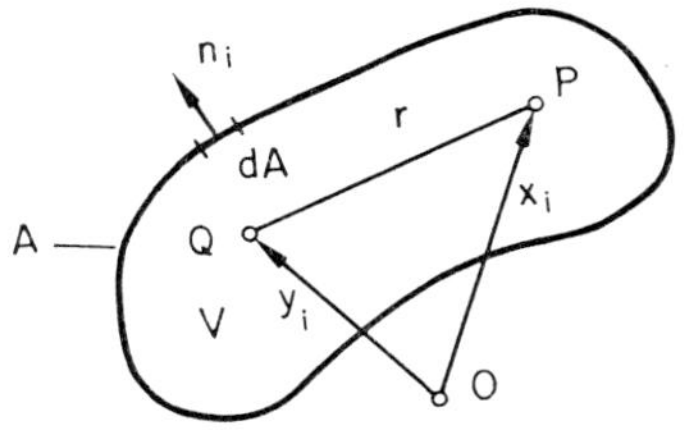

FIG. 1. Sketch illustrating notations.

In the evaluation of $p(x_i,t) - p_0$ in the way prescribed by eq. (12), one may distinguish three typical cases (Figs. 1a, b, c):

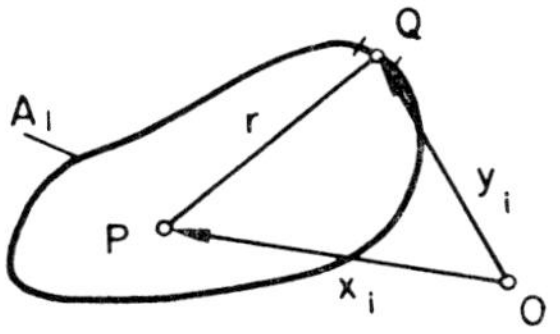

FIG. 1a.

(a) q vanishes interior to a finite V enclosed by A_1. Then a surface integral, over A_1 only, represents all possible contributions of q' exterior to A_1.

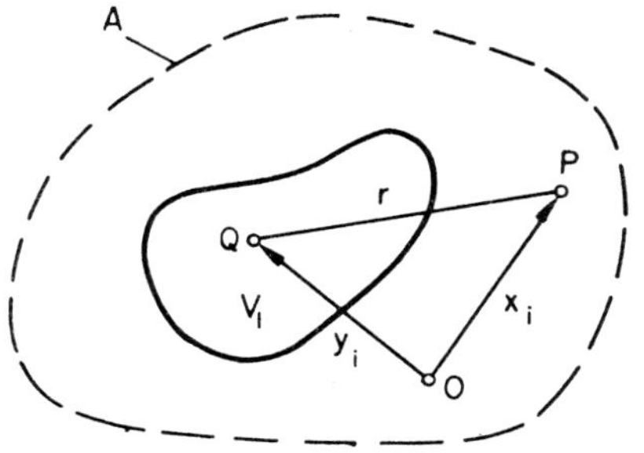

FIG. 1b.

(b) q vanishes exterior to a finite V_1 and the flow is continuous in the sense that no bounding surfaces are present. Then a volume integral over V_1 alone represents all possible contributions. This is because the surface A can under these circumstances be chosen to lie sufficiently far away from the source region V_1, so that the medium may be assumed undisturbed there; or better, disturbances have not yet arrived at A.

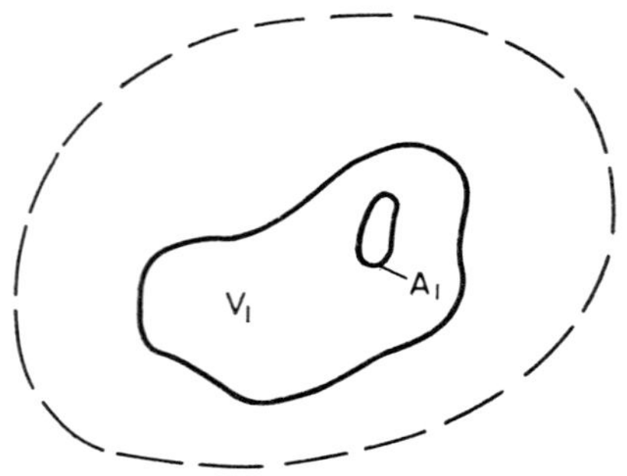

FIG. 1c.

(c) If there are discontinuities (e.g. solid surfaces A_1) present in a limited flow region V_1, then both volume and surface integrals may contribute to the pressure $p - p_0$ interior or exterior to the effective flow region.

In aerodynamic noise theory case (b) is dealt with when considering the noise generated by free shear flows (e.g. jet noise). The more general case (c) is to be considered in the theory of boundary-layer induced noise or all kinds of rotor noise. For the combined effects of flow and solid boundaries a further distinction is made, whether the boundaries are perfectly rigid or compliant.

Without making any such specifications, we shall first consider the surface and volume integrals in eq. (12) separately.

4. DEVELOPMENT OF THE SURFACE INTEGRALS

We introduce the vector of a surface element, $dA_j = n_j\, dA$, with

$$\frac{\partial \cdots}{\partial n}\, dA = \frac{\partial \cdots}{\partial y_j}\, n_j\, dA = \frac{\partial \cdots}{\partial y_j}\, dA_j \tag{15}$$

into the surface integrals (S.I.) of eq. (12):

$$\text{S.I.} = \frac{1}{4\pi}\int_A \left\{\frac{1}{r}\left[\frac{\partial p}{\partial y_j}\right] + \frac{1}{r^2}\frac{\partial r}{\partial y_j}\,[p] + \frac{1}{a_0 r}\frac{\partial r}{\partial y_j}\left[\frac{\partial p}{\partial t}\right]\right\} dA_j. \tag{16}$$

When using the trivial identities

$$\frac{\partial r}{\partial y_j} = -\frac{\partial r}{\partial x_j} = -\frac{x_j - y_j}{r} \tag{17}$$

and the convention

$$\frac{\partial \cdots}{\partial x_j} = \frac{\partial \cdots}{\partial x_i}\,\delta_{ij};\quad dA_j = \delta_{ij}\, dA_i, \tag{18}$$

then eq. (16) reads:

$$\text{S.I.} = \frac{1}{4\pi}\int_A \frac{1}{r}\left[\frac{\partial p\,\delta_{ij}}{\partial y_j}\right] dA_i - \frac{1}{4\pi}\int_A \left\{\frac{1}{r^2}\,[p\delta_{ij}]\,\frac{\partial r}{\partial x_i} + \frac{1}{a_0 r}\left[\frac{\partial p\,\delta_{ij}}{\partial t}\right]\frac{\partial r}{\partial x_i}\right\} dA_j. \tag{19}$$

Now by definition

$$\left[\frac{\partial\cdots}{\partial t}\right] = \frac{\partial\cdots}{\partial t}\bigg|_{t\,\text{repl. by}\,t^+} = \frac{\partial[\cdots]}{\partial t^+} \tag{20}$$

and from eq. (13):

$$\frac{\partial t^+}{\partial x_i} = -\frac{1}{a_0}\frac{\partial r}{\partial x_i} = \frac{1}{a_0}\frac{\partial r}{\partial y_i} = -\frac{\partial t^+}{\partial y_i}. \tag{21}$$

Identities (20) and (21) are used in the final transformation:

$$\frac{\partial}{\partial x_i}\left(\frac{1}{r}[\cdots]\right) = [\cdots]\frac{\partial}{\partial x_i}\frac{1}{r} + \frac{1}{r}\frac{\partial[\cdots]}{\partial t^+}\frac{\partial t^+}{\partial x_i} = -\frac{1}{r^2}[\cdots]\frac{\partial r}{\partial x_i} - \frac{1}{a_0 r}\left[\frac{\partial\cdots}{\partial t}\right]\frac{\partial r}{\partial x_i} \tag{22}$$

to obtain the surface integral (19) in the alternative form:

$$\text{S.I.} = \frac{1}{4\pi}\int_A \frac{1}{r}\left[\frac{\partial p\,\delta_{ij}}{\partial y_j}\right] dA_i + \frac{1}{4\pi}\int_A \frac{\partial}{\partial x_i}\left(\frac{1}{r}[p\delta_{ij}]\right) dA_j. \tag{23}$$

Since the integration over dA_j depends only on the coordinates y_i and not on the coordinates of the field point x_i, which are fixed during the integration, one may retain $\partial/\partial x_i$ outside the second integral:

$$\text{S.I.} = \frac{1}{4\pi}\int_A \frac{1}{r}\left[\frac{\partial p\,\delta_{ij}}{\partial y_j}\right] dA_i + \frac{1}{4\pi}\frac{\partial}{\partial x_i}\int_A \frac{1}{r}[p\delta_{ij}]\, dA_j. \tag{24}$$

This rather lengthy, though extremely simple reformulation of the surface integrals in eq. (12) will be succeeded by similar step-by-step derivations of other transformations to follow. This is done because these notes are not meant as just a review of a well-accepted theory, but as a toolbox for actual working in the field of flow noise.

5. CONSIDERATION OF THE VOLUME INTEGRALS

In this section we shall discuss the various contributions to the pressure field $p - p_0$ caused by the volume integrals of three fundamentally different components in the original source function q:

$$q = q_1 + q_2 + q_3$$

$$q_1 = \frac{\partial}{\partial t}\left\{m + \frac{1}{a_0^2}\frac{\partial}{\partial t}(p - a_0^2\rho)\right\} = \frac{\partial M}{\partial t}$$

$$q_2 = -\frac{\partial}{\partial x_i}(f_i + mc_i) = -\frac{\partial F_i}{\partial x_i}$$

$$q_3 = \frac{\partial^2}{\partial x_i\,\partial x_j}(\rho c_i c_j - \tau_{ij}) = \frac{\partial^2 T_{ij}}{\partial x_i\,\partial x_j}. \tag{25}$$

This will help to illustrate how one can draw a formal analogy of the sound generated aerodynamically to sound fields originating from ordinary acoustic point sources like monopoles, dipoles and quadrupoles. The distinction of three types of simple sources corresponds to three different ways in which one can cause kinetic energy to be converted into the acoustic energy of fluctuating longitudinal motions in a compressible medium:

1. By forcing the mass in a fixed region of space to fluctuate as with a loudspeaker diaphragm embedded in a very large baffle.
2. By forcing the momentum in a fixed region of space to fluctuate or, which is the same thing, forcing the rates of mass flux across fixed surfaces to vary; both these occur when a perfectly solid object vibrates without volume changes.
3. By forcing the rate of momentum flux across a fixed surface to vary. This is the typical feature when sound is generated by fluctuating shearing motions in a turbulent fluid flow.

This, after Lighthill[1], is a linear sequence of methods of energy conversion in that each of the described mechanisms prove less efficient than the preceding one. The individual character of the three types of sources mainly involved in aerodynamic noise will be discussed next by considering the acoustical equivalent of a volume element of q_1, q_2, q_3, in this order.

5.1. *Source Element as Equivalent Monopole Emitter*

The pressure variations caused by the volume integral over $[q_1]/4\pi r$ are:

$$p - p_0 = \frac{1}{4\pi} \int_V \frac{1}{r} \left[\frac{\partial M}{\partial t}\right] dV. \tag{26}$$

We now confine to a very small portion out of the continuous source distribution $q_1(y_i)$ as attached to an infinitesimal volume element. r may then be assumed constant during the integration:

$$p_M - p_0 = \frac{1}{4\pi r} \lim_{V\to 0} \int_V \left[\frac{\partial M}{\partial t}\right] dV. \tag{27}$$

The subscript M is chosen because of the expected monopole character of the source element. In order to relate this to ordinary acoustic terms, we may rewrite eq. (27) as

$$p_M - p_0 = \frac{1}{4\pi r} \rho_0 \left[\frac{\partial Q}{\partial t}\right] \tag{28}$$

by assuming that $[\partial M/\partial t]$ approaches infinity such that the result of the integration in (27) remains finite while $V \to 0$.

By setting

$$\left[\frac{dQ}{\partial t}\right] = \omega \hat{Q} f\left(t - \frac{r}{a_0}\right) \tag{29}$$

where ω is dimensionally a frequency, we find $p_M - p_0$ in exactly the form of a monopole-radiation field:

$$p_M - p_0 = \frac{\omega \rho_0 \hat{Q}}{4\pi} \frac{f\left(t - \dfrac{r}{a_0}\right)}{r}. \tag{30}$$

$\hat{Q}$ is a direct measure of the equivalent monopole-source strength at y_i in the flow. (One may compare eq. (30), for instance, with eqs. (63.6) and (73.7) in ref. 6.)

From our analysis $\hat{Q}$ is found finite only when $p - a_0^2\rho$ varies considerably with time or when the mass addition m possesses a fluctuating component. Lighthill[1] has argued that as long as heating of the fluid is kinetic, departure of p from $a_0^2\rho$ will be small and hence will not make a significant contribution to pressure fluctuations radiated from the disturbed fluid. Even if the fluid is artificially heated, as in the case of a hot jet issuing into a cold atmosphere, sound radiation from the second term in q_1 is thought to be negligible except, perhaps, under extraordinary circumstances.

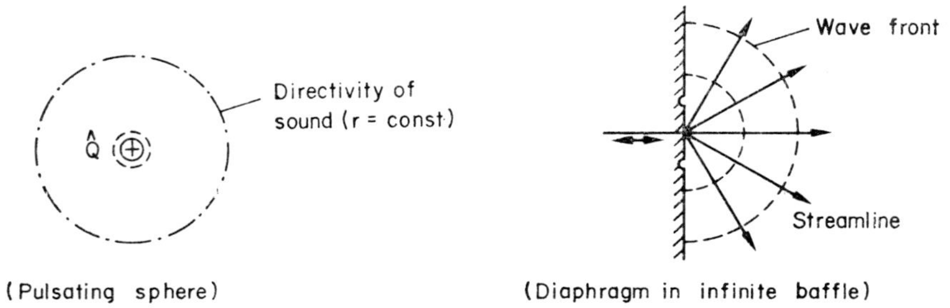

FIG. 2. Monopole-type source.

5.2. *Source Element as Equivalent Dipole Emitter*

The pressure variations caused by the volume integral over $[q_2]/4\pi r$ are:

$$p - p_0 = -\frac{1}{4\pi}\int_V \frac{1}{r}\left[\frac{\partial F_i}{\partial y_i}\right] dV. \tag{31}$$

To transform this integral, we need the following identities which will repeatedly be used in subsequent developments:

$$\frac{\partial[\cdots]}{\partial y_i} = \frac{\partial}{\partial y_i}\bigg|_{t^+=\text{const.}} [\cdots] + \frac{\partial}{\partial t^+}\bigg|_{y_i=\text{const.}} [\cdots]\frac{\partial t^+}{\partial y_i} \tag{32}$$

and with (17), (21) and (20):

$$\frac{\partial}{\partial y_i}\left(\frac{1}{r}[\cdots]\right) = [\cdots]\frac{\partial}{\partial y_i}\frac{1}{r} + \frac{1}{r}\frac{\partial[\cdots]}{\partial y_i} = \frac{1}{r^2}[\cdots]\frac{\partial r}{\partial x_i} + \frac{1}{r}\left[\frac{\partial\cdots}{\partial y_i}\right] + \frac{1}{a_0 r}\left[\frac{\partial\cdots}{\partial t}\right]\frac{\partial r}{\partial x_i}. \tag{33}$$

Finally by adding (22) and (33):

$$\frac{\partial}{\partial x_i}\left(\frac{1}{r}[\cdots]\right) + \frac{\partial}{\partial y_i}\left(\frac{1}{r}[\cdots]\right) = \frac{1}{r}\left[\frac{\partial\cdots}{\partial y_i}\right]. \tag{34}$$

When inserting (34) into eq. (31), and applying the divergence theorem

$$\int_V \frac{\partial\cdots}{\partial y_i}\, dV = \int_A \cdots\, dA_i, \tag{35}$$

we recognize that the pressure field may be thought of as the result of a volume integral over V and a surface integral over A enclosing V:

$$p - p_0 = -\frac{1}{4\pi}\frac{\partial}{\partial x_i}\int_V \frac{1}{r}[F_i]\,dV - \frac{1}{4\pi}\int_A \frac{1}{r}[F_i]\,dA_i. \tag{36}$$

We now proceed with only the volume integral in showing that a very small portion out of a continuous source distribution $q_2(y_i)$, as attached to an infinitesimal volume element, acts upon the surrounding fluid just like an acoustic dipole. In a procedure similar to that in subsection 5.1:

$$p_D - p_0 = -\frac{1}{4\pi}\frac{\partial}{\partial x_i}\left(\frac{1}{r}\lim_{V\to 0}\int_V [F_i]\,dV\right) = -\frac{1}{4\pi}\frac{\partial}{\partial x_i}\left(\frac{1}{r}[K_i]\right). \tag{37}$$

For simplicity, we assume the vector K_i, dimensionally a force, to be parallel to the x_1-direction:

$$K_i = \{K_1; 0; 0\}$$

$$\cos\vartheta = \frac{x_1 - y_1}{r} \tag{38}$$

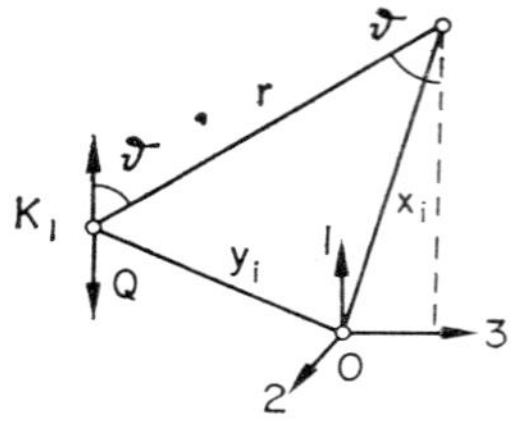

and set

$$[K_1] = \hat{K}_1 f\left(t - \frac{r}{a_0}\right). \tag{39}$$

Under these assumptions we find $p_D - p_0$ in the following form:

$$p_D - p_0 = -\frac{1}{4\pi}\frac{\partial r}{\partial x_1}\frac{\partial}{\partial r}\left(\frac{1}{r}[K_1]\right) = -\frac{\hat{K}_1}{4\pi}\cos\vartheta\frac{\partial}{\partial r}\frac{f\left(t-\frac{r}{a_0}\right)}{r} \tag{40}$$

where identity (17) has been used. This closely resembles, as we will show next, the superposition of the pressure fields of two monopoles, a very short distance h apart, pulsating in exact antiphase:

$$p - p_0 = \frac{\omega\rho_0}{4\pi}\left(\frac{\hat{Q}_1 f\left(t-\frac{r_1}{a_0}\right)}{r_1} + \frac{\hat{Q}_2 f\left(t-\frac{r_2}{a_0}\right)}{r_2}\right)$$

$$\hat{Q}_1 = -\hat{Q}_2 = -\hat{Q}. \tag{41}$$

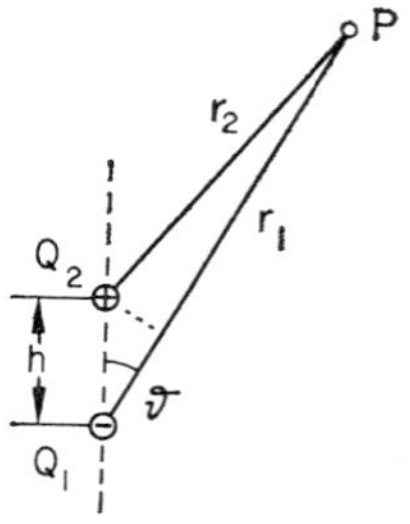

We now let $\hat{Q}$ become very big and simultaneously h become very small such that in the following procedure the product $h\hat{Q}$ remains constant at a certain value. Hence, one may write for the radii

$$r_1 \to r$$

$$r_2 \to r + \Delta r;\quad \Delta r = -h\cos\vartheta \quad \text{for } h \to 0.$$

For compactness, we set:

$$\frac{f\left(t - \frac{r}{a_0}\right)}{r} = g(t, r) \tag{42}$$

$$p - p_0 = \frac{\omega \rho_0 \hat{Q}}{4\pi} \{g(t, r + \Delta r) - g(t, r)\}$$

and hence we find $p - p_0$ in exactly the form of a dipole-radiation field:

$$p_D - p_0 = -\frac{\omega \rho_0 \hat{Q}\, h \cos \vartheta}{4\pi} \lim_{\Delta r \to 0} \frac{g(t, r + \Delta r) - g(t, r)}{\Delta r} = -\frac{\omega \rho_0 (h\hat{Q})}{4\pi} \cos \vartheta \frac{\partial}{\partial r} g(t, r) \tag{43}$$

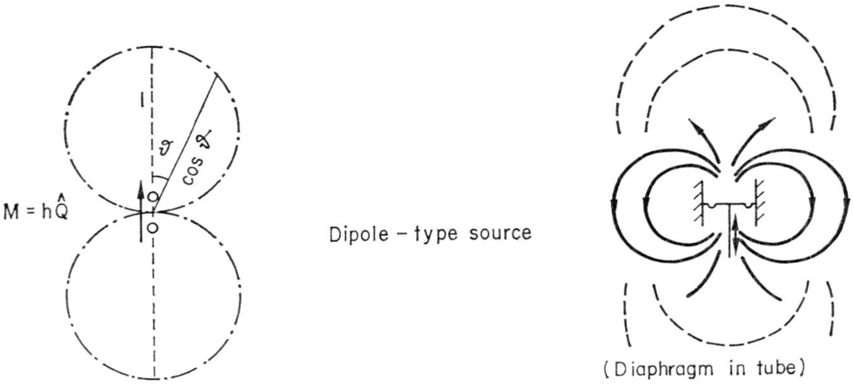

FIG. 3. Dipole-type source.

Thus, an isolated element of the source distribution q_2 would indeed act upon the surrounding fluid in just the same way as a dipole with

$$\hat{K}_1 \text{ equivalent to } \omega \rho_0 (h\hat{Q}) = \omega \rho_0 \hat{M}$$

being a measure of the source strength at y_i in the flow.

5.3. *Source Element as Equivalent Quadrupole Emitter*

The pressure variations caused by the volume integral over $[q_3]/4\pi r$ are finally considered:

$$p - p_0 = \frac{1}{4\pi} \int_V \frac{1}{r} \left[\frac{\partial^2 T_{ij}}{\partial y_i\, \partial y_j}\right] dV. \tag{44}$$

To transform this integral, we first apply relation (34) to T_{ij}:

$$\frac{1}{r} \left[\frac{\partial T_{ij}}{\partial y_j}\right] = \frac{\partial}{\partial x_j} \left(\frac{1}{r} [T_{ij}]\right) + \frac{\partial}{\partial y_j} \left(\frac{1}{r} [T_{ij}]\right) \tag{45}$$

and then apply the relation (34) to $\partial T_{ij}/\partial y_j$:

$$\frac{1}{r}\left[\frac{\partial^2 T_{ij}}{\partial y_i \, \partial y_j}\right] = \frac{\partial}{\partial x_i}\left(\frac{1}{r}\left[\frac{\partial T_{ij}}{\partial y_j}\right]\right) + \frac{\partial}{\partial y_i}\left(\frac{1}{r}\left[\frac{\partial T_{ij}}{\partial y_j}\right]\right). \tag{46}$$

We then insert (45) into the first term on the right of eq. (46):

$$\frac{1}{r}\left[\frac{\partial^2 T_{ij}}{\partial y_i \, \partial y_j}\right] = \frac{\partial^2}{\partial x_i \, \partial x_j}\left(\frac{1}{r}[T_{ij}]\right) + \frac{\partial^2}{\partial x_i \, \partial y_j}\left(\frac{1}{r}[T_{ij}]\right) + \frac{\partial}{\partial y_i}\left(\frac{1}{r}\left[\frac{\partial T_{ij}}{\partial y_j}\right]\right). \tag{47}$$

By using the divergence theorem (35) twice, we may split the volume integral in (44) into a volume plus two surface integrals:

$$p - p_0 = \frac{1}{4\pi}\frac{\partial^2}{\partial x_i \, \partial x_j}\int_V \frac{1}{r}[T_{ij}]\, dV + \frac{1}{4\pi}\frac{\partial}{\partial x_i}\int_A \frac{1}{r}[T_{ij}]\, dA_j + \frac{1}{4\pi}\int_A \frac{1}{r}\left[\frac{\partial T_{ij}}{\partial y_j}\right] dA_i. \tag{48}$$

We consider the volume integral in (48) in order to show that a very small portion of $q_3(y_i)$, as attached to an infinitesimal volume element, acts just like an acoustic quadrupole:

$$p_Q - p_0 = \frac{1}{4\pi}\frac{\partial^2}{\partial x_i \, \partial x_j}\left(\frac{1}{r}\lim_{V\to 0}\int_V [T_{ij}]\, dV\right) = \frac{1}{4\pi}\frac{\partial^2}{\partial x_i \, \partial x_j}\left(\frac{1}{r}[\theta_{ij}]\right). \tag{49}$$

Again for simplicity, we assume the tensor θ_{ij} to consist only of one component, θ_{11},

$$\cos\vartheta = \frac{x_1 - y_1}{r}$$

and set

$$[\theta_{11}] = \hat{\theta}_{11} f\left(t - \frac{r}{a_0}\right) \tag{50}$$

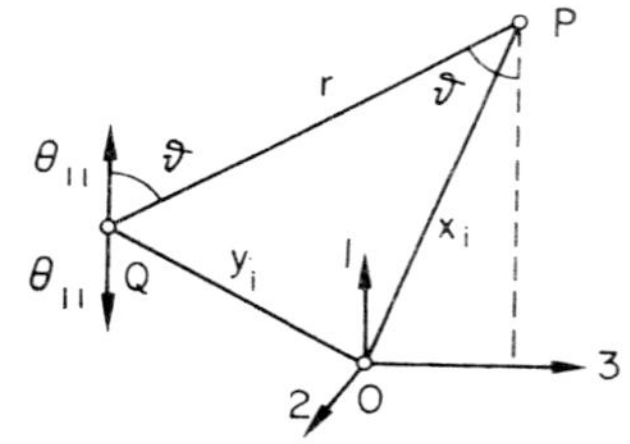

For such a source element we find with identity (17) and abbreviation (42):

$$p_Q - p_0 = \frac{\hat{\theta}_{11}}{4\pi}\frac{\partial}{\partial x_1}\left(\frac{\partial r}{\partial x_1}\frac{\partial g}{\partial r}\right) = \frac{\hat{\theta}_{11}}{4\pi}\frac{\partial r}{\partial x_1}\frac{\partial}{\partial r}\left(\frac{\partial r}{\partial x_1}\frac{\partial g}{\partial r}\right) = \frac{\hat{\theta}_{11}}{4\pi}\cos\vartheta\frac{\partial}{\partial r}\left(\cos\vartheta\frac{\partial g}{\partial r}\right). \tag{51}$$

This closely resembles, as we will finally show, the superposition of the pressure fields originating from two dipoles in line, a very short distance l apart, and oscillating in exact anti-phase:

$$p - p_0 = -\frac{\omega\rho_0}{4\pi}\left\{\hat{M}_1\cos\varphi_1\frac{\partial g_1}{\partial r_1} + \hat{M}_2\cos\varphi_2\frac{\partial g_2}{\partial r_2}\right\} \tag{52}$$

$$\hat{M}_1 = -\hat{M}_2 = -\hat{M}$$

$$\cos\varphi_1 = \frac{x_1 - y_1}{r_1} = \cos\vartheta$$

$$\cos\varphi_2 = \frac{(x_1 - l) - y_1}{r_2}.$$

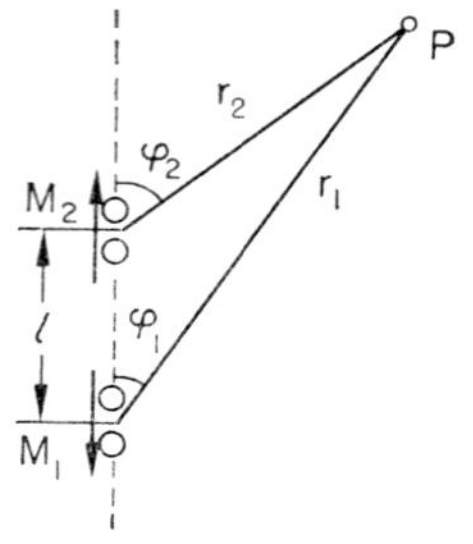

In a similar procedure as in subsections 5.1 and 5.2, now keeping the product $l\hat{M}$ constant, we have:

$$r_1 \to r$$

$$r_2 \to r + \Delta r; \quad \Delta r = -l \cos \vartheta \text{ for } l \to 0$$

$$p_Q - p_0 = \frac{\omega \rho_0 \hat{M} l \cos \vartheta}{4\pi} \lim_{\Delta r \to 0} \frac{1}{\Delta r} \left(\frac{x_1 - l - y_1}{r + \Delta r} \frac{\partial}{\partial (r + \Delta r)} g(t, r + \Delta r) \right.$$

$$\left. - \frac{x_1 - y_1}{r} \frac{\partial}{\partial r} g(t, r) \right) = \frac{\omega \rho_0 (l\hat{M})}{4\pi} \cos \vartheta \frac{\partial}{\partial r} \left(\cos \vartheta \frac{\partial g}{\partial r} \right). \tag{53}$$

From the equivalence of eqs. (53) and (51) it is inferred that the considered source element would act like a quadrupole with

$$\hat{\theta}_{11} \text{ equivalent to } \omega \rho_0 (l\hat{M})$$

as a measure of the source strength present at y_i in the flow. It may be shown by far-field considerations (see section 4 of Part II) that the directional distributions of the radiated

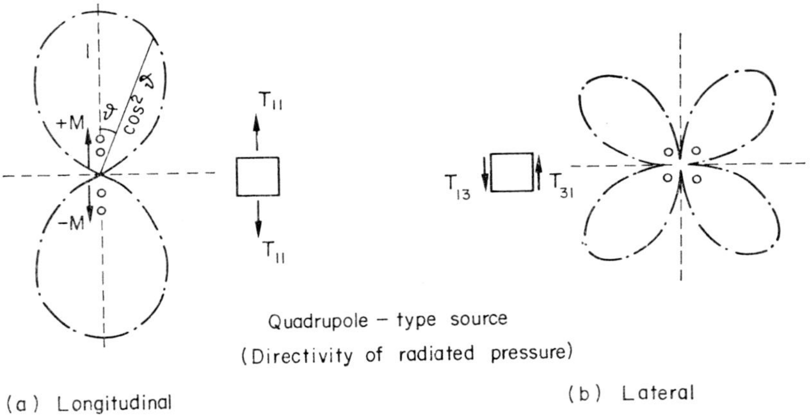

FIG. 4. Quadrupole-type source.

sound pressure is dominated by a term $\cos^2\vartheta$ for the longitudinal (T_{11}) and a term $\cos \vartheta \sin \vartheta$ for the lateral (T_{13}) quadrupole. Figure 4 illustrates this.

At the end of this section, the author feels that he owes the reader an explanation as to why he treated the acoustic analogy to aerodynamic noise in such detail. The reason simply is that one is, without fail, confronted with acoustic terminology in almost any paper on jet noise, be it theoretical or experimental. In many cases, however, these acoustic analogies prove to be dispensable and rather evoke various misconceptions. We may accept a formal relationship of, say, a volume element of turbulent fluid and an acoustic quadrupole; and yet it does not really help us in the calculation of the sound radiation from the flow as a whole. As long as we are not in a position as to describe the source function q in some detail and estimate its integrals, we are equally unable to develop an analogue acoustic source model for the sound generated.

6. FINAL FORM OF GENERAL SOLUTION OF THE WAVE EQUATION

We will now forget about simple-source analogies as exhibited in the preceding sections. We want to recall, however, some of the straightforward mathematical treatments of the original surface and volume integrals in eq. (12) of section 3. These are eqs. (24), (26), (36) and (48). In summing up all possible contributions to the pressure fluctuations generated by a general fluid flow with arbitrary boundaries, one may logically regroup them with respect to their mathematical appearance:

$$(p - p_0)\, 4\pi = \frac{\partial^2}{\partial x_i\, \partial x_j} \int_V \frac{1}{r} [T_{ij}]\, dV - \frac{\partial}{\partial x_i} \int_V \frac{1}{r} [F_i]\, dV + \frac{\partial}{\partial x_i} \int_A \frac{1}{r} [T_{ij} + p\delta_{ij}]\, dA_j + \int_V \frac{1}{r} \left[\frac{\partial M}{\partial t}\right] dV + \int_A \frac{1}{r} \left[\frac{\partial}{\partial y_j} (T_{ij} + p\delta_{ij}) - F_i\right] dA_i. \tag{54}$$

The composition of T_{ij}, F_i and M follows from eq. (25) where these abbreviations were first introduced. Hence,

$$T_{ij} + p\delta_{ij} = \rho c_i c_j + p\delta_{ij} - \tau_{ij} \tag{55}$$

and with the momentum equation in the form (4):

$$\frac{\partial}{\partial y_j} (\rho c_i\, c_j + p\delta_{ij} - \tau_{ij}) - F_i = - \frac{\partial \rho c_i}{\partial t} \tag{56}$$

we may write eq. (54) in an alternative form:

$$(p - p_0)\, 4\pi = \frac{\partial^2}{\partial x_i\, \partial x_j} \int_V \frac{1}{r} [T_{ij}]\, dV - \frac{\partial}{\partial x_i} \int_V \frac{1}{r} [F_i]\, dV + \frac{\partial}{\partial x_i} \int_A \frac{1}{r} [\rho c_i\, c_j + p\delta_{ij} - \tau_{ij}]\, dA_j + \int_V \frac{1}{r} \left[\frac{\partial M}{\partial t}\right] dV - \int_A \frac{1}{r} \left[\frac{\partial \rho c_i}{\partial t}\right] dA_i. \tag{57}$$

An almost identical integral form as in (57) is finally preferred, since it allows a most vivid interpretation of the surface integrals:

$$(p - p_0)\, 4\pi = \frac{\partial^2}{\partial x_i\, \partial x_j} \int_V \frac{1}{r} [T_{ij}]\, dV - \frac{\partial}{\partial x_i} \int_V \frac{1}{r} [F_i]\, dV + \int_V \frac{1}{r} \left[\frac{\partial M}{\partial t}\right] dV + \frac{\partial}{\partial x_i} \int_A \frac{1}{r} [P_i]\, dA + \frac{\partial}{\partial x_i} \int_A \frac{1}{r} [\rho c_i\, c_n]\, dA - \int_A \frac{1}{r} \left[\frac{\partial \rho c_n}{\partial t}\right] dA. \tag{58}$$

Here c_n denotes the (fluctuating) flow velocity as normal to the bounding surface A,

$$c_i \, dA_i = c_i \, n_i \, dA = c_n \, dA. \tag{59}$$

Also the stress vector P_i has been introduced. It represents the resultant force per unit area acting upon the surface adjacent to the flow:

$$P_i = n_j \, (p\delta_{ij} - \tau_{ij}). \tag{60}$$

Equation (58) is a most general integral of the wave equation (9). For a majority of practical applications, one may considerably reduce the number of integral terms contributing to the perturbation pressure field. The first integral with T_{ij} is the celebrated Lighthill integral for an unbounded flow, as equivalent to a quadrupole distribution. The surface integral with P_i basically represents the back-reaction of a solid body to the contiguous flow. This dipole-type source distribution was first discussed by Curle[7] in his extension to the aerodynamic sound theory. It includes forces (i) due to the impact of sound waves from the quadrupole distribution on the solid surface, and (ii) due to the hydrodynamic flow itself, i.e. its turbulent motion. The surface integrals with the fluid normal velocity c_n only enter when compliant boundaries are present. Especially the last integral can be a most effective source term due to its monopole character. It is recognized as the classical Rayleigh integral for calculating the sound from extended acoustic sources, e.g. an array of loudspeakers, radiating into a medium otherwise at rest. That reminds us that the surface vibrations indicated must not necessarily be flow-induced.

7. TREATMENT OF EXTERNAL BOUNDARIES

All the source terms listed in the preceding section were treated as if the integrands were completely known at any single volume or surface element. Such an approach may seem unrealistic with respect to practical applications; it is, however, consistent at least, provided all bounding surfaces are wholly immersed in the noise generating source flow. The pressure at the wall beneath a turbulent boundary layer, for instance, can be made accessible experimentally by pressure transducers suitably mounted flush with the wall. But is this known-integrand approach a realistic one where the surfaces lie exterior to the disturbed flow region V_1?

Powell[8], in his development of the reflection principle in aerodynamic noise, first pointed out that in the presence of external boundaries the surface integrals over A in eq. (24) should reasonably be split as follows:

Figure 5 exhibits a generalized case (c) of section 3. Provided the field point $P(x_i)$ lies within the region enclosed by the total surface $A = A_1 + A_2 + A_3$, one can always choose A_3 so that both p and $\partial p/\partial y_i$ vanish on it (one may fancy a surface A_3 so far away as to not have been reached, yet, by the sound waves emitted):

$$(\text{S.I.})_{A3} = 0. \tag{61}$$

The contribution of A_1 (in contact with the flow region itself) is conveniently kept in the same form as described by the surface integrals in eq. (58):

$$(\text{S.I.})_{A1} = \frac{\partial}{\partial x_i} \int_{A_1} \frac{1}{r} [P_i] \, dA + \frac{\partial}{\partial x_i} \int_{A_1} \frac{1}{r} [\rho c_i c_n] \, dA - \int_{A_1} \frac{1}{r} \left[\frac{\partial \rho c_n}{\partial t} \right] dA. \tag{62}$$

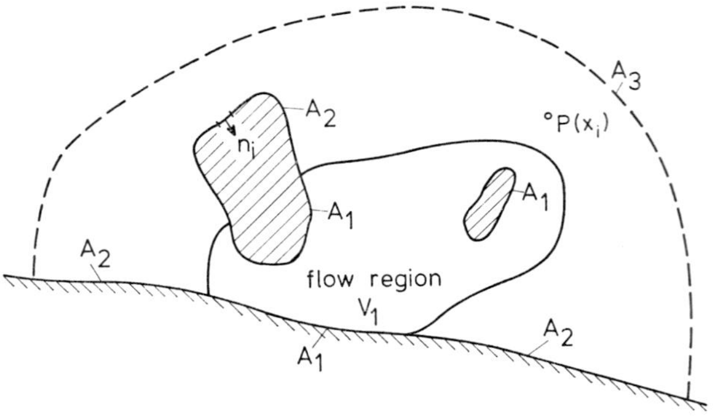

FIG. 5. Sketch of possible boundaries in aerodynamic noise problems.

The remaining pieces of the closed surface, denoted by A_2, contribute to the disturbances at $P(x_i)$ in a way that suggests a notation similar to the original one in eq. (24):

$$(\text{S.I.})_{A2} = \frac{\partial}{\partial x_i} \int_{A_2} \frac{1}{r} [p] \, dA_i + \int_{A_2} \frac{1}{r} \left[\frac{\partial p}{\partial y_j} \right] dA_j.$$

The second of these integrals may preferably be written in terms of the fluctuating velocity, which in a region far from the flow is related to the pressure gradient by the inviscid and linearized momentum equation:

$$\frac{\partial p}{\partial y_j} = -\rho_0 \frac{\partial c_j}{\partial t} \tag{63}$$

$$(\text{S.I.})_{A2} = \frac{\partial}{\partial x_i} \int_{A_2} \frac{1}{r} [p] \, dA_i - \rho_0 \int_{A_2} \frac{1}{r} \left[\frac{\partial c_n}{\partial t} \right] dA. \tag{64}$$

In contrast to the integrands in eq. (62), which are presumed to be given by the flow parameters, the surface distributions over A_2 obviously involve derived quantities. The specification of the latter would presume a detailed knowledge of the radiation field beforehand. This example probably shows most vividly the limitations inherent in the wave-equation approach to aerodynamic noise theory: The source functions cannot in general be assumed known independent of the quantities to be calculated.

The above equations are appropriate for treating reflection phenomena of surfaces adjacent to a turbulent flow. In this context Powell[8] reached the following conclusion which is rather surprising: "The pressure dipole distribution on a plane, infinite and rigid surface accounts for the reflection in that surface of the volume distribution of acoustic quadrupole generators of a contiguous inviscid fluid flow and for nothing more." For this and some other interesting features of boundary-layer induced noise, the reader may refer to ref. 8.

Powell's result was extended by Ffowcs-Williams (cited in ref. 3) to homogeneous, plane, compliant surfaces. He was able to show that the effect of surface vibration is merely to change the phase of the image-system quadrupoles. Thus, for a plane and infinite boundary the sound remains of quadrupole origin if the flow itself can be replaced by a quadrupole distribution, i.e. provided $F_i = 0$ and $\partial M/\partial t = 0$ in the most general integral formulation (58).

This section, concluding the more general aspects of the aerodynamic noise theory as based on the solution of an inhomogeneous wave equation, yielded a possible variant to the surface integrals in (58). In starting from this extremely general integral, Part II of this review paper will discuss certain approximations and specifications, which have proved valid and useful to almost any application to topical aircraft noise problems. Those important specifications relevant to jet noise in particular will be treated separately in Parts III and IV of this article.

PART II: APPROXIMATIVE PROCEDURES IN AERODYNAMIC NOISE THEORY

1. SIMPLIFICATIONS FOR THE RADIATED FAR-FIELD

The integral representation of the pressure, as developed in Part I (e.g., eq. (58)), can be simplified at field points far enough from the noise-generating flow regions. This can still be done on a fairly general basis without making any specifications for the source functions involved in the respective integrands. The only restriction to be made is that the spacial extent of the source region itself, and that of the incorporated boundaries, is limited.

For an acoustic point source (be it a monopole, dipole or quadrupole) one usually talks of a far-field, where the distance r from the concentrated source is very large compared with the wavelength λ of the sound radiated:

$$r \gg \frac{\lambda}{2\pi} = \frac{a_0}{\omega} \text{ (acoustic far-field).} \tag{1}$$

For expanded sources, as is usually the case in aerodynamic noise generation, the acoustic far-field approximation may be extended by an additional postulate:

$$r \gg \Lambda \text{ (geometric far-field)} \tag{2}$$

i.e. for points at distances large compared with the dimensions of the flow (see Fig. 6).

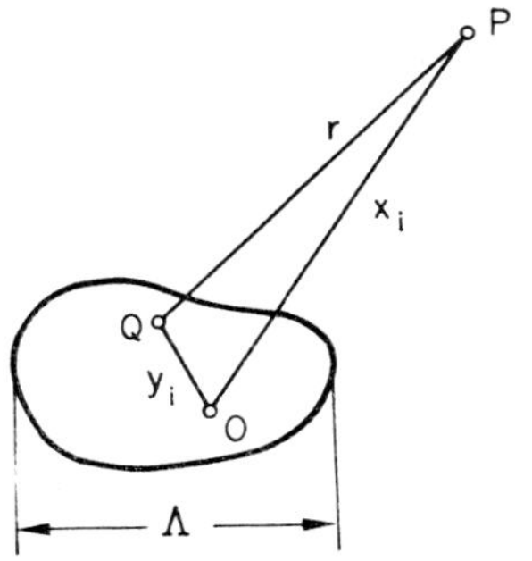

FIG. 6.

A generalized form of condition (1), as applicable to not necessarily harmonic fluctuations of an arbitrary source function, reads:

$$\frac{1}{a_0}\left[\frac{\partial\cdots}{\partial t}\right] \gg \frac{1}{r}\,[\cdots]. \tag{1a}$$

If now the origin O of the coordinates is conveniently chosen to lie within the flow region, a practical consequence of condition (2) for aerodynamic noise is the following approximation:

$$\frac{x_i - y_i}{r}\,[\cdots] \approx \frac{x_i}{|\,x_i\,|}\,[\cdots]_{t^+ = t - \frac{|x_i - y_i|}{a_0}}. \tag{2a}$$

That means y_i must be retained in the retarded-time prescription to account for the cancelling effect of phase differences in the sound rays reaching $P(x_i)$ simultaneously.

Approximations (1a) and (2a) will repeatedly be used in the following sections. They are sufficient for a formal treatment of general source functions. If one tries to specify for a given flow configuration, as will be done for a jet in Parts III and IV, one will find that still another criterion is significant for the characteristics of the radiated sound field. It is the ratio of the acoustic wavelength and a length scale characteristic of the source flow, which is easily seen to have some bearing on whether or not one may neglect retarded-time differences when integrating the contributions from different points of emission. The limiting case

$$\lambda \gg \Lambda \text{ (acoustically compact source)} \tag{3}$$

will hardly be satisfied for a significant range of acoustic wavelengths, as long as Λ stands for a typical dimension of the source region as a whole, e.g. that of a jet. It may, however, occasionally be possible to identify some other length scale as relevant to the turbulence in the flow for which a condition similar to (3) is more likely to be fulfilled. This idea of compact-source elements present in acoustically noncompact flows will be discussed in section 6 of Part II.

2. NEAR- AND FAR-FIELD OF MONOPOLE-TYPE SOURCES

The sound field described by the volume and surface integrals in the third column of eq. (58) of Part I has the simplest characteristics. In the geometric far-field, where a condition like (2a) holds, the amplitude of the pressure fluctuations p' varies like $1/r$, no matter how big λ is:

$$p' = \frac{1}{4\pi}\frac{1}{|\,x_i\,|}\left\{\int_V\left[\frac{\partial M}{\partial t}\right]dV - \int_A\left[\frac{\partial \rho c_n}{\partial t}\right]dA\right\}. \tag{4}$$

The situation is different for the fluctuating particle velocity c_i'. This is evident from the inviscid and linearized momentum equation (compare eq. (63) of Part I)

$$\frac{\partial c_i'}{\partial t} = -\frac{1}{\rho_0}\frac{\partial p'}{\partial x_i} \tag{5}$$

when one combines identities (22) and (17) of Part I:

$$\frac{\partial}{\partial x_i}\left(\frac{1}{r}\,[\cdots]\right) = -\left(\frac{1}{a_0}\left[\frac{\partial\cdots}{\partial t}\right] + \frac{1}{r}\,[\cdots]\right)\frac{x_i - y_i}{r^2}. \tag{6}$$

It is only in the acoustic far-field (condition (1a)) that the particle velocity varies like $1/r$ in amplitude like the pressure:

$$c_i' = \frac{1}{4\pi\, \rho_0\, a_0} \frac{x_i}{|\, x_i\,|^2} \left\{ \int_V \left[\frac{\partial M}{\partial t} \right] dV - \int_A \left[\frac{\partial \rho c_n}{\partial t} \right] dA \right\}. \tag{7}$$

Both prove to be proportional and in phase, the constant of proportionality being $\rho_0\, a_0$, the wave impedance of the medium

$$\frac{p'}{c_n'} = \rho_0\, a_0 \tag{8}$$

where c_n' denotes the velocity component normal to wave front.

The far-field relationship (8) is equally valid for any arbitrary source distribution, provided the source region is sufficiently far away. In the near-field of more complex sources, however, more intricate relations between the field quantities will be found.

3. NEAR- AND FAR-FIELD OF DIPOLE-TYPE SOURCES

We consider now a sound field as described by the three integrals in the second column of eq. (58) of Part I. In view of the identity (6), one notices that for such a dipole distribution the pressure field itself consists of two parts, which may be associated with an acoustic near-field and far-field, respectively. The one dominating in the far-field, where both conditions (1a) and (2a) are satisfied, may be written in the form:

$$p' = \frac{1}{4\pi\, a_0} \frac{x_i}{|\, x_i\,|^2} \left\{ \int_V \left[\frac{\partial F_i}{\partial t} \right] dV - \int_A \left[\frac{\partial}{\partial t} (P_i + \rho c_i\, c_n) \right] dA \right\}. \tag{9}$$

For the particle velocity in a dipole field we need the same manipulation as later for the derivation of the pressure in a quadrupole field. It is therefore performed here in combining relation (6) with identities (21) and (20) of Part I:

$$\begin{aligned}
\frac{\partial^2}{\partial x_i\, \partial x_j} \left(\frac{1}{r} [\cdots] \right) &= - \frac{\partial}{\partial x_j} \left\{ \left(\frac{1}{a_0} \left[\frac{\partial \cdots}{\partial t} \right] + \frac{1}{r} [\cdots] \right) \frac{1}{r} \frac{\partial r}{\partial x_i} \right\} \\
&= - \left(\frac{1}{a_0} \left[\frac{\partial \cdots}{\partial t} \right] + \frac{1}{r} [\cdots] \right) \left(\frac{1}{r} \frac{\partial^2 r}{\partial x_i\, \partial x_j} - \frac{1}{r^2} \frac{\partial r}{\partial x_j} \frac{\partial r}{\partial x_i} \right) \\
&\quad - \frac{1}{r} \frac{\partial r}{\partial x_i} \left\{ \frac{1}{a_0} \left[\frac{\partial^2 \cdots}{\partial t^2} \right] \left(- \frac{1}{a_0} \frac{\partial r}{\partial x_j} \right) - \left(\frac{1}{a_0} \left[\frac{\partial \cdots}{\partial t} \right] + \frac{1}{r} [\cdots] \right) \frac{1}{r} \frac{\partial r}{\partial x_j} \right\}.
\end{aligned}$$

If herein we use the fact that for the second derivative of r, due to eqs. (17), (18) of Part I,

$$\frac{\partial^2 r}{\partial x_i\, \partial x_j} = \frac{\partial}{\partial x_i} \left(\frac{x_j - y_j}{r} \right) = \frac{1}{r} \delta_{ij} - \frac{1}{r} \frac{\partial r}{\partial x_i} \frac{\partial r}{\partial x_j}, \tag{10}$$

then we can finally write:

$$\begin{aligned}
\frac{\partial^2}{\partial x_i\, \partial x_j} \left(\frac{1}{r} [\cdots] \right) &= \left(\frac{1}{a_0^2} \left[\frac{\partial^2 \cdots}{\partial t^2} \right] + \frac{2}{a_0 r} \left[\frac{\partial \cdots}{\partial t} \right] + \frac{2}{r^2} [\cdots] \right) \frac{1}{r} \frac{\partial r}{\partial x_i} \frac{\partial r}{\partial x_j} \\
&\quad - \left(\frac{1}{a_0} \left[\frac{\partial \cdots}{\partial t} \right] + \frac{1}{r} [\cdots] \right) \frac{1}{r} \frac{\partial^2 r}{\partial x_i\, \partial x_j}
\end{aligned}$$

$$\frac{\partial^2}{\partial x_i x_j}\left(\frac{1}{r}[\cdots]\right) = \left(\frac{1}{a_0^2}\left[\frac{\partial^2\cdots}{\partial t^2}\right] + \frac{3}{a_0 r}\left[\frac{\partial\cdots}{\partial t}\right] + \frac{3}{r^2}[\cdots]\right)\frac{1}{r}\frac{\partial r}{\partial x_i}\frac{\partial r}{\partial x_j} - \left(\frac{1}{a_0 r}\left[\frac{\partial\cdots}{\partial t}\right] + \frac{1}{r^2}[\cdots]\right)\frac{1}{r}\delta_{ij}. \tag{11}$$

This inserted into the expression for $-\rho_0{}^{-1}\partial p'/\partial x_j$, according to eqs. (5) and (58) of Part I, yields a fairly complicated velocity field for the dipole distribution. Under conditions (1a) and (2a) this would reduce to

$$c_j' = \frac{1}{4\pi\,\rho_0\,a_0^2}\,\frac{x_i\,x_j}{|\,x_i\,|^3}\left\{\int_V\left[\frac{\partial F_i}{\partial t}\right]dV - \int_A\left[\frac{\partial}{\partial t}(P_i + \rho c_i\,c_n)\right]\right\}dA. \tag{12}$$

When comparing eqs. (9) and (12), we see once again that a direct consequence of assumptions (1a) and (2a) is to obtain p' and c_j' in exactly the same form. One may therefore invert the argument by saying that a proper far-field is reached at a distance where one finds pressure and velocity in phase. This distance would then vary for different frequency components. The final decrease in amplitude like $1/r$ in the far-field is the same for all types of sources.

4. NEAR- AND FAR-FIELD OF QUADRUPOLE-TYPE SOURCES

The first integral in eq. (58) of Part I, representing a quadrupole distribution, is finally considered. In view of identity (11), we could easily write down the complete pressure field. Very close to the source region those terms would dominate that go like $1/r^3$ in amplitude. This compares with the near-field terms in the corresponding dipole and monopole fields, which go like $1/r^2$ and $1/r$, respectively. One may say that each $\partial/\partial x_i$ increases the maximum negative exponent of r by one. Thus, the amplitude of the velocity always decays a bit faster in the very near field of an arbitrary source.

We only write down the far-field approximation for the pressure in a quadrupole field:

$$p' = \frac{1}{4\pi\,a_0^2}\,\frac{x_i\,x_j}{|\,x_i\,|^3}\int_V\left[\frac{\partial^2\,T_{ij}}{\partial t^2}\right]dV. \tag{13}$$

From this it is evident that the directional characteristics of a longitudinal quadrupole (T_{11}) would go like

$$\left(\frac{x_1}{r}\right)^2 = \cos^2\vartheta$$

and that of a lateral quadrupole (T_{13}) like

$$\frac{x_1}{r}\,\frac{x_3}{r} = \cos\vartheta\,\sin\vartheta$$

as earlier discussed in subsection 5.3 of Part I.

For a real flow, where any field property is more likely the integral effect of a continuous source distribution, the directivity patterns of radiated sound entirely depend on the characteristics of this source distribution. For a completely random orientation of the source elements, as in assumed isotropic turbulence, one would rather expect a uniform spherical directional distribution in the far-field, even if any single volume element were believed to act like a quadrupole.

5. INTENSITY OF THE RADIATED SOUND

So far, we have only considered the instantaneous sound pressure p' in the far-field. This would be sufficient from an experimenter's viewpoint since he could use any pressure-measuring device to depict the corresponding mean-square or root-mean-square value of the fluctuating pressure. In a purely analytical approach, however, evaluation for such an averaged quantity means a considerably increased complexity in the integral formulation.

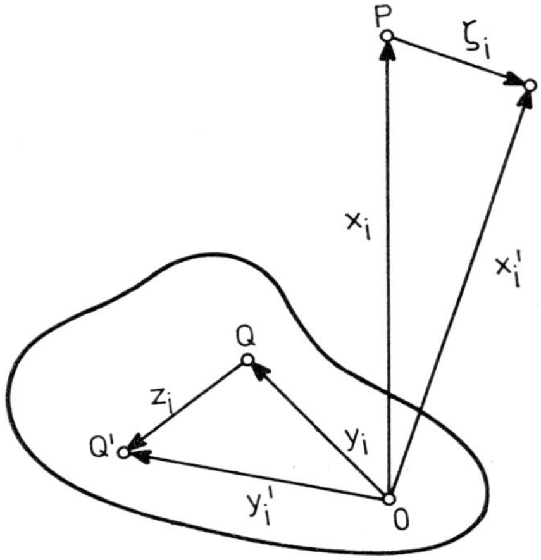

FIG. 7. Sketch showing displacements in source region (z_i) and field region (ζ_i).

To begin with, we define a more general mean product, the so-called covariance of two simultaneous pressures at field points x_i and x_i' displaced by $\zeta_i = x_i' - x_i$:

$$C_p\,(x_i, \zeta_i) = \overline{p'(x_i, t)\, p'(x_i', t)} \tag{14}$$
$$x_i' = x_i + \zeta_i$$
$$y_i' = y_i + z_i.$$

For x_i and x_i' in the same far-field, e.g. that of a volume distribution of monopoles, according to eq. (4):

$$C_p = \frac{1}{16\pi^2}\,\frac{1}{|\,x_i\,|\,|\,x_i'\,|}\,\overline{\int_V \left[\frac{\partial M}{\partial t}\right] dV\,(y_i) \int_V \left[\frac{\partial M}{\partial t}\right]' dV'\,(y_i')}. \tag{15}$$

The square brackets in eq. (15) should be read as follows:

$$\left[\frac{\partial M}{\partial t}\right] = \frac{\partial M}{\partial t}\left(y_i,\, t - \frac{|\,x_i - y_i\,|}{a_0}\right) = \frac{\partial M}{\partial t}\,(y_i,\, t^+) = \left[\frac{\partial M}{\partial t}\right]_{t^+}$$

$$\left[\frac{\partial M}{\partial t}\right]' = \frac{\partial M}{\partial t}\left(y_i',\, t - \frac{|\,x_i' - y_i'\,|}{a_0}\right) = \frac{\partial M}{\partial t}\,(y_i', t^+ + \tau) = \left[\frac{\partial M}{\partial t}\right]'_{t^+ + \tau}$$

Since time is kept constant when integrating over V, one may first time-average and then integrate, i.e. write a double integral over the covariances of the integrands:

$$C_p = \frac{1}{16\pi^2} \frac{1}{|x_i||x_i'|} \int\limits_{V(y_i)} \int\limits_{V(y_i')} \overline{\left[\frac{\partial M}{\partial t}\right]_{t+} \left[\frac{\partial M}{\partial t}\right]'_{t++\tau}} \, dV \, dV' \tag{16}$$

It is noted that in (16) not only is $[\partial M/\partial t]'$ to be taken at another locus than $[\partial M/\partial t]$ but also at another instant of time. The important difference in the retarded times, τ, depends on the coordinates in the following manner:

$$\begin{aligned} a_0\tau &= |x_i - y_i| - |x_i' - y_i'| \\ &= \sqrt{x_i^2 - 2x_i y_i + y_i^2} - \sqrt{(x_i + \zeta_i)^2 - 2(x_i + \zeta_i)(y_i + z_i) + (y_i + z_i)^2}. \end{aligned} \tag{17}$$

Herein, one may, for the geometric far-field (see section 1),

$$|y_i|, |z_i| \ll |x_i|,$$

and with the additional assumption

$$|\zeta_i| \ll |x_i|,$$

neglect quantities of second order of smallness, yielding approximately

$$a_0\tau = |x_i| - \frac{x_i y_i}{|x_i|} - |x_i + \zeta_i| + \frac{(x_i + \zeta_i)(y_i + z_i)}{|x_i + \zeta_i|} = \frac{x_i}{|x_i|}(z_i - \zeta_i). \tag{18}$$

This approximative τ can never be neglected, unless one assumes a condition like (3) for an acoustically compact source flow.

A most significant measure of the radiation is the local covariance for $\zeta_i = 0$, i.e. the mean-square of the pressure. In the far-field, where pressure and velocity are proportional, this is related to the sound intensity by

$$I = \overline{p' v_n'} = \frac{\overline{p'^2}}{\rho_0 a_0}. \tag{19}$$

With all this in mind, we may now write down the intensity I for the different types of sources considered earlier:

(a) Sound intensity originating from volumetric monopole distribution, according to eq. (16):

$$I_M = \frac{1}{16\pi^2 \rho_0 a_0} \frac{1}{|x_i|^2} \int\limits_V \int\limits_V \overline{\left[\frac{\partial M}{\partial t}\right]_{t+} \left[\frac{\partial M}{\partial t}\right]'_{t++\tau}} \, dV \, dV' \tag{20}$$

where here and in subsequent equations:

$$a_0\tau = \frac{x_i z_i}{|x_i|}.$$

The corresponding formula for a surface distribution of monopoles (see eq. (4)) is found analogous to (20):

$$I_M = \frac{1}{16\pi^2 \rho_0 a_0} \frac{1}{|x_i|^2} \int\limits_A \int\limits_A \overline{\left[\frac{\partial \rho c_n}{\partial t}\right]_{t+} \left[\frac{\partial \rho c_n}{\partial t}\right]'_{t++\tau}} \, dA \, dA'. \tag{21}$$

(b) Sound intensity originating from dipole distributions, according to eq. (9):

$$I_D = \frac{1}{16\pi^2 \rho_0 a_0^3} \frac{x_i x_j}{|x_i|^4} \int\limits_V \int\limits_V \overline{\left[\frac{\partial F_i}{\partial t}\right]_{t+} \left[\frac{\partial F_j}{\partial t}\right]'_{t++\tau}} dV\, dV' \tag{22}$$

$$I_D = \frac{1}{16\pi^2 \rho_0 a_0^3} \frac{x_i x_j}{|x_i|^4} \int\limits_A \int\limits_A \overline{\left[\frac{\partial}{\partial t}(P_i + \rho c_i c_n)\right]_{t+} \left[\frac{\partial}{\partial t}(P_j + \rho c_j c_n)\right]'_{t++\tau}} dA\, dA'. \tag{23}$$

(c) Sound intensity originating from a volumetric quadrupole distribution, according to eq. (13):

$$I_Q = \frac{1}{16\pi^2 \rho_0 a_0^5} \frac{x_i x_j x_k x_l}{|x_i|^6} \int\limits_V \int\limits_V \overline{\left[\frac{\partial^2 T_{ij}}{\partial t^2}\right]_{t+} \left[\frac{\partial^2 T_{kl}}{\partial t^2}\right]'_{t++\tau}} dV\, dV'. \tag{24}$$

Note that the subscripts of the second vector or tensor integrands in eqs. (22) to (24) had to be changed due to the twofold integration. Apart from this, the treatment of the higher-order acoustic sources is much the same as for the monopole.

6. IDEALIZED MODEL OF TURBULENCE STRUCTURE

The reader will agree that the integral representations for the intensity in the far-field of arbitrary aerodynamic sources, as derived in section 5, still look rather hopeless, at least from an experimenter's point of view. Once again it becomes evident that the utility of most of the theoretical results depends entirely on the ability of observers to estimate certain, up to now, unmeasured quantities in the flow and at the boundaries. At present, the evaluation of the integrals creates almost insurmountable problems, unless one condescends to a number of simplifying assumptions.

The only really straightforward theoretical model that provides a complete set of conditions by means of which the whole problem of sound generation can be solved analytically, is the model of homogeneous and isotropic turbulence in an incompressible and inviscid fluid. This was treated by Proudman[9] and represents one of the earliest applications of Lighthill's theory.

Unfortunately, turbulence is far from being isotropic in such flows where the sound production is strongest, as for instance in the mixing region of a jet. From correlation measurements in actual flows we can only surmise that at a fixed point in the flow a fluctuating quantity varies at random with time. When correlating overall signals, recorded with

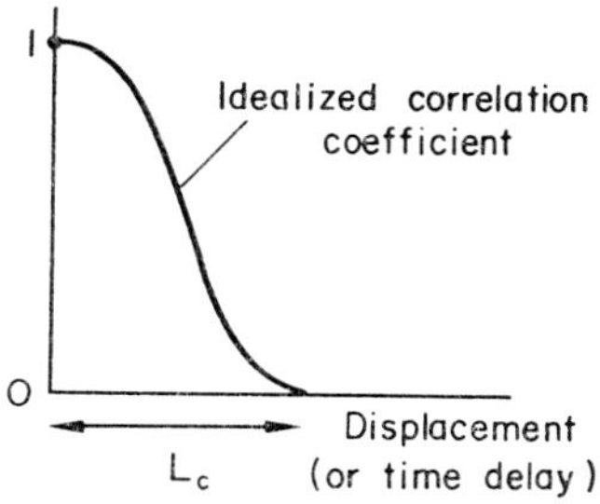

FIG. 8. Idealized correlation coefficient.

varying time delay, one finds a nearly monotonical decay of the normalized correlation function R with increasing time delay. Similarly, at a given instant of time any disturbance shows a highly chaotic variation with position throughout a turbulent flow; and although its values at two points which are fairly close together show some statistical correlation, which of course tends to unity as the points tend to coincide, it is well established that there is a distance, say L_c beyond which the overall signals are effectively uncorrelated (compare Fig. 8).

6.1 "*Correlation Volumes*" and "*Correlation Areas*"

The last-mentioned experimental evidence of a typical length scale L_c in turbulence suggests a preliminary definition of a volumetric scale with respect to the correlation integrals involved in expressions of the form:

$$C_p \sim \int\limits_{V(y_i)} \left(\int\limits_{V(y_i')} \overline{[\cdots]_{y_i,t+} [\cdots]_{y_i',t++\tau}} \, dV' \right) dV. \tag{25}$$

It is indeed tempting to try and get rid of the double integral by setting

$$\frac{\int\limits_{V(y_i')} \overline{[\cdots]_{y_i,t+} [\cdots]_{y_i',t++\tau}} \, dV'}{\overline{[\cdots]^2_{y_i}}} = V_c\left(y_i, \frac{x_i}{|x_i|}\right) \tag{26}$$

where V_c has the dimensions of a length cubed. One hestitates, however, to call this a correlation volume as characteristic of the turbulence, because V_c does not only depend on y_i but also, via the retarded-time differences τ, on the field point x_i, i.e. on the radiation angle. Such a definition would only be a fictitious reduction in complexity of the volume integrals in section 5. V_c and a correlation area A_c, defined in a like manner by

$$\frac{\int\limits_{A(y_i')} \overline{[\cdots]_{y_i,t+} [\cdots]_{y_i',t++\tau}} \, dA'}{\overline{[\cdots]^2_{y_i}}} = A_c\left(y_i, \frac{x_i}{|x_i|}\right) \tag{27}$$

could only attain some physical significance if one were allowed to introduce a second approximation which is discussed in the ensuing subsection.

6.2. *Neglect of Retarded-time Differences*

The effect of differences in retarded times at different points of emission on the instantaneous pressure in the far-field was seen to be a fairly complicated one. None the less, if the source function, e.g. the $\partial^2 T_{ij}/\partial t^2$ in the integral of eq. (13), were known, then the evaluation of the integral could principally be done by taking into account the proper phases of any single source element. If now we consider the intensity or a more general pressure covariance in the far-field, we may regard the inner integral in an expression like (25) as a new source function, which is to be integrated over the whole source region V. We notice that in this case the local magnitude of the source function can no longer be regarded as a property given by the flow characteristics only, since the covariance involved is a function of the distant field point too.

This difficulty could be vigorously overcome by ignoring retarded-time effects right from the beginning. Such an unrealistic assumption would suppose that the source region as a whole could be regarded as acoustically compact (cf. section 1). It is at this stage where the concept of limited correlation volumes and areas proves to be helpful.

Let us assume tentatively that τ can be neglected within a limited flow region which is the effective volume over which the integration of the covariance in (25) yields a finite contribution. If the dimensions of the volume, thus suitably defined by eq. (26) with $\tau = 0$, appears to be very small compared with the wave length of the radiated sound, then one may establish an acoustically compact source element as attached to V_c, although the source region as a whole may be acoustically noncompact.

Under these circumstances we may write equation (25) in the following form:

$$C_p \sim \int_{V(y_i)} \overline{[\cdots]_{y_i}^2}\, V_c(y_i)\, dV. \tag{28}$$

Here, both the mean square of the respective source function, e.g. that of $\partial M/\partial t$ in eq. (20), and the correlation volume as defined by an integral over the normalized simultaneous covariance

$$V_c(y_i) = \frac{\int_{V(y_i')} \overline{[\cdots]_{y_i}[\cdots]_{y'}}\, dV'}{\overline{[\cdots]_{y_i}^2}} \tag{28a}$$

depend only on the statistical properties of the turbulence at a given point y_i in the flow.

For source functions given as surface distributions, as in eqs. (21) and (23), the analogous approximation for acoustically compact surface elements yields the following form of the double integrals:

$$C_p \sim \int_{A(y_i)} \overline{[\cdots]_{y_i}^2}\, A_c(y_i)\, dA \tag{29}$$

and

$$A_c(y_i) = \frac{\int_{A(y_i')} \overline{[\cdots]_{y_i}[\cdots]_{y'}}\, dA'}{\overline{[\cdots]_{y_i}^2}}. \tag{29a}$$

6.3. *Acoustic Intensity from Unit Volume of Turbulence*

After having defined physically meaningful correlation volumes and areas in the flow and on the boundaries, we could rewrite eqs. (20) to (24) for the sound intensity in the respective far-fields in exactly the way prescribed by expressions (28) and (29). Instead of actually doing so, we make use of still another fact suggested by the statistical character of our simplified turbulence model.

Imagine the source region divided up into a set of non-overlapping subregions (with n centres) which fill it completely, each with a volume or area equal to the local correlation volume or area as defined in the preceding section. (Figure 9, taken from ref. 19, is to illustrate this assumption.) Then, by hypothesis, all the resulting source elements, which

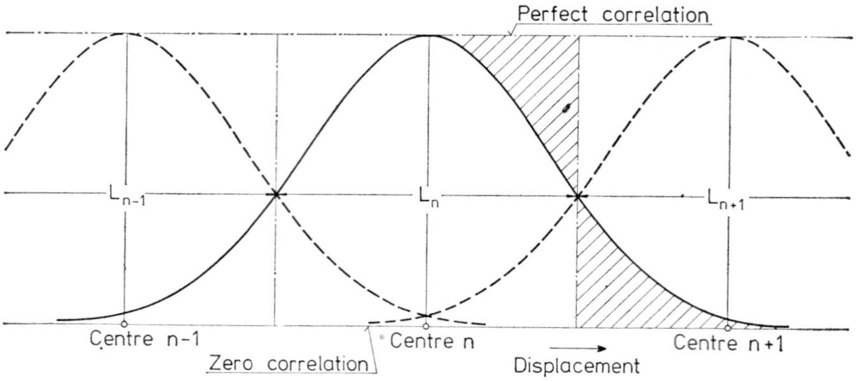

FIG. 9. One-dimensional schematic of three adjacent "correlation volumes" V_{n-1}, V_n, V_{n+1} in quasi-isotropic turbulence.

sometimes are thought of as eddies, will fluctuate in a completely uncorrelated way. As a result, the intensity field due to the sum of the turbulent elements equals the sum of the intensity fields of each element separately. We may, therefore, introduce an acoustic intensity per unit volume or area of turbulence, J and J', and define it sensibly as follows:

$$I(x_i) = \int_V J(x_i, y_i)\, dV(y_i) \approx \sum_n J(x_i, y_{i,n})\, V_c(y_{i,n}) \tag{30}$$

and

$$I(x_i) = \int_A J'(x_i, y_i)\, dA(y_i) \approx \sum_n J'(x_i, y_{i,n})\, A_c(y_{i,n}). \tag{31}$$

$V_c(y_{i,n})$ or $A_c(y_{i,n})$ denote the n correlation volumes or areas into which the flow or the boundaries were divided, each of them being interpreted as a coherently radiating entity with its geometric centre located at $y_{i,n}$.

Now, according to eqs. (20) to (24) with all the assumptions introduced for the idealized model of the turbulence:

$$J_M = \frac{1}{16\pi^2\, \rho_0\, a_0}\, \frac{1}{|\, x_i\,|^2}\, \overline{\left(\frac{\partial M}{\partial t}\right)^2}\, V_c \tag{20a}$$

$$J'_M = \frac{1}{16\pi^2\, \rho_0\, a_0}\, \frac{1}{|\, x_i\,|^2}\, \overline{\left(\frac{\partial \rho c_n}{\partial t}\right)^2}\, A_c \tag{21a}$$

$$J_D = \frac{1}{16\pi^2\, \rho_0\, a_0^3}\, \frac{x_i\, x_j}{|\, x_i\,|^4}\, \overline{\frac{\partial F_i}{\partial t}\, \frac{\partial F_j}{\partial t}}\, V_c \tag{22a}$$

$$J'_D = \frac{1}{16\pi^2\, \rho_0\, a_0^3}\, \frac{x_i\, x_j}{|\, x_i\,|^4}\, \overline{\frac{\partial}{\partial t}(P_i + \rho c_i\, c_n)\, \frac{\partial}{\partial t}(P_j + \rho c_j\, c_n)}\, A_c \tag{23a}$$

and

$$J_Q = \frac{1}{16\pi^2\, \rho_0\, a_0^5}\, \frac{x_i\, x_j\, x_k\, x_l}{|\, x_i\,|^6}\, \overline{\frac{\partial^2 T_{ij}}{\partial t^2}\, \frac{\partial^2 T_{kl}}{\partial t^2}}\, V_c. \tag{24a}$$

7. DIMENSIONAL ANALYSIS OF AERODYNAMIC SOUND PRODUCTION

Only very few deductions of direct practical applicability can be made from the theory at this stage despite the many simplifications already made. The simplest one and probably the most useful, too, is an analysis of the dependence of the sound field, for geometrically similar mechanisms of flow production, on a typical velocity U in the flow and a typical linear dimension L, and also on constants of the gas such as ρ_0 and a_0. Only in the light of such an analysis can a variety of experiments with similar flow configurations be co-ordinated.

The dimensional analysis is carried out for the total sound power radiated into the far-field:

$$P = \int_S I\,dS, \tag{32}$$

where the closed surface S coincides with the acoustic wave front. If a sphere, with the origin coinciding with the origin of the coordinates x_i, is chosen as the surface of integration:

$$P \sim |x_i|^2 I. \tag{33}$$

I is taken as the contribution of an element of the source distributions as described by the equations on page 256. The time derivatives $\partial/\partial t$ are replaced by a typical radian frequency $\omega = 2\pi f$. We note that this analysis will not reveal the spectral characteristics of the radiated sound. Furthermore, we now confine the discussion to flows whose fluctuations are internally generated by certain flow instabilities rather than by any direct external cause. The dominant frequency f is then found experimentally to vary with the other constants of the flow like

$$f \sim \frac{U}{L} \text{ and hence the time scale } T \sim \frac{L}{U}. \tag{34}$$

A correlation volume, area or length is assumed to scale with L^3, L^2 and L, respectively:

$$V_c \sim L^3;\ A_c \sim L^2;\ L_c \sim L. \tag{35}$$

An assumption that has important consequences is that in our simplified turbulence model we assumed that retarded-time differences might be neglected within a correlation volume or area, i.e. we regarded a turbulence element as acoustically compact ($L_c \ll \lambda$). This means that for a given frequency according to (34)

$$\frac{L_c}{\lambda} \sim \frac{L}{\lambda} \sim \frac{U}{\lambda f} = \frac{U}{a_0} \ll 1. \tag{36}$$

(a) For monopole-like vibrations of a bounding surface excited by the flow the analysis yields with eq. (21a):

$$P_M \sim \frac{1}{\rho_0 a_0} \overline{\left(\frac{\partial \rho c_n}{\partial t}\right)^2} A_c^2 \tag{37}$$

and with $\rho c_n \sim \rho_0 U$:

$$P_M \sim \frac{1}{\rho_0 a_0} \left(\frac{U}{L}\right)^2 (\rho_0 U)^2 (L^2)^2$$

$$P_M \sim \frac{\rho_0 L^2 U^4}{a_0} = \rho_0 L^2 U^3 \left(\frac{U}{a_0}\right). \tag{38}$$

(b) For dipole radiation from pressure fluctuations caused by the flow at a solid surface bounding the flow, with eq. (23a):

$$P_D \sim \frac{1}{\rho_0 a_0^3} \overline{\frac{\partial P}{\partial t} \frac{\partial P}{\partial t}} A_c^2, \tag{39}$$

where

$$P = \frac{x_i}{|x_i|} P_i \sim \rho_0 U^2$$

and hence:

$$P_D \sim \frac{1}{\rho_0 a_0^3} \left(\frac{U}{L}\right)^2 (\rho_0 U^2)^2 (L^2)^2$$

$$P_D \sim \frac{\rho_0 L^2 U^6}{a_0^3} = \rho_0 L^2 U^3 \left(\frac{U}{a_0}\right)^3. \tag{40}$$

(c) Finally, the dimensional analysis yields for the quadrupole-like radiation from a turbulent flow according to eq. (24a):

$$P_Q \sim \frac{1}{\rho_0 a_0^5} \overline{\left(\frac{\partial^2 T}{\partial t}\right)^2} V_c^2 \tag{41}$$

and with

$$T = \frac{x_i x_j}{|x_i|^2} T_{ij} \sim \rho_0 U^2$$

$$P_Q \sim \frac{1}{\rho_0 a_0^5} \left(\frac{U}{L}\right)^4 (\rho_0 U^2)^2 (L^3)^2$$

$$P_Q \sim \frac{\rho_0 L^2 U^8}{a_0^5} = \rho_0 L^2 U^3 \left(\frac{U}{a_0}\right)^5. \tag{42}$$

For free jets, both model and full-scale tests, this approximate prediction (42) is borne out surprisingly well by a great number of experiments, provided the Mach number of the flow U/a_0 is not too small and does not exceed 1. This is emphasized because some of the assumptions made in the analysis, e.g. $T \sim \rho_0 U^2$, can only be considered as very rough approximations. Lighthill himself suggested that in careful experimental studies one would be able rather to detect departures from laws such as (42).

The celebrated "U^8-law" is only too frequently regarded as the inevitable outcome of any theory of sound generated by turbulent flows. It should be clear, however, that this is rather a very special result obtained by applying the theory to a highly simplified model of turbulence.

In a steadily maintained flow the energy per unit volume will be roughly proportional to $\rho_0 U^2$, and the total rate of supply of energy to

$$(\rho_0 U^2)\,(UL^2).$$

Hence, the ratio of the total acoustic power output to the supply of mechanical power in the flow, e.g. the power of a jet engine, will satisfy

$$\eta_Q \sim \left(\frac{U}{a_0}\right)^5. \tag{43}$$

The ratio η can be described as the efficiency of aerodynamic sound production. Equation (43) makes it clear that turbulence at low Mach numbers is quite an exceptionally inefficient producer of sound, the constant of proportionality involved being of the order 10^{-4} from jet experiments. This subsequently justifies a presupposition which was tacitly assumed valid throughout the theory. In the absence of amplifying resonators near the flow, the sound produced is so weak relative to the motions producing it that no significant back-reaction on the flow field itself is to be expected.

PART III: APPLICATION TO JET NOISE

1. INTRODUCTION

In Parts I and II of this paper a fairly general scope was given of the basic equations governing the analysis and general concepts as applied in aerodynamic noise theories. No intrinsic specifications were introduced concerning a particular type of source flow. Lighthill's original ideas have initiated, though, a substantial number of more specific theories on the generation of sound by various flow configurations. In this Part III the subject is specialized to study free turbulence as a source of sound. Application to a topical problem, that of the noise from jets, is obvious. Some results reviewed will also prove useful in a more general context, as for instance the modifications to the theory required in studies of all such flows where turbulence is convected at high speeds (see section 5).

The noise of a free air jet can be regarded as a separate entity in aircraft noise, even when sound is known to emerge to a large extent from the exhaust systems, arising from conditions in the nozzle or upstream of it (e.g. compressor noise). For, whatever these conditions may be, a new, very intense turbulence is created in the shear layer where the jet fluid, after leaving the nozzle, exchanges momentum with the ambient air. The noise which this generates can be thought as an additive to combustion noise and turbomachinery noise. Furthermore, to a good approximation, it is uninfluenced by the presence of the nozzle, and can be visualized as sound radiated into essentially unbounded space by a limited extent of turbulent gas flow.

This acoustic power output is usually regarded as a mere by-product of the main jet flow, a phenomenon which has itself no marked influence on the flow development. By contrast, the near-field includes not only outwardly propagating waves, but also local reciprocating motions and pressure fluctuations, such as may be directly induced by fluctuating vortex movements in the turbulent flow. This near sound field (sometimes termed "pseudo-sound" (refer to section 6)) is of particular concern to the designer of structures to be placed near jet orifices, but the body of knowledge that has been most sought after by aircraft and rocket engineers comprises the acoustic power radiated by a jet, its frequency spectrum and directivity.

2. SIMPLIFYING ASSUMPTIONS COMMON IN JET NOISE THEORY

Before discussing the structure of the sources which dominate in a jet, we may recall the specific presuppositions that can be made in contrast to the more general analysis in Parts I and II:

1. The medium is assumed unbounded, i.e. there are no foreign bodies fully or partly immersed in the flow, nor are there any reflecting solid boundaries or discontinuities in the medium outside the flow. All the surface integrals in eq. (58) of Part I must therefore vanish:

$$\int_A \frac{1}{r} [\cdots]\, dA = 0.$$

2. The rate of fluctuating mass injection is zero throughout the flow:

$$m = 0.$$

3. There is no externally applied fluctuating force that could vary locally. Hence,

$$\frac{\partial}{\partial x_i} \int_V \frac{1}{r} [F_i]\, dV = 0.$$

4. The effects of heat conduction and viscosity are assumed unimportant to the extent that an isentropic approximation can be written in the form

$$\frac{\partial^2}{\partial t^2} (p - a_0^2 \rho) = 0$$

and subsequently in eq. (58) of Part I:

$$\int_V \frac{1}{r} \left[\frac{\partial M}{\partial t}\right] dV = 0.$$

The only one integral to which the general solution (58) thus reduces is

$$p - p_0 = \frac{1}{4\pi} \frac{\partial^2}{\partial x_i\, \partial x_j} \int_V \frac{1}{r} [T_{ij}]\, dV. \tag{1}$$

5. Finally setting

$$\tau_{ij} = 0$$

for the viscous stresses, the Lighthill tensor in eq. (1) is approximated by

$$T_{ij} = \rho c_i\, c_j. \tag{2}$$

6. There has often been some reasoning whether or not one may, at least for moderate Mach numbers, replace ρ by its mean value ρ_0. We therefore write:

$$T_{ij} \approx \rho_0\, c_i\, c_j \tag{3}$$

in order to indicate that the validity of the latter assumption may be more restricted than that of the preceding ones.

These five (or six) assumptions will be presupposed in what follows unless something else is stated. We shall see that definite results for the intensity and characteristics of the radiated sound field can only be arrived at when one accepts certain additional assumptions concerning the main flow characteristics and, above all, the statistical properties of the turbulence which will be discussed in the course of the analysis.

3. "SELF-NOISE" AND "SHEAR-NOISE"

We now and in the following confine to the far-field approximation of (1) (cf. eq. (13) of Part II):

$$p' = \frac{1}{4\pi\, a_0^2} \frac{x_i\, x_j}{|\, x_i \,|^3} \int_V \left[\frac{\partial^2 T_{ij}}{\partial t^2}\right] dV \tag{4}$$

and on the assumptions 5 and 6 above:

$$p' \approx \frac{\rho_0}{4\pi\, a_0^2} \frac{x_i\, x_j}{|\, x_i \,|^3} \int_V \left[\frac{\partial^2 c_i\, c_j}{\partial t^2}\right] dV. \tag{5}$$

This equation was the starting-point of Proudman's[9] application of the theory to isotropic turbulence. He first simplified the notation by carrying out the tensor contractions on the right of eq. (5) before proceeding with the analysis:

$$x_i\, c_i = |\, x_i \,|\, c_x = |\, x_i \,|\, c$$

$$p' \approx \frac{\rho_0}{4\pi\, a_0^2} \frac{1}{|\, x_i \,|} \int_V \left[\frac{\partial^2 c^2}{\partial t^2}\right] dV. \tag{6}$$

The instantaneous velocity component parallel to the direction of x_i, $c_x = c$, may be split into a mean and a fluctuating part:

$$c = \bar{c} + c'$$

$$c^2 = \bar{c}^2 + \bar{c}\, c' + c'^2. \tag{7}$$

If $\bar{c}$ stands for the mean flow and c' for the superimposed turbulence, then it becomes clear from eq. (7) that sound can be generated in two essentially different ways:

(i) by turbulence–turbulence interaction. This may be called "self-noise" since it would exist even when there is no mean flow present as in the theoretical model of isotropic turbulence;

(ii) by turbulence–mean flow interaction. This will be seen to contribute considerably to the noise from shearing flows like jets.

It is tempting to argue from eq. (7) directly that $\rho c_i c_j$ can be a most efficient source function when fluctuations in c_j are multiplied by a large mean value of c_i. This, however, does not apply when $\bar{c}_i$ is uniform in space, because the first time derivative of $\rho c_i c_j$ is then exactly a space derivative, as may be seen from eq. (4) of Part I:

$$\frac{\partial \rho\, \bar{c}_i\, c_j}{\partial t} = -\frac{\partial}{\partial x_k} (\rho \bar{c}_i\, c_j\, c_k + p \bar{c}_j\, \delta_{jk}). \tag{8}$$

The fact that the right-hand side of (8) represents a space derivative indicates that such an amplification by the mere presence of mean flow is not likely to take place, since the corresponding source terms act like an octupole field and must be expected to radiate relatively little sound, especially at the lower Mach numbers.

However, Lighthill[2] has shown that large non-cancelling values of the quadrupole strength density may occur where there is a large mean velocity gradient (causing considerable changes of velocity across a single eddy). To make the argument precise, the time derivative of the momentum flux is now thrown into a form which brings out directly the importance of mean shear. By using the basic equations (4) and (1) of Part I:

$$\begin{aligned}\frac{\partial \rho\, c_i\, c_j}{\partial t} &= \rho c_i \frac{\partial c_j}{\partial t} + c_j \frac{\partial \rho c_i}{\partial t} = c_i \frac{\partial \rho c_j}{\partial t} - c_i\, c_j \frac{\partial \rho}{\partial t} + c_j \frac{\partial \rho c_i}{\partial t} \\ &= -\, c_i \left(\frac{\partial \rho c_j\, c_k}{\partial x_k} + \frac{\partial p}{\partial x_j}\right) + c_i\, c_j \frac{\partial \rho c_k}{\partial x_k} - c_j \left(\frac{\partial \rho c_i\, c_k}{\partial x_k} + \frac{\partial p}{\partial x_i}\right) \\ &= -\frac{\partial \rho c_i\, c_j\, c_k}{\partial x_k} + \rho c_j\, c_k \frac{\partial c_i}{\partial x_k} - \frac{\partial p c_i}{\partial x_j} + p \frac{\partial c_i}{\partial x_j} + c_i\, c_j \frac{\partial \rho c_k}{\partial x_k} \\ &\quad - \rho c_j\, c_k \frac{\partial c_i}{\partial x_k} - c_i\, c_j \frac{\partial \rho c_k}{\partial x_k} - \frac{\partial p c_j}{\partial x_i} + p \frac{\partial c_j}{\partial x_i} \\ &= p \left(\frac{\partial c_i}{\partial x_j} + \frac{\partial c_j}{\partial x_i}\right) - \frac{\partial}{\partial x_k} (\rho c_i\, c_j\, c_k + p c_i\, \delta_{jk} + p c_j\, \delta_{ik}). \end{aligned} \tag{9}$$

When neglecting the last term of (9), which represents an octupole field, we retain

$$\frac{\partial \rho c_i\, c_j}{\partial t} = p \left(\frac{\partial c_i}{\partial x_j} + \frac{\partial c_j}{\partial x_i}\right), \tag{10}$$

i.e. the very compact form of the product of pressure and the rate-of-strain tensor. The quantities on the right of eq. (10) can now be split into mean and fluctuating components, thus revealing again two distinct mechanisms by which sound can be generated according to (4),

$$\frac{\partial^2 T_{ij}}{\partial t^2} = \left(\frac{\partial \bar{c}_i}{\partial x_j} + \frac{\partial \bar{c}_j}{\partial x_i}\right) \frac{\partial p}{\partial t} + \frac{\partial}{\partial t} p \left(\frac{\partial c_i'}{\partial x_j} + \frac{\partial c_j'}{\partial x_i}\right). \tag{11}$$

Lighthill used an argument inferred from the assumption of an isotropic turbulent velocity field in a heavily sheared flow to show that the first term on the right of (11) may presumably dominate in a jet. The pressure fluctuations in the flow would, on this assumption, act as the genuine source function with an amplification by the mean shear. Suppose the mean velocity were given by

$$\begin{aligned}\bar{c}_i &= \{\bar{c}_1;\ 0;\ 0\} \\ \frac{\partial \bar{c}_1}{\partial x_j} &= \left\{0;\ \frac{\partial \bar{c}_1}{\partial x_2};\ 0\right\}.\end{aligned} \tag{12}$$

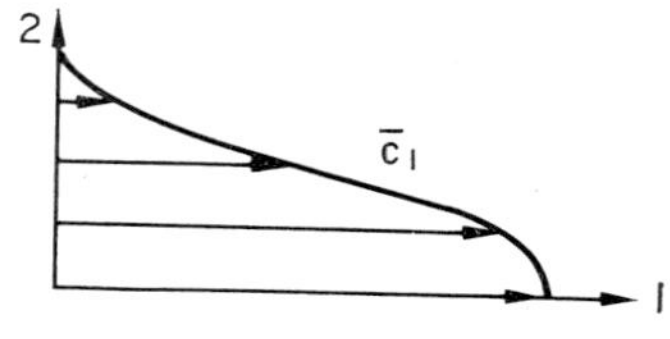

FIG. 10. Idealized two-dimensional shear layer.

The pressure in the far-field then reads:

$$p' = \frac{2}{4\pi a_0^2} \frac{x_1 x_2}{|x_i|^3} \int\limits_V \frac{\partial \bar{c}_1}{\partial y_2} \left[\frac{\partial p}{\partial t}\right] dV \tag{13}$$

and the intensity of what is now termed "shear noise" (cf. eq. (24) of Part II):

$$I = \frac{4}{16\pi^2 \rho_0 a_0^5} \frac{x_1^2 x_2^2}{|x_i|^6} \int\limits_V \int\limits_V \frac{\partial \bar{c}_1}{\partial y_2} \left(\frac{\partial \bar{c}_1}{\partial y_2}\right)' \overline{\left[\frac{\partial p}{\partial t}\right]_{t+} \left[\frac{\partial p}{\partial t}\right]'_{t++\tau}} dV\, dV'. \tag{14}$$

The intensity is thus related to the two-point cross correlation of $\partial p/\partial t$ and its double integral over the whole flow region. In some cases of interest the mean rate of strain, which appears as a weighting function, will be constant over an effective correlation volume as defined in section 6 of Part II, i.e. in a region of nonzero $\partial p/\partial t$ covariance,

$$I = \frac{4}{16\pi^2 \rho_0 a_0^5} \frac{x_1^2 x_2^2}{|x_i|^6} \int\limits_V dV \left(\frac{\partial \bar{c}_1}{\partial y_2}\right)^2 \int\limits_V \overline{\left[\frac{\partial p}{\partial t}\right]_{t+} \left[\frac{\partial p}{\partial t}\right]'_{t++\tau}} dV'. \tag{15}$$

Equation (15) gives the intensity at x_i as a volume integral over V. The integrand can thus be interpreted as the contribution to the intensity produced by unit volume of turbulence at y_i, though it strictly involves integration of the correlation volume

$$dV' = d^3y' = d^3z \text{ (see Fig. 7)} \tag{16}$$

over all space. We denote this contribution to the intensity by J by analogy with the definition in subsection 6.3 of Part II:

$$I = \int J(x_i, y_i)\, dV(y_i). \tag{17}$$

For the "shear-noise" from our two-dimensional shear layer we finally write:

$$J = \frac{4}{16\pi^2 \rho_0 a_0^5} \frac{x_1^2 x_2^2}{|x_i|^6} \bar{\tau}_{12}^2 \int P(z, \tau)\, d^3z \tag{18}$$

where

$$P(z, \tau) = \overline{\left[\frac{\partial p}{\partial t}\right]_{t+} \left[\frac{\partial p}{\partial t}\right]'_{t++\tau}}.$$

The analogue for the "self-noise" is conveniently written as

$$I = \frac{\rho_0}{16\pi^2 a_0^5} \frac{1}{|x_i|^2} \int\limits_V dV \int\limits_V \overline{\left[\frac{\partial^2}{\partial t^2}(c^2 - \overline{c^2})\right]_{t+} \left[\frac{\partial^2}{\partial t^2}(c^2 - \overline{c^2})\right]'_{t++\tau}} dV' \tag{19}$$

and again for the intensity per unit volume of turbulence:

$$J = \frac{\rho_0}{16\pi^2 a_0^5} \frac{1}{|x_i|^2} \int \Pi(z, \tau)\, d^3z \tag{20}$$

where

$$\Pi(z, \tau) = \overline{\left[\frac{\partial^2}{\partial t^2}(c^2 - \overline{c^2})\right]_{t+} \left[\frac{\partial^2}{\partial t^2}(c^2 - \overline{c^2})\right]'_{t++\tau}}.$$

Both "shear-noise" and "self-noise" have been thoroughly treated by different authors on the basis of a theoretical model of isotropic turbulence. The details of their analyses will not be reproduced here. They have both in common that the simultaneous covariances in P and Π respectively were considered only (i.e. $\tau = 0$).

Proudman[9] was able to approximate to the covariance in (20) which is of the fourth degree in the velocities, by using an idea of Batchelor[10] that essentially assumes a normal joint-probability distribution of the velocities and their time derivatives at y_i and y_i'. The evaluation of the integral then yields

$$\int \Pi(z, 0)\, d^3z = \alpha \frac{4\pi(\overline{c^2})^4}{L}; \quad L = \frac{(\overline{c^2})^{3/2}}{\epsilon} \tag{21}$$

where L is the length scale of the turbulence and $\overline{c^2}$ the mean square value of one of the velocity components. ϵ is the mean rate of dissipation of energy per unit mass. The dimensionless constant α depends on the theoretical form chosen for the correlation function $f(z)$ that is usually denoted by this symbol in isotropic turbulence. Proudman calculated α

(i) from published tables of Heisenberg's form of $f(z)$ with the result $\alpha \approx 40$;

(ii) from the kinematically possible form of $f(z)$,

$$f(z) = \exp\left(-\frac{\pi z^2}{4L^2}\right) \tag{22}$$

with the result $\alpha \approx 13$.

With (21) inserted into eq. (20), we find for the intensity per unit volume of isotropic turbulence:

$$J = \frac{1}{4\pi \mid x_i \mid^2} \frac{\rho_0 \alpha}{a_0^5} \frac{(\overline{c^2})^4}{L} = \frac{1}{4\pi \mid x_i \mid^2} \rho_0 \alpha \epsilon \left(\frac{\overline{c^2}}{a_0^2}\right)^{5/2}. \tag{23}$$

From this we deduce that the intensity of the radiated sound decreases like $|x_i|^{-2}$ in any one of the three entirely equivalent directions x_i. This obvious result is due to the random oscillations of the directions of the axes of the quadrupoles in the flow.

Lilley[11] in his estimation of "shear-noise" introduced more or less the same model as Proudman did, in order to calculate the required integral over the $\partial p/\partial t$ covariance in eq. (18). He was able to show that approximately

$$\int P(z, 0)\, d^3z = \frac{16.32\, \rho_0^2\, \bar{\tau}_{12}^4\, L_{11}^5\, \overline{c^2}}{225\pi} \beta \tag{24}$$

where L_{11} is the longitudinal length scale of the turbulence. The derivation of (24) was based on the assumption of incompressible isotropic turbulence with the longitudinal velocity correlation being given by the kinematically possible form (22) with $L = L_{11}$. Anisotropy was only empirically allowed for in the term β, which might be expected to lie within the range 1/2 to 1/3.

With the result (24) inserted into eq. (18), we find for the intensity per unit volume of turbulence superimposed on a heavily sheared mean flow:

$$J = \frac{128}{225\, \pi^3 \mid x_i \mid^2} \frac{x_1^2\, x_2^2}{\mid x_i \mid^4} \frac{\rho_0 \beta}{a_0^5} \bar{\tau}_{12}^6\, L_{11}^5\, \overline{c^2}. \tag{25}$$

On the basis of a detailed study of the specific characteristics of the flow in the different regions of a circular air jet, i.e.

(a) pseudo-laminar shear layer near jet exit
(b) turbulent mixing region
(c) turbulent transition region
(d) fully developed turbulent flow region,

Lilley also gave estimates for the sound production of any of these flow regions separately. Some of his conclusions may serve as a substitute for a more rigorous treatment of his comprehensive paper:

1. The estimate of the strength of the equivalent distribution of acoustic quadrupoles in a low speed circular jet, using measured results of the flow structure, have shown that the central region of the mixing region (where the gradient of the mean velocity is highest) is mainly responsible for the bulk of the noise emitted.

2. In the mixing region of the jet the greatest contribution to the noise emitted is from the interaction between the turbulence and mean shear (something like 80% of the total noise emitted), whereas in the fully developed turbulent flow region, further downstream, the contribution is shared between the effects of shear amplifications and "self-noise".

3. From approximate estimates of the sound spectrum from the turbulence it is found that the high frequency contribution comes from the first 2 to 3 diameters of the jet mixing region, though the low frequency content is not negligible. The bulk of the low frequency noise comes from 4 to 6 diameters from the jet exit. The remainder of the low frequency noise comes from regions further downstream but is of low intensity. In the latter region the equivalent acoustic quadrupole strength falls off proportional to $1/y_1^7$, a result found independently by Ribner.[12]

At the end of this section, it is noted that the concept of an isotropic turbulence structure with a source pattern of independently radiating eddies is physically meaningful, at most, in a system of coordinates convected with the individual source elements. That is, one has to define a kind of "average eddy convection speed", U_i say, and then imagine the source flow divided up into a set of non-overlapping subregions which fill it completely, each with a volume equal to the local (convected) correlation volume as defined in section 6 of Part II. It is only in such a co-moving system that retarded-time differences in the covariances P and Π of eqs. (18) and (20) could be ignored, with some justification, within a correlation volume. Thereafter, the analysis can be carried through on the assumption that the turbulence be isotropic in that convected system. This procedure, however, will be discussed separately in section 5. As a result, we shall find that a certain convection factor, multiplying the non-convected results, takes care of the convection of the turbulence. Thus the estimates (23) and (25), valid for a low-speed jet, can also be adjusted to the case of a high-speed jet.

Another modification to the above results that is immediately evident concerns the mean density ρ in the jet, which may, in cases, deviate considerably from the value ρ_0 in the ambient air. With

$$T_{ij} \approx \bar{\rho} c_i c_j \tag{3a}$$

the intensity per unit volume of turbulence as given, e.g. by eq. (20) for the "self-noise" should be multiplied by a density factor

$$R(y_i) = \left(\frac{\bar{\rho}}{\rho_0}\right)^2 \tag{26}$$

where $\bar{\rho} = \bar{\rho}(y_i)$ is representative of the actual mean density value within a certain correlation volume at y_i. The same modification is to be made to both final results (23) and (25) for "self-noise" and "shear-noise", respectively. The general U^8-law, too, can easily be modified when performing the dimensional analysis in J_Q according to eqs. (24a), (41) and (42) of Part II:

$$J_Q \sim \frac{1}{|x_i|^2} \frac{\rho_0 U^8}{La_0^5} \left(\frac{\bar{\rho}}{\rho_0}\right)^2. \tag{27}$$

4. INTEGRATION OF J OVER "SLICES" OF JET

Independently of Lilley's estimation of the source strength distribution along a jet, Ribner[12] analysed the acoustic power emitted by a "slice" of jet (section between two neighbouring planes normal to the jet axis) as a function of distance x of the slice from the nozzle with diameter D.

Ribner's approach is based on an idealized jet flow model. It presupposes "similar" profiles of the mean and fluctuating quantities.

(i) For the mixing region the assumed similarity laws are:

$$\begin{aligned} U = \bar{c}_1 &= U_0 f_1 \left(\frac{r - D/2}{x}\right) \\ \overline{c_1'^2} &= U_0^2 f_2 \left(\frac{r - D/2}{x}\right) \\ \overline{c_2'^2} &= U_0^2 f_3 \left(\frac{r - D/2}{x}\right) \end{aligned} \tag{28}$$

etc. (r = radial distance from jet axis)

$$L = x f_4 \left(\frac{r - D/2}{x}\right) \quad (L = \text{typical length scale}).$$

(ii) For the fully developed jet the assumed similarity laws are:

$$\begin{aligned} U = \bar{c}_1 &= u\, g_1 \left(\frac{r}{x}\right) \\ \overline{c_1'^2} &= u^2 g_2 \left(\frac{r}{x}\right) \end{aligned} \tag{29}$$

etc. with $u \sim U_0 \dfrac{D}{x}$

$$L = x\, g_4 \left(\frac{r}{x}\right).$$

From eqs. (27) to (29) we can approximate how the acoustic power output P may vary proportional to the typical parameters of the flow, here especially with distance x:

$$dP \sim \frac{\rho_0 U^8}{a_0^5 L} \left(\frac{\bar{\rho}}{\rho_0}\right)^2 d^3y. \tag{30}$$

4.1. *Sound Power from Annular Element in the Mixing Region*

The volume of an annular element of the mixing layer is given in cylindrical coordinates for the axisymmetric case by

$$dV = d^3y \mathrel{\hat{=}} 2\pi r\, dr\, dx. \tag{31}$$

Ribner approximates this as

$$dV = D\, \pi\, dr\, dx$$

in assuming that the errors due to $r < D/2$ and $r > D/2$ will roughly cancel with regard to the noise power output. The expression can be put in the form

$$dV \sim D\, x\, d\left(\frac{r - D/2}{x}\right) dx, \tag{32}$$

since D and x do not depend on r. Inserting (32) and the similarity laws (28) into eq. (30) yields

$$dP \sim \frac{\rho_0 U_0^8 D}{a_0^5} F\left(\frac{r - D/2}{x}\right) d\left(\frac{r - D/2}{x}\right) dx$$

where all the functions of $(r - D/2)/x$ have been lumped into F. On the similarity assumptions made, the integration with respect to $(r - D/2)/x$ yields just a multiplicative constant for any distance x. Therefore,

$$\frac{dP}{dx} \sim \frac{\rho_0 D U_0^8}{a_0^5} \tag{33}$$

states that the acoustic power emitted from a slice of jet a distance x from the nozzle in the initial mixing region is a constant independent of x ("x^0-law").

4.2. *Sound Power from Annular Element in Fully Developed Jet*

We obtain quite a different result when now an annular volume element of the fully developed jet is considered:

$$dV = 2\pi r\, dr\, dx \sim x^2 \left(\frac{r}{x}\right) d\left(\frac{r}{x}\right) dx.$$

Upon substitution of this and the similarity laws (29) into eq. (30), there results:

$$dP \sim \frac{\rho_0 U_0^8 D^8}{a_0^5} \frac{1}{x^7} G\left(\frac{r}{x}\right) d\left(\frac{r}{x}\right) dx.$$

The integration over r/x of all the functions lumped into G again just yields a constant depending only on the type of similar profiles g_n chosen or measured. Thus finally

$$\frac{dP}{dx} \sim \frac{\rho_0 D U_0^8}{a_0^5} \left(\frac{x}{D}\right)^{-7} \quad (\text{"}x^{-7}\text{-law"}). \tag{34}$$

Both the x^0 and the x^{-7}-laws, as exhibited schematically on Fig. 11, needed some comments because of the very rough approximations made in the analysis. For example, one may argue that the jet is far from being self-preserved up to at least 40 diameters from the exit. The question then is: where does dP/dx begin to fall like x^{-7} and what happens in the transition region? Again the reader may refer to Lilley's paper, where the assumptions made under (28) and (29) are discussed in more detail.

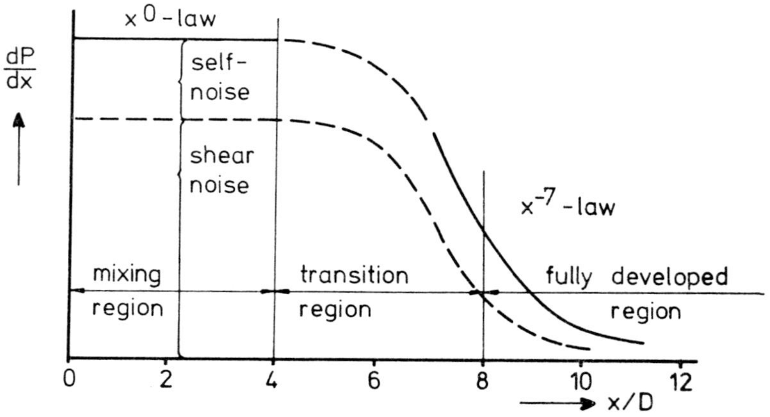

FIG. 11. Idealized strength distribution of noise sources along jet (after Lilley[11] and Ribner[12]).

5. INFLUENCE OF RETARDED TIME ON JET NOISE

In all the preceding sections we did not explicitly account for eddy convection, which is known as a characteristic phenomenon in jet turbulence. The reason is that for a certain class of turbulence models the effect of eddy convection on the radiated sound field results in an additional convection factor multiplying the formulae valid for the non-convected case, especially that for isotropic turbulence.

When proceeding from the subsonic jet to jets at supersonic speeds, major modifications to the theory are obviously required. Apart from the effect of retarded-time differences in the flow, there is also the problem of changes in density at high Mach numbers, i.e. the question whether or not assumption 6 of section 2 remains valid. Abandonment of this simplification led Phillips[13] to a reformulation of the basic wave equation in terms of the pressure such as to eliminate the density in the source term on the right-hand side of his "convected wave equation". His paper was denoted to obtaining asymptotic solutions for large Mach numbers describing the conditions just outside an idealized shear layer. For this, Phillips' equation

$$\left\{\left(\frac{\partial}{\partial t} + \bar{c}_1 \frac{\partial}{\partial x_1}\right)^2 - \frac{\partial}{\partial x_i} \overline{a^2} \frac{\partial}{\partial x_i}\right\} \log \frac{p}{p_0} = \gamma \frac{\partial c_i}{\partial x_j} \frac{\partial c_j}{\partial x_i} \tag{35}$$

reduces considerably, and the radiation may be interpreted as an emission of "eddy Mach waves" which is quite independent of the temporal development of the turbulence that plays such an important role at low speeds.

Ffowcs-Williams,[14] and independently Ribner,[4] worked out an extension of the theory for transonic and supersonic ranges of eddy convection speed based entirely on the now classical Lighthill equation. Thus, starting from an integral equation like (24) of Part II, Ffowcs-Williams regarded the turbulence as represented by a tensor T_{ij}, as usual, but to be specified in a frame of reference moving along with the eddy structure. An attempt is made below to make the details of such a procedure comprehensible to those who are not too familiar with generalized functions, etc.

In analogy to the formulations (18) and (20) for "self-noise" and "shear-noise" respectively, we write the intensity per unit volume of turbulence in the more general form (cf. eq. (24) of Part II):

$$J = \frac{1}{16\pi^2 \rho_0 a_0^5} \frac{1}{|x_i|^2} \int S(z_i, \tau)\, d^3z \tag{36}$$

where

$$S(z_i, \tau) = \frac{x_i x_j x_k x_l}{|x_i|^4} \overline{\left[\frac{\partial^2 T_{ij}}{\partial t^2}\right]_{t^+} \left[\frac{\partial^2 T_{kl}}{\partial t^2}\right]'_{t^+ + \tau}}.$$

Here, according to the far-field approximations made under section 5 of Part II:

$$t^+ = t - \frac{|x_i - y_i|}{a_0} = t - \frac{|x_i|}{a_0} + \frac{x_i y_i}{a_0 |x_i|} \tag{37}$$

$$\tau = \frac{x_i z_i}{a_0 |x_i|}.$$

Before we transform into a system of coordinates moving with the source pattern, we will first rewrite the integrand in (36) in a more convenient form. It essentially represents a two-point cross correlation of a tensor field, with displacement z_i and time delay τ,

$$S(z_i, \tau) = \frac{x_i x_j x_k x_l}{|x_i|^4} \overline{f_{ij}(y_i, t^+) f_{kl}(y_i + z_i, t^+ + \tau)}. \tag{38}$$

Several identical notations for f_{ij} and f_{kl} are feasible (cf. eq. (20) of Part I):

$$f_{ij} = \left[\frac{\partial^2 T_{ij}}{\partial t^2}\right]_{t^+} = \frac{\partial^2 [T_{ij}]}{\partial t^{+2}}$$

$$f_{kl} = \left[\frac{\partial^2 T_{kl}}{\partial t^2}\right]'_{t^+ + \tau} = \frac{\partial^2 [T_{kl}]'}{\partial (t^+ + \tau)^2} = \frac{\partial^2 [T_{kl}]'}{\partial t^{+2}} = \frac{\partial^2 [T_{kl}]'}{\partial \tau^2} \tag{39}$$

or, alternatively, with $t_{ij} = \partial T_{ij}/\partial t$; $t_{kl} = \partial T_{kl}/\partial t$:

$$f_{ij} = \left[\frac{\partial t_{ij}}{\partial t}\right]_{t^+} = \frac{\partial [t_{ij}]}{\partial t^+}$$

$$f_{kl} = \left[\frac{\partial t_{kl}}{\partial t}\right]'_{t^+ + \tau} = \frac{\partial [t_{kl}]'}{\partial (t^+ + \tau)} = \frac{\partial [t_{kl}]'}{\partial t^+} = \frac{\partial [t_{kl}]'}{\partial \tau}. \tag{40}$$

The following identity will prove useful:

$$\frac{\partial[t_{ij}]_{t^+}}{\partial t^+}\frac{\partial[t_{kl}]'_{t^++\tau}}{\partial t^+} = \frac{\partial}{\partial\tau}\left(\frac{\partial[t_{ij}]}{\partial t^+}[t_{kl}]'\right)$$

$$= \frac{\partial}{\partial\tau}\left\{\frac{\partial}{\partial t^+}([t_{ij}]\,[t_{kl}]') - [t_{ij}]\frac{\partial[t_{kl}]'}{\partial t^+}\right\} \tag{41}$$

$$= \frac{\partial}{\partial\tau}\left\{\frac{\partial}{\partial t^+}([t_{ij}]\,[t_{kl}]') - \frac{\partial}{\partial\tau}([t_{ij}]\,[t_{kl}]')\right\}$$

where

$$[t_{ij}]\,[t_{kl}]' = \frac{\partial[T_{ij}]_{t^+}}{\partial t^+}\frac{\partial[T_{kl}]'_{t^++\tau}}{\partial t^+}$$

and hence with the above identity:

$$[t_{ij}]\,[t_{kl}]' = \frac{\partial}{\partial\tau}\left\{\frac{\partial}{\partial t^+}([T_{ij}]\,[T_{kl}]') - \frac{\partial}{\partial\tau}([T_{ij}]\,[T_{kl}]')\right\}. \tag{42}$$

$S(z_i, \tau)$, which is the time or ensemble average of (41) with (42), involves terms like

$$\lim_{T\to\infty}\frac{1}{2T}\int_{-T}^{+T}\frac{\partial}{\partial t^+}([\cdots]\,[\cdots]')\,dt^+$$

which integrate directly

$$\lim_{T\to\infty}\frac{1}{2T}([\cdots]\,[\cdots]')\Big|_{-T}^{+T}$$

and can make no contribution, provided the values of [. . .] and [. . .]′ are limited at any arbitrary $\pm T$, and $2T$ can be chosen sufficiently large. The latter condition implies that the fluctuating processes in the flow field are statistically stationary in time. As a result of these considerations, the integrand in (36) may be written as compact as

$$S(z_i, \tau) = \frac{x_i\,x_j\,x_k\,x_l}{|\,x_i\,|^4}\frac{\partial^4}{\partial\tau^4}\overline{[T_{ij}]\,[T_{kl}]'} = \frac{\partial^4 W}{\partial\tau^4}. \tag{43}$$

Furthermore, we introduce the delta function, which is defined, in general, by the following integral:

$$F(\tau) = \int \delta(\tau' - \tau)\,F(\tau')\,d\tau' \tag{44}$$

where the integral ranges from any number less than τ to any number greater than τ. One important feature of the derivatives of the delta function follows immediately on integration by parts:

$$\int d\tau'\,\frac{\partial^n F(\tau')}{\partial\tau'^n}\,\delta(\tau' - \tau) = (-1)^n\int d\tau'\,F(\tau')\,\frac{d^n}{d\tau'^n}\,\delta(\tau' - \tau)$$

$$= (-1)^n\int dg\,F(\tau')\,\frac{d^n}{dg^n}\,\delta(g) \tag{45}$$

where $$g = \tau' - \tau; \; dg = d\tau'$$

and the integration is taken in the sense g increasing and the integration range including the neighbourhood of $g = 0$.

With the aid of the delta function the integral in (36) can be written as an integral over the four independent variables z_i and τ',

$$\int S(z_i, \tau)\, d^3z = \int d^3z \int dg\; W(y_i, z_i, \tau') \frac{d^4}{dg^4}\, \delta(g). \tag{46}$$

For an evaluation of this integral, Ffowcs-Williams proposed the following transformation of the space-time coordinates:

$$\lambda = \epsilon \mid U_i \mid \tau'; \qquad d\lambda = \epsilon \mid U_i \mid d\tau' \tag{47}$$

$$\eta_i = z_i - U_i\, \tau'; \quad d^3\eta = d^3z \tag{48}$$

Herein λ is simply the time scale variable τ' normalized to an equivalent length variable by multiplying τ' by a velocity scale characteristic of the turbulence. This is the typical turbulence velocity fluctuation level written as

$$\sqrt{\overline{c^2}} = \epsilon \mid U_i \mid \quad \text{or} \quad \sqrt{\overline{c^2}} \sim \mid U_i \mid.$$

Transformation (48) should not be mistaken as an ordinary Galilei transformation of all space variables, e.g. $y_i = \hat{y}_i + U_i t$, and consequently W imagined as $\hat{W}(\hat{y}_i, \hat{z}_i, \tau)$ as perceptible by an observer moving with this new system of coordinates. The physical interpretation of transformation (48) rather is so that the cross-correlation function at a fixed position y_i is to be taken as $W(y_i, z_i, \tau)$ but, for a given $\tau =$ const, as a function of the displacement $z_i = U_i\tau + \eta_i$. If the turbulence were indeed convected with a certain constant speed U_i without decay or amplification, then η_i might be interpreted as a kind of displacement measured in that system attached to the convected eddy structure.

The integral (46) in terms of the new variables reads

$$\int S(z_i, \tau)\, d^3z = \int d^3\eta \int dg\; W(y_i, \eta_i, \lambda)\, \partial^{(4)}(g) \tag{49}$$

$$= \frac{1}{\epsilon \mid U_i \mid} \int d^3\eta \int d\lambda\; W(y_i, \eta_i, \lambda)\, \delta^{(4)}(g),$$

where now

$$g = \tau' - \tau = \frac{\lambda}{\epsilon \mid U_i \mid} - \frac{x_i}{a_0 \mid x_i \mid} (\eta_i + U_i\, \tau') = A_i\, \eta_i + B\lambda \tag{50}$$

with

$$A_i = -\frac{x_i}{a_0 \mid x_i \mid}$$

and

$$B = \frac{1}{\epsilon \mid U_i \mid} \left(1 - \frac{M_i\, x_i}{\mid x_i \mid}\right).$$

Here we have written $M_i = U_i/a_0$ as the Mach number (vector) at which the eddies are being convected through a medium of uniform sound speed a_0.

Consider now the specific features of g as a function of the four variables η_i and λ. grad g is defined as a vector with four components,

$$\text{grad } g = \left\{\frac{\partial g}{\partial \eta_1}; \ \frac{\partial g}{\partial \eta_2}; \ \frac{\partial g}{\partial \eta_3}; \ \frac{\partial g}{\partial \lambda}\right\} \tag{51}$$

which is directed normal to the surface $g = 0$ denoted by S. The surface normal is defined as

$$n_i = \frac{\text{grad } g}{|\text{ grad } g\ |} \tag{52}$$

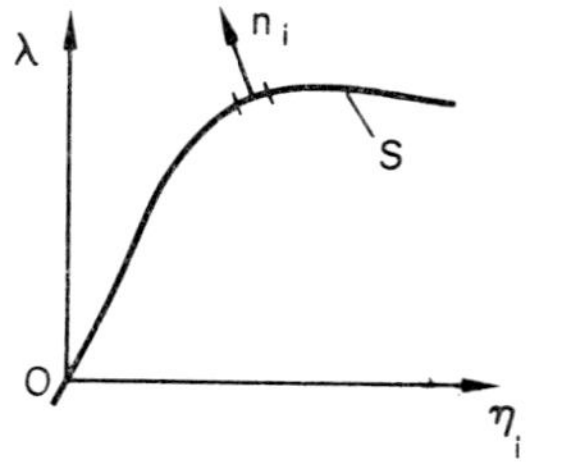

where

$$|\text{ grad } g\ | = \sqrt{\left(\frac{\partial g}{\partial \eta_i}\right)^2 + \left(\frac{\partial g}{\partial \lambda}\right)^2} = \sqrt{A_i^2 + B^2} = C \tag{53}$$

(independent of η_i and λ).

Without loss of generality, we may rotate the space-time system in a way that $\tilde{\eta}_1$ of the new system just lies parallel to A_i:

$$A_i \eta_i = |\ A_i\ |\ \tilde{\eta}_1. \tag{54}$$

This first manipulation, with obviously

$$d^3\tilde{\eta} = d\tilde{\eta}_1\ d\tilde{\eta}_2\ d\tilde{\eta}_3 = d^3\eta \tag{55}$$

$$\tilde{\eta}_i^2 = \tilde{\eta}_1^2 + \tilde{\eta}_2^2 + \tilde{\eta}_3^2 = \eta_i^2$$

yields for the integral (49):

$$\int S(z_i, \tau)\ d^3z = \frac{1}{\epsilon\ |\ U_i\ |}\int d\tilde{\eta}_2\ d\tilde{\eta}_3 \int d\tilde{\eta}_1\ d\lambda\ W(y_i, \tilde{\eta}_i, \lambda)\ \delta^{(4)}(|\ A_i\ |\ \tilde{\eta}_1 + B\lambda). \tag{56}$$

In order to ease the integration in the (η_1,λ)-plane, we may perform a second manipulation by setting for $\tilde{\eta}_1$ and λ new coordinates σ and ζ which are again orthogonal:

$$\sigma = \frac{B\tilde{\eta}_1 - |\ A_i\ |\ \lambda}{C}$$

$$\zeta = \frac{|\ A_i\ |\ \tilde{\eta}_1 + B\lambda}{C} = \frac{g}{|\text{ grad } g\ |}. \tag{57}$$

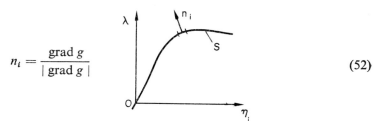

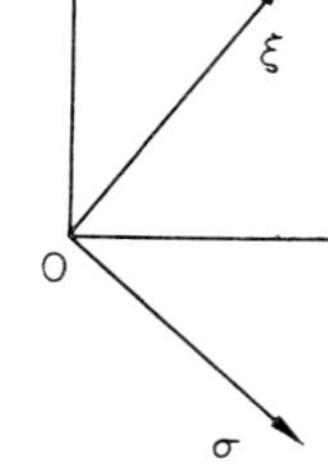

Since again for this simple transformation

$$d\sigma\ d\zeta = d\tilde{\eta}_1\ d\lambda \tag{58}$$

$$\sigma^2 + \zeta^2 = \tilde{\eta}_1^2 + \lambda^2,$$

we may write (56) as

$$\int S(z_i, \tau)\, d^3z = \frac{1}{\epsilon \mid U_i \mid} \int d\tilde{\eta}_2\, d\tilde{\eta}_3 \int d\sigma\, d\zeta\; W(y_i, \tilde{\eta}_2, \tilde{\eta}_3, \sigma, \zeta)\; \delta^{(4)}(g) \tag{59}$$

$$= \frac{1}{\epsilon \mid U_i \mid C} \int d\tilde{\eta}_2\, d\tilde{\eta}_3\, d\sigma \int d(C\zeta)\; W \frac{d^4}{d(C\zeta)^4}\, \delta(C\zeta).$$

Now with identity (45) and definition (44) of the delta function:

$$\int S(z_i, \tau)\, d^3z = \frac{1}{\epsilon \mid U_i \mid C} \int d\tilde{\eta}_2\, d\tilde{\eta}_3\, d\sigma \int d(C\zeta)\, \frac{\partial^4 W}{\partial (C\zeta)^4}\, \delta(C\zeta) \tag{60}$$

$$= \frac{1}{\epsilon \mid U_i \mid C} \int d\tilde{\eta}_2\, d\tilde{\eta}_3\, d\sigma\, \frac{\partial^4 W}{\partial (C\zeta)^4}\bigg|_{g=0}$$

$$= \frac{1}{\epsilon \mid U_i \mid C^5} \int d\tilde{\eta}_2\, d\tilde{\eta}_3\, d\sigma\, \frac{\partial^4 W}{\partial \zeta^4}\bigg|_{\zeta=0}.$$

Progress must cease at this general result unless we can say something more about the character of

$$W = W(y_i, \eta_i, \lambda) = W(y_i, \tilde{\eta}_i, \lambda) = W(y_i, \tilde{\eta}_2, \tilde{\eta}_3, \sigma, \zeta).$$

If we could choose the correlation function W so that it would only depend on the absolute value of the orthogonal coordinates, say η_i^2 and λ^2, then according to relations (55) and (58) W would remain unchanged when transformations (54) and (57) are performed. This condition is exactly met when one assumes W to be isotropic in space-time, i.e.

$$W_{\text{is}} = W\left(y_i, \sqrt{\eta_i^2 + \lambda^2}\right) = W\left(y_i, \sqrt{\tilde{\eta}_i^2 + \lambda^2}\right) = W\left(y_i, \sqrt{\tilde{\eta}_2^2 + \tilde{\eta}_3^2 + \sigma^2 + \zeta^2}\right). \tag{61}$$

All the preceding manipulations in the integral of eq. (36) thus would not have been worthwhile unless one presupposes a specific model for the turbulence in a system convected with the eddy structure.

By inserting W_{is} into (60),

$$\int S(z_i, \tau)\, d^3z = \frac{1}{\epsilon \mid U_i \mid C^5} \int d\tilde{\eta}_2\, d\tilde{\eta}_3\, d\tilde{\eta}_1\, \frac{\partial^4 W_{\text{is}}}{\partial \lambda^4}\bigg|_{\lambda=0} \tag{62}$$

we may finally write the acoustic intensity per unit volume of turbulence as

$$J = \frac{1}{16\pi^2\, \rho_0\, a_0^5}\, \frac{1}{\mid x_i \mid^2}\, \frac{1}{(\epsilon \mid U_i \mid C)^5} \int d^3\eta\, \frac{\partial^4 W_{\text{is}}}{\partial \tau^4}\bigg|_{\tau=0} \tag{63}$$

where a relation like (47) between λ and τ has been used.

We can recover the non-convected result by setting the convection speed U_i to zero and hence from (53) and (50),

$$(\epsilon \mid U_i \mid C)^5 = \left[\left(1 - \frac{M_i\, x_i}{\mid x_i \mid}\right)^2 + \epsilon^2 M_i^2\right]^{5/2}, \tag{64}$$

we see that an important effect of eddy convection may be extracted from the integral (36).

This is the so-called convection or Mach factor. On the assumption of an isotropic eddy structure convected at constant speed as a frozen pattern, which does never begin nor end, this convection factor (64) would be very significant for the sound radiation at high Mach numbers. It would simply multiply Proudman's result (23). If, however, turbulence in a jet happened to be anything but isotropic (in any system of coordinates) then of course the factor (64) could not, beforehand, tell us very much.

Provided the integral in (63) happens to be independent of Mach number, then one would expect Lighthill's eighth-power law (cf. eq. (42) of Part II) to become

$$P_Q \sim \rho_0 L^2 U^3 \frac{M^5}{\left[\left(1 - \frac{M_i x_i}{|x_i|}\right)^2 + \epsilon^2 M_i^2\right]^{5/2}} \tag{65}$$

and for Mach numbers $M \gg 1$ degenerate to

$$P_Q \sim \rho_0 L^2 U^3. \tag{65a}$$

That means the efficiency η_Q of the sound production (refer to (43) of Part II) would not keep increasing like M^5 but reach a certain value which then remains constant with Mach number.

In the investigation of convection effects we started from an equation like (36) or (20) as typical for the intensity of "self-noise". If instead we would consider an integral like (18) ("shear-noise"), the only difference in the analogue development would be that the corresponding covariance of the pressure involves only first time derivatives. Therefore, the second derivative of the delta function would appear and hence the procedure in (60) yield a C^{-3} and consequently in the step from (62) to (63) the convection factor would come out as $(\epsilon|U_i|C)^{-3}$ instead of $(\epsilon|U_i|C)^{-5}$.

6. SOUND AND "PSEUDOSOUND"

When one looks at the radiated acoustic field far away from a source flow, one usually thinks of it in terms of fluctuating pressures caused by wave-like changes in density of a compressible medium. For a complete knowledge of this sound field it is quite sufficient to know all details of the pressure only. Mature microphone measuring techniques are available for a long time to depict and analyse the pressure.

When, on the other hand, one considers the turbulence in a flow at moderate Mach numbers, one tends to think of it in terms of fluctuating velocities caused by disordered (random) motions of fluid particles (eddies) with essentially no changes in density taking part in the process of mutual exchange of momentum. For a complete knowledge of this turbulent near-field it is quite sufficient to know, inversely, all details of the particle velocities. In fact there are, nowadays, no principal difficulties in analysing such velocity fields by means of refined hot-wire techniques.

In the case of an acoustic far-field we have already seen in section 2 of Part II that pressure and velocity are related by

$$\frac{\partial c_i'}{\partial t} = -\frac{1}{\rho_0}\frac{\partial p'}{\partial x_i}. \tag{66}$$

These velocities are so closely connected with the sound pressure that nobody would hit upon the idea to term "pseudoturbulence" in distinction of real turbulent velocities in a general fluid motion.

In the turbulent case, on the other hand, the pressure is related to the velocities by an inhomogeneous Laplace equation:

$$\frac{\partial^2 p}{\partial x_i^2} = -\bar{\rho}\,\frac{\partial^2 c_i\, c_j}{\partial x_i\, \partial x_j}, \tag{67}$$

which is easily derived from eqs. (5) and (6) of Part I with $\partial^2 \rho/\partial t^2 = 0$. This pressure, although so closely connected with the turbulent velocities, is sometimes—rather misleadingly—termed "pseudosound".

Both differential equations (66) and (67) may be integrated straightforwardly. For the radiated far-field of an arbitrary source flow (cf. sections 2 to 4 of Part II):

$$p'(x_i, t) = \rho_0\, a_0\, c_n'(x_i, t). \tag{68}$$

Thus, the instantaneous pressure at a point is exactly proportional to the local velocity normal to wave front. The general integral of Poisson's equation (67) is similar to that of the inhomogeneous wave equation in section 3 of Part I:

$$p(x_i, t) = \frac{\bar{\rho}}{4\pi} \int_V \frac{1}{r} \frac{\partial^2 c_i\, c_j}{\partial y_i\, \partial y_j}\, dV + \bar{\rho} \int_A \frac{1}{r} \frac{\partial p}{\partial n}\, dA + \bar{\rho} \int_A \frac{1}{r^2} \frac{\partial r}{\partial n}\, p\, dA. \tag{69}$$

This shows that even in the absence of any bounding surfaces the pressure at a point x_i is an integral extended over the whole velocity field.

The remote analogy to the radiated sound pressure is obvious. One cannot solve the problem analytically unless one makes decisive assumptions concerning the structure of the turbulence and the decay of the velocity correlation function involved. By assuming isotropy, Proudman[9] and Lilley[11] were able to solve the sound problem analytically for the so-called "self-noise" and "shear-noise" originating from an idealized model of a jet. Batchelor[10] and Kraichnan[15], on the other hand, calculated the pressure fluctuations within flows satisfying the same conditions. In a heavily sheared mean flow one may again distinguish between two different mechanisms generating pressure,

(i) pressure due to turbulence–turbulence interaction (p_{T-T}) and
(ii) pressure due to turbulence–mean shear interaction (p_{T-M}).

The result for both contributions to the pressure can be written in the form:

$$\begin{aligned} \overline{p_{T-T}^2} &= 2\bar{\rho}^2\, (\overline{c^2})^2 \int_0^\infty z \left(\frac{df(z)}{dz}\right)^2 dz \\ \overline{p_{T-M}^2} &= \frac{8}{15}\, \bar{\rho}^2\, \overline{c^2}\, \bar{\tau}_{12}^2 \int_0^\infty z\, f(z)\, dz \end{aligned} \tag{70}$$

where the quantities involved are known already from section 3. It is noted, however, that $\bar{\tau}_{12} = \partial \bar{c}_1/\partial x_2$ is now kept constant throughout the flow, whereas in section 3 it had only been assumed uniform over a typical correlation volume. We will write down the final results for the pressure when three different forms for $f(z)$ are chosen, the first two of which we already know from section 3.

(a) With Heisenberg's form of the correlation function:

$$\overline{p^2_{T-T}} = 0.34\ \bar{\rho}^2\ (\overline{c^2})^2. \tag{71a}$$

(b) With $f(z) = \exp(-\sigma^2 z)$:

$$\overline{p^2_{T-T}} = \bar{\rho}^2\ (\overline{c^2})^2$$

$$\overline{p^2_{T-M}} = \frac{4}{15}\ \frac{\bar{\tau}^2_{12}\ \bar{\rho}^2\ \overline{c^2}}{\sigma^2}. \tag{71b}$$

(c) With $f(z) = \exp(-\sigma z)$:

$$\overline{p^2_{T-T}} = 0.5\ \bar{\rho}^2\ (\overline{c^2})^2$$

$$\overline{p^2_{T-M}} = \frac{8}{15}\ \frac{\bar{\tau}^2_{12}\ \bar{\rho}^2\ \overline{c^2}}{\sigma^2}. \tag{71c}$$

(For the details of the analysis refer to the original papers.)

The reason why some people introduced the term "pseudosound" into jet noise theory might be a historical one. In one of the earliest publications on the acoustics of a non-homogeneous moving medium, Blokhintsev[16] pounced upon the genuine pressure fluctuations in the disturbed flow. This was, however, when he studied feasible background noise sources for purely acoustic measurements with a microphone inserted in a flow on which an *external* sound field is superimposed. It was only in this special case that Blokhintsev found it useful to term one of the possible probe-flow interferences at the sound receiver "pseudo-sound".

7. DIRECT MEASUREMENT OF FLUCTUATING PRESSURE IN A JET

Experiments in turbulence have until recently favoured the measurement of instantaneous velocities and their statistical characteristics. This is mainly due to the early development of reliable hot-wire anemometers whereas pressure probes were suspected of inevitably creating and recording additional spurious pressure fluctuations due to probe-flow interactions.

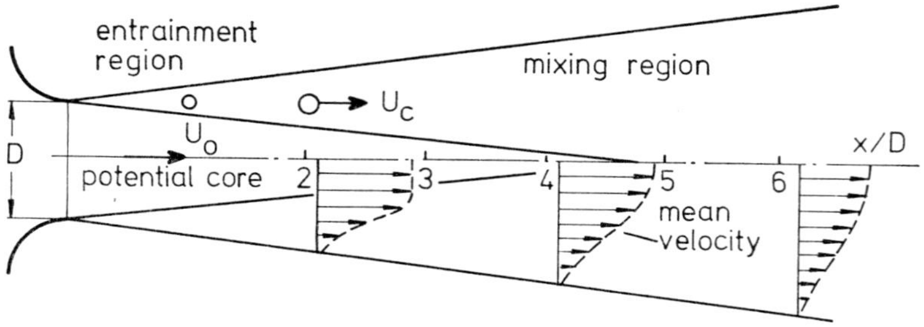

FIG. 12. Circular jet flow configuration.

In the case of a jet, however, the author[17] showed that this problem can be overcome when suitably shaped static-pressure probes are used to detect the original pressure fluctuations in the flow. The characteristics of the pressure field on both sides of the high-intensity mixing layer of a jet was discussed in some detail in the paper cited above. The pressure level, for instance, was found to exceed the values expected for isotropic turbulence (cf. eq. (71)) by a factor up to 10^3 on the jet axis, this being one of the reasons why the pressure could be recorded without noticeable distortions.

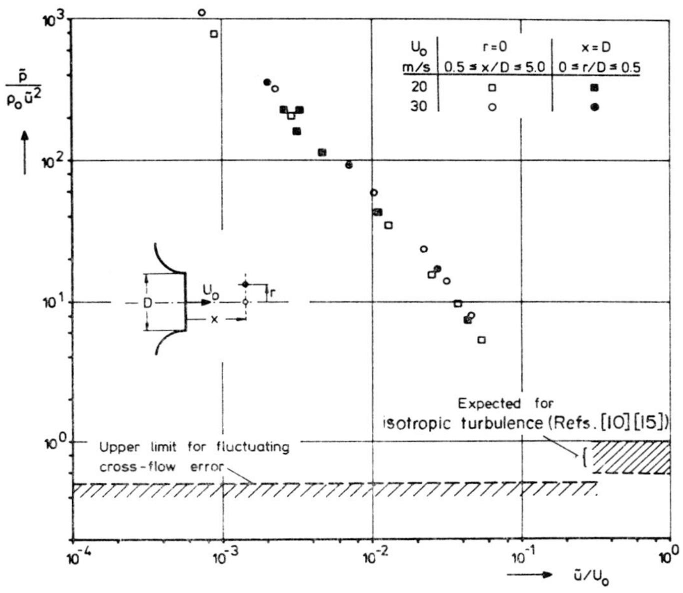

FIG. 13. Ratio $\tilde{p}/\rho_0\,\tilde{u}^2$ against local values $\tilde{u}/U_0$ in jet core. (Measurements by Fuchs[17].)

Part of the measured pressures could be interpreted on the basis of a simple theoretical model. It was assumed that the disturbances, pressure as well as velocities, are induced by a nearly frozen pattern of sources convected in the central regions of the mixing layer.

The convection speed U_c was taken from cross-correlation measurements as roughly uniform and approximately 0.65 of the jet exit speed U_0. On this assumption, the integration of the linearized momentum equation yields the instantaneous pressure fluctuation as proportional to the local axial velocity component u':

$$p' = -\rho_0\,(U - U_c)\,u' \sim \rho_0\,U\,u' \text{ (compare Fig. 14).} \tag{72}$$

This first-order approximation to the conditions in and close to the jet works surprisingly well in parts of the jet flow. The prediction that inside of the mixing layer, where $U_c < U$, the pressure should be in anti-phase with the axial velocity, and outside, where $U_c > U$, it should be in phase with u, could be justified fairly well by cross-correlation measurements with microphone and crossed hot-wire probes reported by Lau, Fuchs and Fisher.[18]

Relation (72) would hold for any type of frozen source pattern for points sufficiently close to the source region. The phase of simultaneously measured velocities in the core and in the entrainment region indicated, however, that the sources, if they are present, must be of higher order, i.e. dipoles or quadrupoles. Further experimental and theoretical studies are under way. In particular, it was felt that the signals, before correlating them, should be passed through narrow-band frequency analysers in order to receive more information about the statistical behaviour of specific components in the spectra.

Relation (72) implies that the disturbances in the jet are propagated in the downstream direction in a wave-like manner. Comparison of the spectra in the mixing region and on both sides of it suggests that it is only a limited range of frequencies which are associated with such an orderly structure in the turbulence. Spectra on the axis exhibit a pronounced peak in the basically broad-band distribution at a Strouhal number

$$\frac{f_{\text{peak}}\, D}{U_0} = 0.4\text{–}0.5 \tag{73}$$

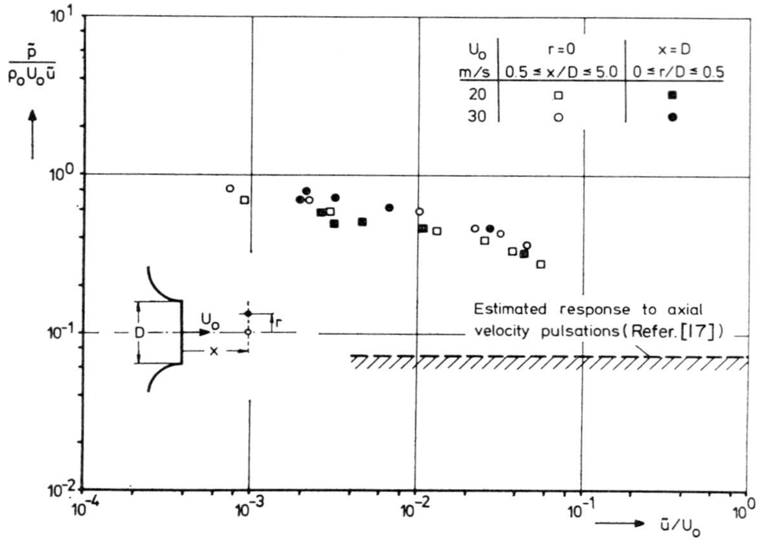

FIG. 14. Ratio $\tilde{p}/\rho_0\, U_0\, \tilde{u}$ against local values $\tilde{u}/U_0$ in jet care.

A typical pressure spectrum is shown in Fig. 15. There is already experimental evidence that the frequency components around this peak are indeed propagating down-stream with practically no decay over a distance of a few jet diameters.

In a radial direction these spectral components of the pressure seem to be well correlated over much of the jet. Hence, the simplified physical picture is that of an almost axisymmetric disturbance propagating coherently downstream. For details of recently obtained correlation results (a sample of which is depicted on Fig. 16) please refer to reference 19.

Some of the new results are obviously incompatible with the assumption of an isotropic turbulence structure (be it convected or not) on which, as we have seen in the preceding sections, jet noise theory was, hitherto, mainly based. A wide range of spectral components on both sides of f_{peak} may be less coherent in the longitudinal and lateral directions; still, the wave-like components are found to incorporate a marked portion of the fluctuating energy in the near-field of a jet.

The admittedly oversimplified model discussed in the foregoing must, surely, be modified

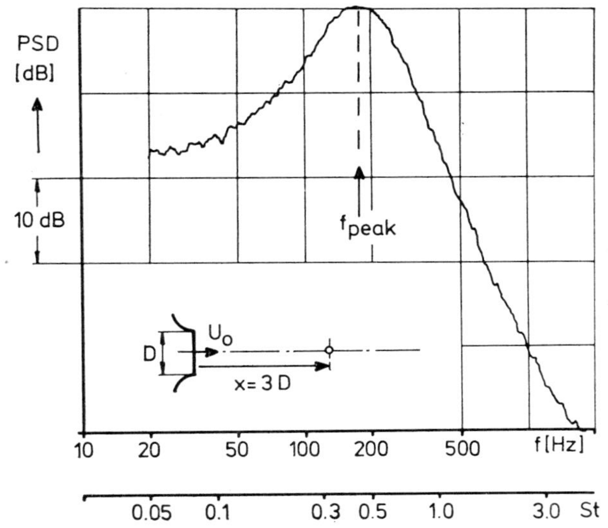

FIG. 15. Pressure probe on jet axis: shape of power spectral density (arbitrary scale) $U_0 = 40$ m/s, bandwidth 31.6 Hz.

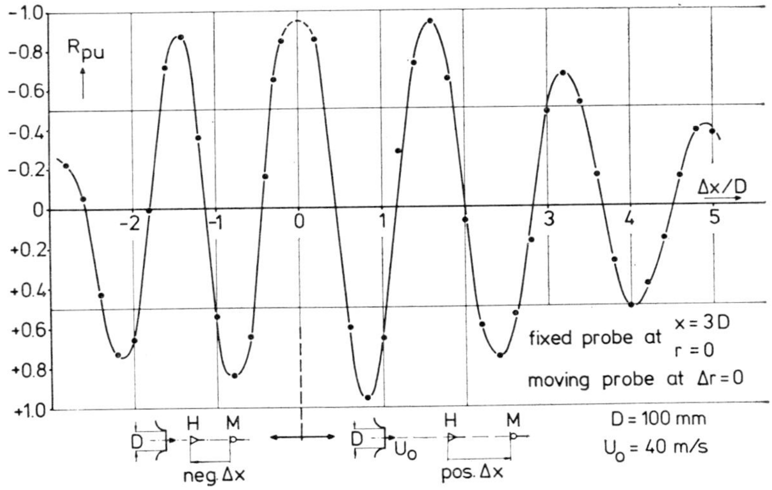

FIG. 16. Longitudinal space correlation coefficient on jet axis, microphone (M) and hot wire (H). Signals filtered at centre frequency 200 Hz, bandwidth 10 Hz.

when estimating the effect of wave-like disturbances on the sound radiated by a jet. The finite extent of the source region, for instance, and the local intensity variations within this region have to be considered. At this stage, we only note that it may be misleading to imagine the source flow as being divided into numerous small subregions (thought of as acoustically compact eddies), with all of them fluctuating and radiating in a completely uncorrelated way. This is because for a wave-like structure even the definition of a correlation volume or area would become irrelevant.

8. ALTERNATIVE APPROACHES TO JET NOISE

When in section 4 of Part II we sought for a solution of the inhomogeneous wave equation, we did not really do anything but transform the original differential equation into an integral equation. On the assumptions made in section 2 the two identical formulations read:

$$p - p_0 = \frac{1}{4\pi} \int \frac{1}{r} \left[\frac{\partial^2 \rho\, c_i\, c_j}{\partial y_i\, \partial y_j} \right] dV \tag{74}$$

$$\frac{\partial^2 \rho\, c_i\, c_j}{\partial x_i\, \partial x_j} = \frac{1}{a_0^2} \frac{\partial^2 p}{\partial t^2} - \frac{\partial^2 p}{\partial x_i^2}. \tag{74a}$$

To be sure, if we were really given $\partial^2 \rho c_i c_j / \partial y_i \partial y_j$ in some closed analytical form, we could certainly use the integral identity (74) to determine the radiation field. The same would be true if we were able to give the pressure in the flow an analytical representation. The latter would essentially mean that we replaced Lighthill's source term, which is basically a non-linear function of fluctuating quantities, by an equivalent source term which is linear in the pressure. This would considerably reduce the complexity in the rest of the analysis.

Our present knowledge of the fluctuating pressure field in a jet is still incomplete so that we cannot yet use it as a known source function in the theory. There are, however, certain characteristic features, typical for any circular jet with and without the assumption of wave-like disturbances, that can already be incorporated in a theoretical model for the pressure or even for the source term q in general.

Michalke[20] has only recently put forward such a new model for the sound generated by circular jets. In Part IV of this article he will also cover methods by which the spectral components of jet noise can be treated individually. Fourier analysis of the source function and solving for the spectral components of the radiated pressure appears to be indispensable, if one wishes to predict directivity distributions in the far-field of a jet. This is the reason why overall directivities deduced from isotropic turbulence models were not explicitly treated here.

This introduction into jet noise would be incomplete if we did not mention Ribner's[4] simple-source theory as an alternative to Lighthill's quadrupole theory. A similar idea as that mentioned above led Ribner (and prior to him Meecham and Ford[5]) to replace Lighthill's T_{ij} by a source term in which the pressure in the flow appears as the driving function for the sound generation. But he did not use the exact relationship (74a). Instead, he introduced a kind of incompressible-flow approximation to the pressure with

$$\frac{\partial^2 p^0}{\partial x_i^2} = - \frac{\partial^2 \rho c_i c_j}{\partial x_i\, \partial x_j}. \tag{75}$$

After Ribner, the true pressure p in the flow then consists of a linear superposition of the "pseudosound" p^0 and real sound p',

$$p = p^0 + p'. \tag{76}$$

Furthermore, he rewrote eq. (74a) as:

$$\begin{aligned} \frac{1}{a_0^2} \frac{\partial^2 p'}{\partial t^2} - \frac{\partial^2 p'}{\partial x_i^2} &= - \frac{1}{a_0^2} \frac{\partial^2 p^0}{\partial t^2} + \frac{\partial^2 p^0}{\partial x_i^2} + \frac{\partial^2 \rho\, c_i\, c_j}{\partial x_i\, \partial x_j} \\ &= - \frac{1}{a_0^2} \frac{\partial^2 p^0}{\partial t^2}. \end{aligned} \tag{77}$$

Again following Ribner's belief, this "dilatation equation" relates the "real" sound p' to the "pseudosound" p^0.

Expression (77) has the form of a wave equation for a medium at rest with a virtual source distribution $-\,\partial^2 p^0/a_0{}^2\partial t^2$. By isentropy Ribner found the following relationship to be justified:

$$-\frac{1}{a_0^2}\frac{\partial^2 p^0}{\partial t^2} \approx -\frac{\partial^2 \rho^0}{\partial t^2}, \tag{78}$$

which would associate the "pseudosound" with changes in density, the latter being associated with corresponding changes in volume. It was on this basis that Ribner developed his theory on aerodynamic sound from fluid dilatations, in which he re-derived many of the results relevant to jet noise in terms of simple sources instead of quadrupoles.

An alternative theoretical approach to the jet noise problem was already alluded to in section 5. It was made by Phillips[13] and concentrated on high-supersonic speeds, where an important issue is the refraction of sound in travelling from a point of emission in the flow through velocity and temperature gradients into the atmosphere beyond. In connexion with refraction effects in jet noise the work of Csanady[21] remains to be mentioned, who also started from a "convected wave equation". For completeness, we may briefly allude to two further alternatives to Lighthill's acoustic analogy.

The one is based on the method of matched asymptotic expansions (refer, e.g., to Möhring, Müller and Obermeier[22] for the corresponding literature). It emerges from this general scheme that the Lighthill analogy appears as only the first term of a Taylor expansion of a more general integral equation of the fluid density. In an application of this new method to a two-dimensional problem one was able to calculate the sound radiation from a pair of spinning vortices. The total power radiated was found to be proportional to U^7 in contrast to the eighth-power law valid for three-dimensional flows.

The last approach which was put forward, among others, by C. L. Morfey is as yet little more than a suggestion based on a general energy-balance equation of the form

$$\frac{\partial E}{\partial t} + \frac{\partial N_i}{\partial x_i} = P\,(x_i, t) \tag{79}$$

where E stands for an energy density and N_i for the vector of an energy-flux density. Similar to q in the inhomogeneous wave equation (eq. (10) of Part I), P represents a potential energy source (or sink) which can again be deduced from the basic flow equations by fundamental manipulations. Unlike q, however, P strongly depends on how one defines E and N_i. Outside the flow P should vanish and eq. (79) reduce to the ordinary balance equation for the acoustic energy.

In ref. 23 the functions E and N_i were chosen to be

$$E = \frac{1}{\bar{\rho} a^2}\, p'^2 + \frac{\bar{\rho}}{2}\, c_i'^2 + \frac{\bar{c}_i}{a^2}\, p'\, c_i' \tag{80}$$

and

$$N_i = p'\, c_i' + \frac{\bar{c}_i}{\bar{\rho} a^2}\, p'^2 + \frac{\bar{c}_i\, \bar{c}_j}{a^2}\, p'\, c_j' + \bar{\rho}\bar{c}_j\, c_j'\, c_i'. \tag{81}$$

In a homogeneous and perfect fluid at rest, which may surround the turbulent flow, these reduce to the acoustic energy and energy flux, respectively:

$$E = \frac{1}{\bar{\rho} a^2} p'^2 + \frac{\bar{\rho}}{2} c_i'^2 \tag{80a}$$

$$N_i = p' c_i'. \tag{81a}$$

Consider now the volume integral of the time-averaged eq. (79):

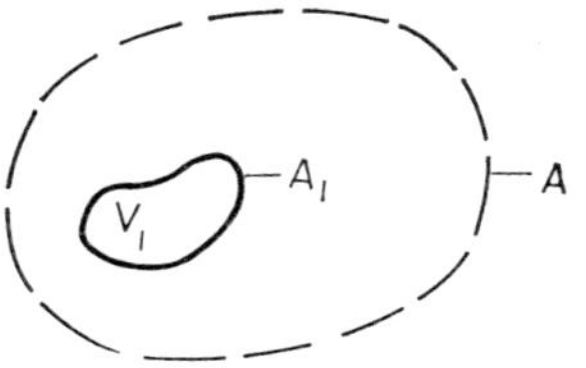

$$\int_A \overline{N_i}\, dA_i = \int_V \overline{P}\, dV. \tag{82}$$

It is clear that P must vanish outside the sound producing flow provided there are no sound sources or dissipative forces present in the fluid surrounding the flow. From eqs. (82) and (81a) then follows that P stands for a sound-source density which is non-zero only in the volume occupied by the fluctuating flow.

When one surmises isentropic conditions, i.e. an inviscid and non-conducting fluid with no external sources of mass, momentum or heat being involved, then the source terms on the right of (79) can be shown to be

$$P = \bar{\rho}\bar{c}_i \left(\frac{\partial c_i'}{\partial x_j} - \frac{\partial c_j'}{\partial x_i}\right) c_j' - \left(\frac{\bar{c}_j}{a^2}\frac{\partial \bar{c}_j}{\partial x_i} + \frac{1}{\bar{\rho}}\frac{\partial \bar{\rho}}{\partial x_i}\right) p' c_i' + \frac{1}{\bar{\rho} a^2}\left(\frac{\partial \bar{c}_j}{\partial x_j} - \frac{\bar{c}_i \bar{c}_j}{a^2}\frac{\partial \bar{c}_j}{\partial x_i}\right) p'^2. \tag{83}$$

The first term in (83) would be zero only if the velocity fluctuations were irrotational. It may be compared with what is called "self-noise" in Lighthill's theory. The remainder of the source terms depends on the magnitude of the gradients of the mean quantities, c_i and ρ_0. In this they resemble a phenomenon which was termed "shear-noise" in section 3.

The generation of sound by a fluctuating flow can thus be related to the properties in the flow. But in contrast to the wave equation approach, the energy balance does not reveal all details about the structure of the radiated field, i.e. its directivity, etc. One advantage of the latter is obvious; it is the local co-variance of the fluctuating quantities that is needed for an estimation of the total sound power emitted. The mutual interference of locally displaced equivalent acoustic sources, which is mainly responsible for the directional characteristics of the sound field, is buried in the covariances and hence automatically taken into account by integration. The unnecessarily powerful procedure of calculating the instantaneous sound pressure prior to a computation of intensity is thus eluded and the retarded-time effects apparently eliminated.

APPENDIX: SOME RECENT DEVELOPMENTS IN JET NOISE RESEARCH
(a note added in proof)

After these introductory notes on the concepts common in aerodynamic noise theory had been sent on to the publishers, another torrent of papers on jet engine noise has been released, mainly from research groups in the United States. Apart from thirteen (!) volumes (NASA, CR 1705–1717) on fan-compressor noise reductions of DC-8 Nacelle, there appeared also quite voluminous interim reports of the General Electric Comp. (AFAPL, TR 72–52, 800 pages approximately) and of the Lockheed Georgia Comp. (AFAPL, TR 72–53, more than 1700 pages). The reported activities are part of a large-scale U.S. Air Force supersonic jet noise investigation program in which Bolt Beranek and Newman Inc., too, participated during the first phase (AFAPL, TR 72–00). There is hope that these obviously large efforts will considerably further our understanding of jet noise both subsonic and supersonic.

On the theoretical side, one seems to have abandoned Lighthill's original acoustic analogy in favour of a unified aerodynamic sound theory which is no longer based on an ordinary-wave-equation approach. Some theorists now dislike the physically unsatisfying formulation which lumps together all those difficult matters as generation, propagation, convection, refraction or scattering of sound and treats them as "equivalent acoustic sources" in a homogeneous medium at rest. P. E. Doak in AFAPL, TR 72–53, Vol. III, for instance, does not deny that (p. 37) "Lighthill's formulation, being mathematically exact, must give the right answers, if only the exact measured value of his source term quantities are used. However, the source term quantities concerned simply cannot, in practice, be measured that exactly, nor can they be deduced exactly enough from presently available theory." And he concludes (p. 63): "Despite its beguiling simplicity the acoustic analogy approach cannot be regarded now as more than a helpful and essential first step towards an explicit theory of aerodynamic sound generation and propagation. In this explicit theory, all 'equivalent source terms' that locally, in the real flow, directly involve the dependent field variable to be predicted must appear not as source terms at all but as 'propagation-type' terms, leaving as source terms only those that can be regarded unambiguously as responsible for local conversion of the energy of the 'non-acoustic' types of motion present into 'acoustic' energy." Doak's paper is recommended as an extraordinary critical comment on almost everything so far known as aerodynamic noise theory. The only attempt which passed Doak's examination is incorporated in AFAPL, TR 72–53: in Vol. IV G. M. Lilley derives a new "convected wave equation" which, for a two-dimensional inviscid mean shear flow (as described here in Fig. 10 and eq. (12) on p. 262) reads

$$\left\{\left(\frac{\partial}{\partial t}+\bar{c}_1\frac{\partial}{\partial x_1}\right)^3-\left(\frac{\partial}{\partial t}+\bar{c}_1\frac{\partial}{\partial x_1}\right)^2\frac{\partial}{\partial x_i}\,\overline{a^2}\,\frac{\partial}{\partial x_i}+2\,\frac{\partial\bar{c}_1}{\partial x_2}\,\frac{\partial}{\partial x_1}\,\overline{a^2}\,\frac{\partial}{\partial x_2}\right\}\log\frac{p}{p_0}=\Lambda(x_i,t) \tag{84}$$

and is of third order in contrast to Phillips' second-order wave equation (eq. (35) of this paper). The right-hand side, $\Lambda(x_i, t)$, contains products and higher-order terms of fluctuating quantities involving all kinds of interactions between velocity, pressure and temperature fluctuations (refer to eq. (4.13) of the original paper).

Relatively little is said in the Lockheed, General Electric and BBN reports about the turbulence models to be introduced into any of the proposed source functions, regardless of the mathematical structure of the equations. A quasi-isotropic convected turbulence structure seems still to prevail (see, however, the wave-model described by Lilley, pp. 36–38 of the above cited report).

On the experimental side, various remote sensing devices (Laser–Doppler Velocimeter, Crossed-beam Schlieren, Pulsed-Laser Interferometer) are currently developed for studying the conjectural source functions in the turbulent flow regions. In addition, the BBN and General Electric groups recommend fluctuating static pressure measurements with inserted microphones to study turbulent large-scale structures in jets.

PART IV: SPECTRAL METHODS IN JET NOISE THEORY

1. INTRODUCTION

Theoretical investigations of jet noise mostly started from the inhomogeneous acoustic wave equation which has been derived by Lighthill.[1,2] He showed that the source term of this equation can be interpreted as a distribution of acoustic quadrupoles in a medium otherwise at rest. Therefore, the sound field produced by the jet will have a directivity. This directivity can be derived from the integral solution of the Lighthill equation.

It is known that the sound intensity far from the jet can be calculated, if the correlation function of the turbulent stress tensor is given. For a turbulent jet this correlation function will, however, depend on the jet Mach number and on the structure of jet turbulence, for instance. Unfortunately, the correlation function is mostly unknown. Therefore, assumptions with respect to the structure of jet turbulence have to be made. A general accepted assumption is that the jet turbulence consists of turbulent eddies which are convected with a certain convection speed. In a frame of reference moving with the convection speed, the structure of turbulence is believed to be approximately isotropic and independent of the Mach number. This model of jet turbulence leads in the fixed frame of reference to a directivity of sound radiation caused by the convection of the turbulent eddies. In this way the directivity pattern has been deduced, among others, by Ffowcs-Williams[14] and by Ribner.[4] It was, however, found that the theoretical directivities did not completely fit the measured values. Especially at angles near the jet axis, differences have often been observed (cf. Jones[24]). Therefore, Mollo-Christensen[25] stated "that the theory of aerodynamic noise could never be checked experimentally in detail".

With respect to noise emission Mollo-Christensen, Kolpin and Martuccelli[26] and Krishnappa and Csanady[27] found experimentally that spectral sound components of low and high frequencies have quite different directivities. Furthermore, the directivities measured in different jets for the same frequency bands can have different shapes. It may be supposed that a method based only on overall correlations can hardly reveal these properties of the mechanism of sound generation, since there is a strong loss of information by using overall correlations. Therefore, it is reasonable to suppose that spectral methods possibly lead to a deeper insight into the problem of noise generation. The radiation of single frequency components of sound can be investigated theoretically by the application of a Fourier transformation with respect to time. In the following, we shall discuss two different spectral methods which have been developed recently.

2. THE SPECTRAL THEORY OF PAO AND LOWSON

Pao and Lowson[28,29] started with the integral solution of the Lighthill equation. In the absence of bounding surfaces this is

$$p(x_i, t) = \frac{1}{4\pi} \int dy_i \, \frac{1}{r_0} \, q(y_i, t - r_0/a_0). \tag{1}$$

Here p is the pressure and q the source term of the Lighthill equation. x_i are the cartesian coordinates of the measuring point, y_i the cartesian coordinates of the source point and dy_i is the volume element of the source region. a_0 is the constant speed of sound in the medium at rest surrounding the jet flow. r_0 is the distance between the source point and the measuring point defined by

$$r_0 = \sqrt{(y_i - x_i)^2} = | \, y_i - x_i \, |. \tag{2}$$

A geometric far-field approximation for large distances $\tilde{r} = | \, x_i \, | \gg | \, y_i \, |$ yields from (2):

$$r_0 \approx \tilde{r} - y_i \, x_i/\tilde{r}. \tag{3}$$

This approximation can be used for calculating the sound field, if the source region is of a restricted extension. This is surely true for a turbulent jet flow.

2.1. *A Solution of the Lighthill Equation for Frequency Components*

It is frequently believed that the jet turbulence is random in time and space. Therefore, Pao and Lowson[28,29] assumed that the turbulence can approximately be described by a superposition of plane waves whose wave numbers and frequencies are arbitrary. Hence, they concluded that the source term q can be presented by a four-dimensional Fourier integral

$$q(y_i, t) = \int dk_i \int d\omega \; \hat{Q}(k_i, \omega) \exp\,[i(k_i \, y_i - \omega t)], \tag{4}$$

where k_i are the components of the wave-number vector with respect to cartesian coordinates and ω is the cyclic frequency. dk_i is the volume element in the wave-number space and $\hat{Q}$ is the wave-number frequency spectral function of q. Introducing (4) and (3) in (1), one obtains to $O\,(\tilde{r}^{-1})$:

$$p(x_i, t) = \frac{1}{4\pi} \int d\omega \, \frac{1}{\tilde{r}} \exp\,[i(k\tilde{r} - \omega t)] \int dk_i \int dy_i \; \hat{Q}(k_i, \omega) \exp\,[i(k_i - kx_i/\tilde{r}) \, y_i] \tag{5}$$

where

$$k = \omega/a_0 \tag{6}$$

is the wave number of sound. Now we can use the properties of the Dirac delta function δ defined by

$$\delta(x_i) = \int \exp\,(ix_i \, y_i) \, dy_i \tag{7}$$

and

$$\int \delta(x_i - A_i)\, G(x_i)\, dx_i = G(A_i). \tag{8}$$

Hence, the solution (5) for the sound pressure becomes with (7) and (8):

$$p(x_i, t) = \frac{1}{4\pi} \int d\omega \, \frac{1}{\tilde{r}}\, e^{i(k\tilde{r} - \omega t)} \int dk_i\, \delta(k_i - kx_i/\tilde{r})\, \hat{Q}(k_i, \omega) \tag{9}$$

$$= \frac{1}{4\pi} \int d\omega \, \frac{1}{\tilde{r}}\, e^{i(k\tilde{r} - \omega t)}\, \hat{Q}(kx_i/\tilde{r}, \omega).$$

We see from (9) that the sound pressure p can be written as a Fourier integral

$$p(x_i, t) = \int d\omega\, e^{-i\omega t}\, P_\omega(x_i). \tag{10}$$

Then with (10) a spectral component of the sound pressure is due to (9):

$$P_\omega(x_i) = \frac{1}{4\pi} \frac{e^{ik\tilde{r}}}{\tilde{r}}\, \hat{Q}(kx_i/\tilde{r}, \omega). \tag{11}$$

The interesting feature of this solution (11) is that there is no integral. As a consequence, a single frequency component $P_\omega(x_i)$ of sound is merely determined by the wave-number spectral function $\hat{Q}_\omega(k_i)$ of the source term at the wave-number vector $k_i = kx_i/\tilde{r}$. It should be noted that the solution (11) is very general, since no use is made of the special configuration of a circular jet.

The sound intensity of a single frequency component is for $k\tilde{r} \gg 1$ proportional to the squared amount of P_ω which is given by

$$| P_\omega(x_i) |^2 = (4\pi\, \tilde{r})^{-2}\, | \hat{Q}_\omega(kx_i/\tilde{r}) |^2. \tag{12}$$

Only this second order spectrum leads to a meaningful result for stationary random functions.

2.2. *Calculation of the Directivity of Sound Intensity*

Pao and Lowson[28] then proceeded to analyze the source term q and its wave-number spectral function $\hat{Q}_\omega$ in more detail. By introducing mean values and fluctuations of the velocity vector, the source term was separated in a shear-noise term and in a self-noise term. The most important quantities of both terms have been evaluated for a jet.

On the other hand, the wave-number–frequency spectral function $\hat{Q}$ is generally unknown. Hence, a simple isotropic turbulence structure defined in the convected frame of reference has been chosen by Pao and Lowson. For this turbulence model the directivities of sound intensity have been derived. The advantage of this method is that the theoretical results can quantitatively be compared with measurements. This has been done using the results of measurements by Mangiarotti, Cuadra, Bowie, Large and Holman.[30] In Fig. 17 the directivities for the overall-noise and low frequency components are shown. We see that the agreement in the overall-noise is within 5 dB, but of less accuracy for the low spectral components. The directivities for higher frequencies are shown in Fig. 18. Pao and Lowson[28] admitted that here "the predictions are very poor". The reduction of sound intensity at small

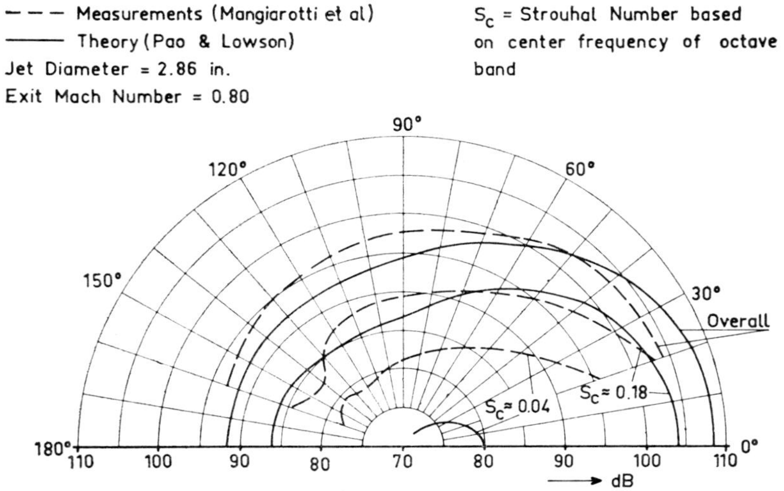

FIG. 17. Comparison of measured and calculated directivities of a circular jet for the overall-value and low frequency components.[28]

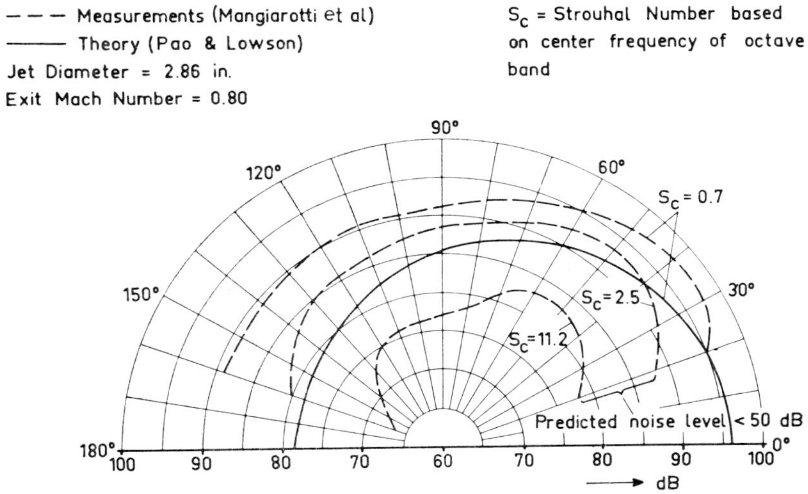

FIG. 18. Comparison of measured and calculated directivities of a circular jet for higher frequency components.[28]

jet angles, found also in other experiments, is not present in the theoretical curve. This reduction of sound intensity has been explained by the refraction of sound waves in the jet.

It should, however, be noted that, for instance, concerning the shear-noise term the circular shape of the jet has been neglected by Pao and Lowson. Thus, the shear-noise intensity depends on the azimuthal angle φ. In order to eliminate this unrealistic dependence, Pao and Lowson simply take the average in the variable φ. Therefore, it may be supposed that the moderate accuracy of the spectral components is possibly due to this crude procedure.

A very interesting property of the solution (11) which, however, has not been mentioned by Pao and Lowson, can be found by introducing polar coordinates ($\tilde{r}$, θ, φ) for the measuring point x_i. If the x_1-direction is identical with the jet axis, we can put

$$x_1 = \tilde{r} \cos \theta; \quad x_2 = \tilde{r} \sin \theta \cos \varphi; \quad x_3 = \tilde{r} \sin \theta \sin \varphi \tag{13}$$

where φ is the azimuthal angle and θ the angle with respect to the jet axis. With (13), eq. (11) becomes

$$P_\omega(\tilde{r}, \theta, \varphi) = (4\pi\tilde{r})^{-1} e^{ik\tilde{r}} \hat{Q}_\omega(k \cos \theta, k \sin \theta \cos \varphi, k \sin \theta \sin \varphi). \tag{14}$$

We see that the wave-number spectral function $\hat{Q}_\omega$ is a periodic function of the azimuthal angle φ. It follows that $\hat{Q}_\omega$ can be expanded as a Fourier series of φ with the period 2π.

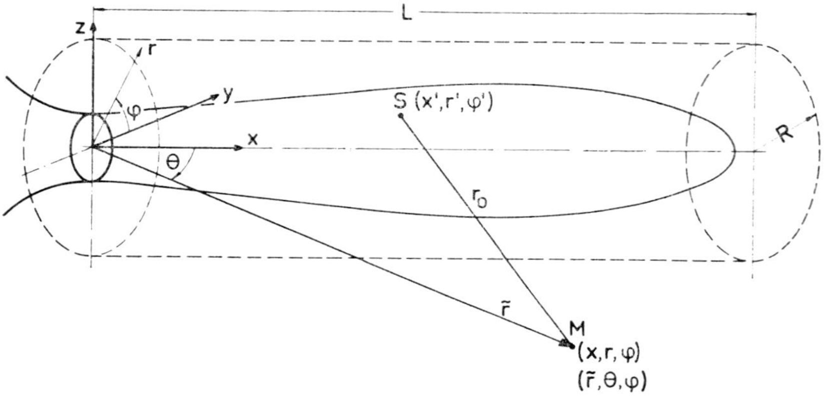

FIG. 19. A circular jet with the coordinate systems and notations used in ref. 20.

On the other hand, the sound intensity of a single frequency component has to be independent of φ for a circular jet. Hence, it follows from (12) that the wave-number spectrum $|Q_\omega|^2$ has to be a function of θ only. A method, which takes care of this property for a circular jet from the beginning, will be discussed in the next section.

3. THE SPECTRAL THEORY OF MICHALKE

Michalke[20,31] investigated the sound generation of circular jets. Consequently, a cylindrical coordinate system (x, r, φ) has been used for the source region as shown in Fig. 19. The jet axis coincides with the x-direction. As usual, the measuring point is described by polar coordinates ($\tilde{r}$, θ, φ). The distance between the source point and the measuring point is again denoted by r_0. Then the integral solution of the Lighthill equation is given by

$$p(\tilde{r}, \theta, \varphi, t) = \frac{1}{4\pi} \iiint_V q(x', r', \varphi', t - r_0/a_0) \frac{r'}{r_0} \, dx' \, dr' \, d\varphi' \tag{15}$$

where

$$r_0 = \sqrt{\tilde{r}^2 - 2x' \tilde{r} \cos \theta - 2r' \tilde{r} \sin \theta \cos (\varphi' - \varphi) + x'^2 + r'^2}. \tag{16}$$

Generally, the source term q vanishes or is negligibly small outside of the flow. Hence, the integration can be restricted to a circular cylinder volume V with a radius R and a length L outside of which $q \equiv 0$. Then we have $0 \leqq \varphi' \leqq 2\pi$, $0 \leqq x' \leqq L$ and $0 \leqq r' \leqq R$.

3.1. *Fourier Analysis of the Source Term and Integral Solution*

The advantage of using a cylindrical coordinate system for the source region is that we can easily take the periodicity of the jet turbulence with respect to φ into account. This periodicity in φ is self-evident, since the instantaneous turbulent fluctuations must have unique values for fixed time at each point of the jet flow independent of any multiple of 2π in circumferential direction. As a consequence, the turbulence as well as the source term q can be expanded in a Fourier series of φ with the period 2π:

$$q(x', r', \varphi', t) = \sum_{m=0}^{\infty} [q_m(x', r', t)\, e^{im\varphi'} + \overline{q_m}(x', r', t)\, e^{-im\varphi'}]. \tag{17}$$

The integer value of m is the azimuthal wave number. The functions q_m are complex and $\overline{q_m}$ is the conjugated complex value of q_m. Each q_m is random in time and uncorrelated with any other one. Hence, the rms-value of the Fourier series is independent of φ'. The source term q has to be bounded for $r' \to 0$. Then it can be shown that for $r' \to 0$

$$q_m = 0(r'^m). \tag{18}$$

With (17) it follows from (15) and (16) that the pressure p can also be expanded in a Fourier series:

$$p(\tilde{r}, \theta, \varphi, t) = \sum_{m=0}^{\infty} [p_m(\tilde{r}, \theta, t)\, e^{im\varphi} + \overline{p_m}(\tilde{r}, \theta, t)\, e^{-im\varphi}]. \tag{19}$$

With (17) and (19) we find from (15) that the mth component of the pressure, random in time, is given by

$$p_m(\tilde{r}, \theta, t) = \frac{1}{4\pi} \iiint_V q_m(x', r', t - r_0/a_0)\, e^{im\chi} \frac{r'}{r_0}\, dx'\, dr'\, d\chi \tag{20}$$

where

$$\chi = \varphi' - \varphi. \tag{21}$$

The second step is now to consider only single azimuth-frequency components. These components can be obtained by means of a Fourier transform with respect to time

$$P_{m\omega}(\tilde{r}, \theta) = \int_{-\infty}^{\infty} p_m(\tilde{r}, \theta, t)\, e^{i\omega t}\, dt$$

$$\tag{22}$$

$$Q_{m\omega}(x', r') = \int_{-\infty}^{\infty} q_m(x', r', t)\, e^{i\omega t}\, dt.$$

Then the Fourier transform of the retarded source term component q_m is

$$\int_{-\infty}^{\infty} q_m(x', r', t - r_0/a_0)\, e^{i\omega t}\, dt = \exp(ikr_0)\, Q_{m\omega}(x', r') \tag{23}$$

where

$$k = \omega/a_0 \tag{24}$$

is the wave number of sound. A single azimuth-frequency component of pressure becomes from (20):

$$P_{m\omega}(\tilde{r}, \theta) = \frac{1}{4\pi} \int_0^L \int_0^R dx'\, dr'\, r'\, Q_{m\omega}(x', r') \int_0^{2\pi} d\chi\, \frac{1}{r_0} \exp[i(kr_0 + m\chi)] \tag{25}$$

where r_0 is given by

$$r_0 = \sqrt{\tilde{r}^2 - 2x'\,\tilde{r}\cos\theta - 2r'\,\tilde{r}\sin\theta\cos\chi + x'^2 + r'^2}. \tag{26}$$

3.2. *The Sound Pressure Components far from the Source Region*

Since we are only interested in the far-field of sound, we introduce a geometric far-field approximation for $\tilde{r} \gg L$, $\tilde{r} \gg R$. Then the expansion of (26) yields

$$r_0 \approx \tilde{r} - x'\cos\theta - r'\sin\theta\cos\chi. \tag{27}$$

With (27) the last integral of (25) can be evaluated to $O(\tilde{r}^{-1})$:

$$\int_0^{2\pi} d\chi\, \frac{1}{r_0} \exp[i(kr_0 + m\chi)] = \frac{e^{ik(\tilde{r} - x'\cos\theta)}}{\tilde{r}} \int_0^{2\pi} d\chi\, e^{i(m\chi - kr'\sin\theta\cos\chi)}$$
$$= \tilde{r}^{-1}\, e^{ik(\tilde{r} - x'\cos\theta)}\, 2\pi i^{-m} J_m(kr'\sin\theta). \tag{28}$$

Here J_m is the Bessel function of first kind and order m. Denoting the phase of the complex $Q_{m\omega}$ by $F_{m\omega}$, we have

$$Q_{m\omega}(x', r') = |\, Q_{m\omega}(x', r')\, |\, \exp[iF_{m\omega}(x', r')]. \tag{29}$$

From (22) it follows that $F_{m\omega}$ is a measure for the convection of a $Q_{m\omega}$-component. With (28) and (29) the sound pressure component far from the source region becomes, due to (25):

$$P_{m\omega}(\tilde{r}, \theta) = i^{-m}\, \frac{e^{ik\tilde{r}}}{\tilde{r}}\, I_{m\omega}(\theta) \tag{30}$$

where

$$I_{m\omega}(\theta) = \tfrac{1}{2} \int_0^L \int_0^R dx'\, dr'\, |\, Q_{m\omega}(x', r')\, |\, r'\, J_m(kr'\sin\theta) \exp[i(F_{m\omega}(x', r') - kx'\cos\theta)]. \tag{31}$$

The sound intensity of a single azimuth-frequency component for $k\tilde{r} \gg 1$ is essentially given by

$$| P_{m\omega} |^2 = \tilde{r}^{-2} | I_{m\omega}(\theta) |^2. \tag{32}$$

If we denote the Fourier transform of the total pressure p in (19) by P_ω, then the power spectral density of sound is given by

$$| P_\omega |^2 = \sum_{m=0}^{\infty} | P_{m\omega} |^2 = \tilde{r}^{-2} \sum_{m=0}^{\infty} | I_{m\omega}(\theta) |^2 \tag{33}$$

since the different functions p_m in the Fourier series (19) have been assumed to be uncorrelated. (33) means that the directivity of a spectral sound component is a superposition of those of the various azimuth-frequency components. $|P_\omega|$ is a quantity which could be compared with measurements and which would approximately agree with the rms-value of a narrow-band-width signal of sound pressure.

It should be noted that up to now nothing has been assumed concerning the structure of jet turbulence. Nevertheless, some essential conclusions may be drawn by means of the solution (30), even if the amplitude and phase function of the source term components are not known in detail.

We see from (31) that the sound pressure of a single azimuth-frequency component is essentially influenced by two oscillating functions which are contained in the integrand of $I_{m\omega}$ and which correspond to two different acoustic mechanisms. The first one is due to the radial interference of sound sources in the jet and is expressed by the Bessel function. This term takes the effect of non-zero acoustic jet thickness into account. The second mechanism is due to the axial interference and convection of sound sources in the jet. It is expressed by the exponential term with imaginary argument in (31).

Let us first discuss the influence of the radial interference function. The Bessel function J_m can be expanded, for small arguments, in

$$J_m(kr' \sin \theta) = \frac{(kR \sin \theta)^m}{m!} \left(\frac{r'}{2R}\right)^m \{1 - 0[(kR \sin \theta)^2]\}. \tag{34}$$

Here, $kR \sin \theta$ is a jet thickness parameter. kR, which is sometimes called Helmholtz number, is proportional to the product of Strouhal number times Mach number. It denotes essentially the ratio of the radius of the source region to sound wavelength. The influence of the jet thickness parameter is symmetric around $\theta = 90°$. We see from (34) and (31) that for fixed $Q_{m\omega}$ and $kR \sin \theta \ll 1$ the value of $I_{m\omega}$ is much smaller for $m \geqq 1$ than for $m = 0$. This means that in this case an axisymmetric source component with $m = 0$ radiates much more sound than a non-axisymmetric source component with $m \geqq 1$. Then, the spectral density of sound (33) is approximately given by $|P_\omega|^2 \approx |P_{0\omega}|^2$. In other words, for sound wavelengths very large compared with the jet diameter, sound will prevailingly be radiated by axisymmetric source components. On the other hand, for $m = 0$ and increasing kR, the radiated sound of axisymmetric source components will become smaller normal to the jet ($\theta = 90°$) since the Bessel function J_0 decreases with increasing argument and can change the sign. This will occur, when the radius R and the sound wavelength will be comparable. Contrary to this, non-axisymmetric source components with $m \geqq 1$ will be important only for higher frequencies, or more precisely, for larger values of kR. They have no radiation in the direction of the jet axis ($\theta = 0°$ and $\theta = 180°$). Therefore, any sound measured at these angles is produced only by axisymmetric source components. It follows that the directivity of sound is quite different for axisymmetric and non-axisymmetric source components.

The influence of convection is not so simple to explain, since the imaginary argument of the exponential term in (31) depends on the unknown phase function $F_{m\omega}$. For a convection of turbulence in the direction of the jet flow, we must have $F_{m\omega} > 0$. It follows that the value of $F_{m\omega} - kx' \cos\theta$ is smaller for $\theta = 0°$ than for $\theta = 180°$. Therefore, the reducing influence of the exponential term is smaller for $\theta = 0°$ than for $\theta = 180°$. As a consequence, the convection effect leads to an intensified sound radiation in jet flow direction with a directivity asymmetric around $\theta = 90°$, as is well known from previous theories. In order to study the convection effect in more detail, we have to introduce a suitable model of turbulence.

3.3. *Sound Generation by Wave-type Jet Turbulence*

In his experimental investigation of jet turbulence, Mollo-Christensen[25] came to the conclusion "that turbulence may be more regular than we think it is. The experimental data are telling us that turbulence comes in packages containing components of all frequencies, and that different frequency components preserve their phase relationship over a few jet diameters."

Wave-type phenomena of jet turbulence were also found by Lau, Fuchs and Fisher[18]. Recently, Crow and Champagne[32] found that a large-scale orderly pattern may exist in the noise-producing region of the jet. They observed spatially growing waves which lead to axisymmetric vortex trains even in the turbulent jet boundary layer.

On the basis of these experimental results, it seems to be reasonable to assume that each azimuth-frequency component of turbulence is of wave-type in jet flow direction. Then each source term component is of the special type

$$Q_{m\omega}(x', r') = \tilde{Q}_{m\omega}(x', r')\, e^{i\alpha x'} \tag{35}$$

where the phase of the complex $\tilde{Q}_{m\omega}$ can only be a function of r' for wave character in x'-direction. $\tilde{Q}_{m\omega}$ and the axial wave-number α will generally depend on m and ω as well as on the flow parameters, e.g. the Mach number.

From (31) it follows with (35) that

$$I_{m\omega} = \tfrac{1}{2} \int_0^L \int_0^R dx'\, dr'\, \tilde{Q}_{m\omega}(x', r')\, r'\, J_m\left(kR \sin\theta \frac{r'}{R}\right) \exp\left[i\alpha L(1 - M_c \cos\theta)\frac{x'}{L}\right] \tag{36}$$

where

$$M_c = k/\alpha = (\omega/\alpha)/a_0 = c_{\text{ph}}/a_0 \tag{37}$$

is the ratio of the constant axial phase velocity to sound speed, i.e. the convection Mach number of a source component. It is obvious from (36) that the convection parameter $\alpha L\,(1 - M_c \cos\theta)$ is responsible for the intensified sound radiation in the direction of the jet ($\theta = 0°$) for $0 < M_c \leqq 1$ as mentioned above. For a convection Mach number $M_c > 1$ the sound pressure component $P_{m\omega}$ can reach a maximum value for a certain angle θ_c defined by

$$1 - M_c \cos\theta_c = 0. \tag{38}$$

Then the reducing influence of the exponential term drops out, and the sound sources at different x'-positions in the jet do not interfere. This effect has been called "Mach wave radiation" in the literature. The influence of the jet thickness parameter $kR \sin \theta$ on $P_{m\omega}$ remains unchanged in (36) as discussed in subsection 3.2.

It may be noted that the directivity of a sound pressure component $P_{m\omega}$ depends strongly on the jet thickness and on the convection parameter as well as on m, but seems to be not very sensitive to the special amplitude distribution $\tilde{Q}_{m\omega}$ of the source term component. This was checked by Michalke[20]. He compared the directivities calculated for various suitably chosen distributions $\tilde{Q}_{m\omega}(x', r') = \tilde{Q}(r')\, g(x')$ which were assumed to be independent of m and ω. The normalized directivity of a rms-sound pressure component is then proportional to the product of a jet thickness factor $|\hat{I}_r|$ and of a convection factor $|\hat{I}_x|$.

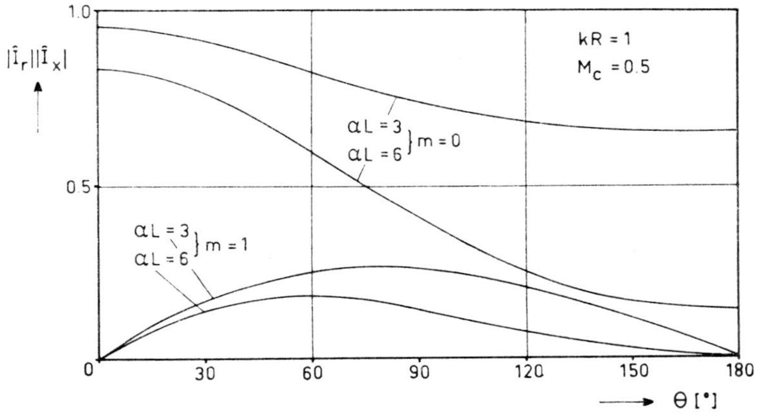

FIG. 20. The rms-sound pressure-directivities of single azimuth-frequency components for the axisymmetric and the first azimuthal source component at a low frequency.[20]

For fixed $\tilde{Q}(r')$- and $g(x')$-distributions, these directivities are shown in Fig. 20 for low frequencies, i.e. for $kR = 1$, for a convection Mach number $M_c = 0.5$ and for values $\alpha L = 3$ and $\alpha L = 6$. We see quite clearly on the curves for $m = 0$ and $m = 1$ that the shape of the directivity produced either by axisymmetric or by the first azimuthal source component is very different. For $m = 0$ there is a maximum of sound pressure in jet flow direction ($\theta = 0°$) and a minimum at $\theta = 180°$. Contrary for $m = 1$ the sound pressure is zero at $\theta = 0°$ and $\theta = 180°$ with a maximum in between, which is, however, still small compared with the sound pressure of the axisymmetric component.

For the same parameters, but for higher frequency ($kR = 4$), the directivities are shown in Fig. 21. For $m = 0$ we have again a maximum at $\theta = 0°$, but very small sound radiation for $\theta = 60°$ up to $120°$. The first azimuthal sound component ($m = 1$) is larger in this region and has a maximum at roughly $\theta = 40°$.

These theoretical results can qualitatively be compared with directivities measured by Mollo-Christensen, Kolpin and Martuccelli[26] for certain frequency bands. Their far-field measurements were concerned with subsonic jets. In Fig. 22 the rms-sound pressure directivities are shown for frequency components with Strouhal numbers $S < 0.5$. As mentioned in subsection 3.2, these curves should approximately agree with the directivity of $|P_\omega|$ defined by (33). We see that there is apparently a maximum of sound pressure at $\theta = 0°$. All

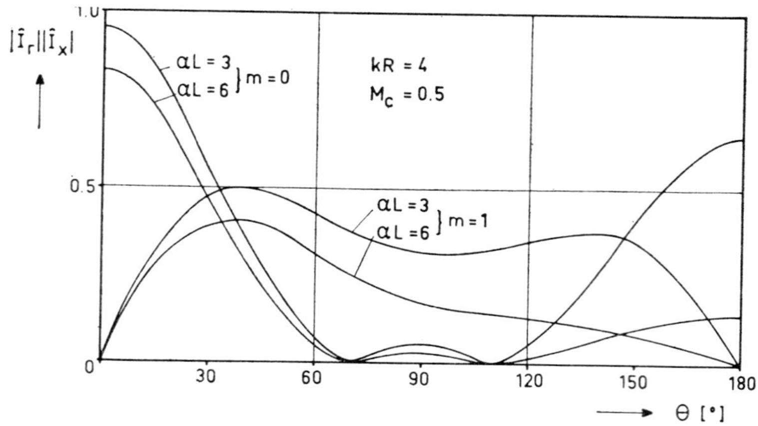

FIG. 21. The rms-sound pressure-directivities of single azimuth-frequency components for the axisymmetric and the first azimuthal source component at a high frequency.[20]

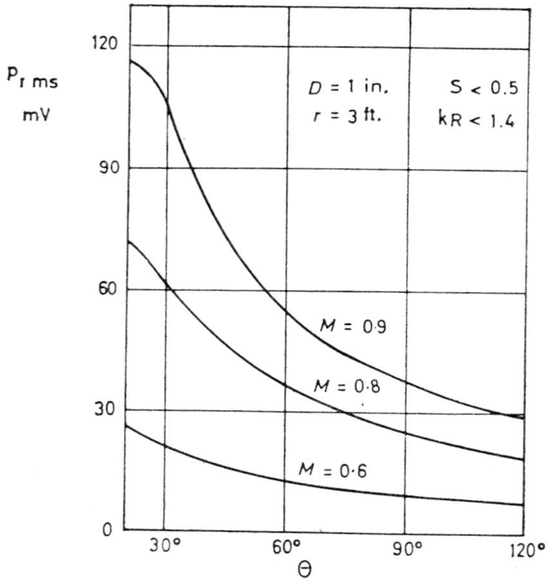

FIG. 22. The rms-sound pressure-directivities for low frequencies (Strouhal numbers $S < 0.5$) and various Mach numbers M measured by Mollo-Christensen, Kolpin and Martuccelli.[26]

curves for Mach numbers $M < 0.9$ show the same tendency as the theoretical directivity for $m = 0$ and $kR = 1$ shown in Fig. 20. In fact, assuming that the jet diameter is roughly twice the radius R of the source region, the value of kR is for $S < 0.5$ and $M < 0.9$ smaller than 1.4. Hence, we can conclude from the theoretical results that the measured directivities are mainly produced by axisymmetric source components. This is in agreement with the results of Crow and Champagne[32] who found strong axisymmetric wave components in the turbulent jet for Strouhal numbers $S \leq 0.6$.

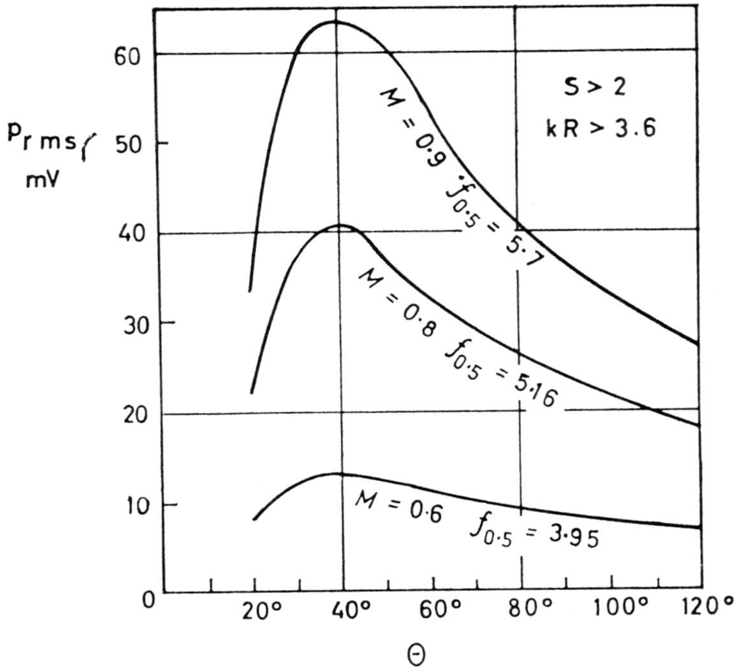

FIG. 23. The rms-sound pressure-directivities for high frequencies (Strouhal numbers $S > 2$) and various Mach numbers M measured by Mollo-Christensen, Kolpin and Martuccelli.[26]

In Fig. 23 the measured directivities are shown for frequency components with Strouhal numbers $S > 2$. Here the directivities have a remarkable peak at roughly $\theta = 40°$ and apparently tend to zero for $\theta = 0°$. Previously, this type of directivity has been explained by means of the refraction effect. Evidently, there is also an alternative explanation. The measured curves have the same tendency as the theoretical directivitiy of the first azimuthal component with $m = 1$ and $kR = 4$ shown in Fig. 21. Hence, in the view of the theoretical results, it can be concluded that these directivities can only be a consequence of a dominating sound radiation from non-axisymmetric source components with $m \geqq 1$. This can occur for larger values of kR. In fact, for $S > 2$ and $M > 0.6$ we find $kR > 3.6$. Furthermore, the shape of the measured directivities for $\theta \to 0$ suggests that the contribution of axisymmetric source components is small compared with that of non-axisymmetric ones at these Strouhal numbers $S > 2$.

One should, however, realize that the specific form of the Lighthill source term—the double divergence of the Reynolds stress tensor—leads to azimuth-frequency components $Q_{m\omega}$ which consist of different terms some of which have first and second derivatives with respect to x' and r'. Then, by partial integration, $I_{m\omega}$ (31) includes some integrals the integrands of which contain the Bessel functions J_m, J_{m-1} or J_{m-2}. As mentioned above, the most important terms for sound radiation at low frequencies ($kR \ll 1$) are those containing J_0. Hence, for sound wavelengths large compared with the radius of the source region, the most effective sound emitters are the azimuthal components of turbulence with $m = 0, 1, 2$. As a consequence, a suppression of the axisymmetric, first and second azimuthal components of jet turbulence should lead to a remarkable reduction of jet noise, at least for low frequencies.

The results due to the above-discussed "expansion scheme for jet noise" have been compared in ref. 33 with those of previous methods. The turbulence model used, for instance, by Ribner[4], Ffowcs-Williams[14] and Pao and Lowson[28] is characterized by the assumption that the turbulent eddies can be described by overall-correlation functions of the convected Gaussian form. This, however, leads to a very special type of turbulence. It is found in ref. 33 that in this case each azimuthal component of the cross-spectral density:

1. Is of wave-type in jet flow direction with a wave-number α independent of m.
2. Does not change its phase in radial direction.
3. Has a radial distribution independent of the frequency ω.
4. Has a spectrum independent of m.
5. Has a coherence in jet flow direction which is larger compared with that of the overall-correlation function, but it independent of ω and m.

The experimental results of Fuchs[19] have, however, shown that some of these assumptions do not hold true, especially in the frequency range most important for jet noise.

The most severe restriction with respect to the radiated noise made in refs. 4, 14 and 28 is believed to be the assumption of small correlation volumes in which the rms-values of the correlated quantities can be treated as constant. Correlation measurements of pressure fluctuations in a circular jet by Fuchs[19] have shown that the axial coherence of certain azimuth-frequency components of jet turbulence can extend over several jet diameters, and the radial coherence can cover the whole cross-section of the jet. For these azimuth-frequency components of turbulence the above-mentioned assumptions appear to be invalidated. Therefore some theoretical results for jet noise derived under these assumptions may be used only with precaution.

4. CONCLUSION

The spectral theory of Pao and Lowson is based on the wave-number frequency spectral function of jet turbulence. A simple universal formula for the sound intensity of a single frequency component has been derived which contains no integral. For a simple structure of turbulence based on the eddy-model, quantitative predictions of shear-noise and self-noise intensities have been made. Without the use of disposable parameters the predictions with respect to the overall-directivity agree within 5 dB with experimental results, while the directivities of frequency components were not predicted well.

The spectral theory of Michalke is not suitable to predict noise-levels quantitatively in a simple way, but seems to reveal some new aspects of the noise generation mechanism. The expansion of the source term in a Fourier series with respect to the azimuthal angle and the application of a Fourier transform with respect to time leads to an integral solution for single-azimuth-frequency components of sound pressure. This solution shows quite clearly the influence of convection of sound sources and of finite acoustic jet thickness. It was found that for low frequencies the sound pressure is generated only by a small number of azimuthal components of turbulence. The directivities produced by axisymmetric and non-axisymmetric source components were found to be quite different. Using a special wave model of jet turbulence, the calculated directivities seem to be not very sensitive with respect to the special amplitude distribution of the source term components. By the results the tendency of measured directivities could be explained. Re-examining the results of previous methods, it is found that these were derived under very restrictive assumptions.

REFERENCES

1. M. J. LIGHTHILL, On sound generated aerodynamically. I. General theory. *Proc. Roy. Soc. London*, **211**, Ser. A (1952), pp. 564–587.
2. M. J. LIGHTHILL, On sound generated aerodynamically. II. Turbulence as a source of sound. *Proc. Roy. Soc. London*, **222**, Ser. A (1954), pp. 1–32.
3. J. E. FFOWCS-WILLIAMS, Hydrodynamic noise. In *Ann. Rev. of Fluid Mech.* **1** (1969), pp. 197–222. Palo Alto: Annual Review, 1969.
4. H. S. RIBNER, Aerodynamic sound from fluid dilatations. Univ. Toronto, UTIAS Rep. 86 (1962).
5. W. C. MEECHAM and G. W. FORD, Acoustic radiation from isotropic turbulence. *J. Acoust. Soc. Amer.* **30** (1958), pp. 318–322.
6. L. D. LANDAU and E. M. LIFSHITZ, *Fluid Mechanics*. Oxford: Pergamon Press, 1963.
7. N. CURLE, The influence of solid boundaries upon aerodynamic sound. *Proc. Roy. Soc. London*, Ser. A, **231** (1955), pp. 505–513.
8. A. POWELL, Aerodynamic noise and the plane boundary. *J. Acoust. Soc. Amer.* **32** (1960), pp. 982–990.
9. I. PROUDMAN, The generation of noise by isotropic turbulence. *Proc. Roy. Soc. London*, Ser. A, **214** (1952), pp. 119–132.
10. G. K. BATCHELOR, Pressure fluctuations in isotropic turbulence. *Proc. Cambr. Phil. Soc.* **47** (1951), pp. 359–374.
11. G. M. LILLEY, On the noise from air jets. A.R.C. 20, 376/N 40/F.M. 2724 (1958).
12. H. S. RIBNER, On the strength distribution of noise sources along a jet. Univ. Toronto, UTIAS Rep. 51 (1958).
13. O. M. PHILLIPS, On the generation of sound by supersonic turbulent shear layers. *J. Fluid Mech.* **9** (1960), pp. 1–28.
14. J. E. FFOWCS-WILLIAMS, The noise from turbulence convected at high speed. *Phil. Trans. Roy. Soc. London*, Ser. A, **255** (1963), pp. 469–503.
15. R. H. KRAICHNAN, Pressure field within homogeneous unisotropic turbulence. *J. Acoust. Soc. Amer.* **28** (1956), pp. 64–73.
16. D. I. BLOKHINTSEV, Acoustics of a nonhomogeneous moving medium. NACA TM 1399 (1956) (Transl.).
17. H. V. FUCHS, Über die Messung von Druckschwankungen mit umströmten Mikrofonen im Freistrahl. DLR FB 70-22 (1970). See also: Measurement of pressure fluctuations within subsonic turbulent jets. *J. Sound Vibration* **22** (3) (1972), pp. 361–378.
18. J. C. LAU, H. V. FUCHS and M. J. FISHER, A study of pressure and velocity fluctuations associated with jet flows. Univ. Southampton, ISVR, Techn. Rep. 28 (1970). See also: The intrinsic structure of turbulent jets. *J. Sound Vibration* **22** (4) (1972), pp. 379–406.
19. H. V. FUCHS, Space correlations of the fluctuating pressure in subsonic turbulent jets. *J. Sound Vibration* **23** (1) (1972), pp. 77–99.
20. A. MICHALKE, A wave model for sound generation in circular jets. DLR FB 70–57 (1970).
21. G. T. CSANADY, The effect of mean velocity variations on jet noise. *J. Fluid Mech.* **26** (1966), pp. 183–197.
22. W. F. MÖHRING, E.-A. MÜLLER and F. F. OBERMEIER, Schallerzeugung durch instationäre Strömung als singuläres Störungsproblem. *Acustica* **21** (1969), pp. 184–188.
23. H. V. FUCHS, Energy balance for small fluctuations in a moving medium. Univ. Southampton, ISVR Techn. Rep. 18 (1969).
24. I. S. F. JONES, Aerodynamic noise dependent on mean shear. *J. Fluid Mech.* **33** (1968), pp. 65–72.
25. E. MOLLO-CHRISTENSEN, Jet noise and shear flow instability seen from an experimenter's viewpoint. *ASME J. Appl. Mech.* Ser. E, **89** (1967), pp. 1–7.
26. E. MOLLO-CHRISTENSEN, M. A. KOLPIN and J. R. MARTUCCELLI, Experiments in jet flows and jet noise far-field spectra and directivity patterns. *J. Fluid Mech.* **18** (1964), pp. 285–301.
27. G. KRISHNAPPA and G. T. CSANADY, An experimental investigation of the composition of jet noise. *J. Fluid Mech.* **37** (1969), pp. 149–159.
28. S. P. PAO and M. V. LOWSON, Spectral technique in jet noise theory. Wyle Laboratories Rep. WR 68-21 (1969).
29. S. P. PAO and M. V. LOWSON, Some applications of jet noise theory. AIAA Paper No. 70-233 (1970).
30. R. A. MANGIAROTTI, D. E. CUADRA, G. E. BOWIE, J. B. LARGE and F. S. HOLMAN, Acoustic and thrust characteristics of the jet efflux from model scale sound suppressors. Part I, Unheated jets. The Boeing Company, Rep. No. D 6-15071, Vol. II (1966).
31. A. MICHALKE, Eine Bemerkung zur Schallerzeugung durch turbulente, runde Freistrahlen. *Z. Flugwiss.* **18** (1970), pp. 479–480.
32. S. C. CROW and F. H. CHAMPAGNE, Orderly structure in jet turbulence. *J. Fluid Mech.* **48** (1971), pp. 547–591.
33. A. MICHALKE, An expansion scheme for the noise from circular jets. *Z. Flugwiss.* **20** (1972), pp. 229–237.

AUTHOR INDEX

SUBJECT INDEX

CONTENTS OF PREVIOUS VOLUMES